# FLEXIS™ and COLDFIRE® V1 MICROCONTROLLERS

*Munir Bannoura*

*Rudan Bettelheim*

*Richard Soja*

**AMT Publishing**

**www.amtpublishing.com**

# FLEXIS™ and COLDFIRE® V1 MICROCONTROLLERS

Published
by
AMT Publishing
Farmington Hills, Michigan
**www.amtpublishing.com**

## First Edition

Printed in the United States of America
ISBN 0-9762973-2-9

Cover art and design by Mike Cipolla, MC3 Designs

To order other books as well as this book, contact

AMT Publishing at www.amtpublishing.com

To order ColdFire® development tools, contact

Freescale Semiconductor at www.freescale.com

Additional Web Links:

ColdFire® product home page:

http://www.Freescale.com/ColdFire

ColdFire® discussion group hosted by Freescale:

http://forums.freescale.com/

ColdFire® independent mailing discussion group:

http://www.wildrice.com/ColdFire

# *About the Authors*

Munir Bannoura was born and raised in the city of Bethlehem in the West Bank of the Jordan River. He graduated in 1974 with a bachelor's degree of science. He immediately joined Burroughs Corporation as a product engineer, working for the corporation in both, Michigan and Scotland, UK. In 1978, he became Professor of Electrical Engineering at the National Institute of Electricity and Electronics in Algeria, North Africa. He joined Motorola Technical Training Operations in 1984 as a customer trainer and course developer. He is now employed by Freescale Semiconductor as a world-wide trainer for advanced microprocessor and microcontroller product lines. Munir has written a book with Amy Dyson on how to microcode the Time Processor Unit titled "TPU Microcoding for Beginners" a second book with Richard Soja titled "MPC5554/MPC5553 Revealed", a third book with Margaret Frances titled "eTPU Programming Made Easy" and a fourth book with Rudan Bettelheim and Richard Soja titled "ColdFire Microprocessors and Microcontrollers". Munir lives in Farmington Hills, Michigan

Richard Soja was born and educated in Scotland, graduating from Aberdeen University in 1974 with a degree in Engineering Science. By pure coincidence, his first job was working in the same Burroughs factory in Scotland where Munir had an assignment, though their paths were not to cross for another ten years or so. After 5 years at Hughes Microelectronics, Richard joined Motorola, East Kilbride in 1984, mainly providing microcontroller application engineering support for European automotive and industrial customers. He has spent the last 16 years in Austin, Texas working initially for Motorola Semiconductor Sector, and latterly Freescale Semiconductor. He is currently responsible for the system definition and architecture of new SoC products, and contributed to the architecture and design of the MPC5500 family. He is widely published in the US and Europe and has been awarded 4 patents associated with the design of microcontroller products.

Rudan Bettelheim was born in Prague the Czech Republic. Rudan learned English during a two-year stay in India, then completed school in London and graduated with a bachelor's degree in Electrical and Electronic Engineering at the Polytechnic of Central London (now Westminster University). Electronics was a hobby throughout and Rudan built many electronic devices before and after graduating. After graduation in 1979 he joined Motorola Semiconductor in East Kilbride Scotland as a systems engineer, developing many early microcontroller applications for Motorola's customers. In 1984 Rudan relocated to Munich Germany, and in 1990 to Austin Texas, always within the Microcontroller business of Motorola, now Freescale. Over the years Rudan has worked in many different roles from design (the first generation of the MPC500 microcontroller family), to applications engineering, systems engineering, marketing, and tools product management. At the time of writing this book Rudan is a Product Manager working on the definition of new 32-bit microcontroller products in the Industrial and Multi-Market Operation within Freescale's Microcontroller Systems Division

# *Acknowledgements*

The authors would like to thank Freescale Semiconductor for supporting this book and for giving them permission to include some of the figures, tables and other information from existing Freescale documents. Special thanks to Jeff Bock for his continuous support of the book. We would like to thank John Bodnar and Joe Circello for their superb support and for answering frequent questions. We would also like to acknowledge the excellent technical help and support provided by Melissa Hunter, Ruth Rhoades, Michael Norman, Eric Southers, Matt Peters and David Niewolny.

Many thanks to Lydia Ziegler and Rebeca Delgado for reviewing the S08 chapter and the MSCAN chapter, respectively. Thanks also to Joanna Bettelheim for thoroughly proof reading the book. A final thank you is extended to all other hard working Freescale Semiconductor managers and engineers who made the products described in this book possible. To anyone we have missed, please forgive us.

# Note from the Authors

In the late summer of 2003, Richard and Munir traveled together to visit a number of customers in Europe to teach the MPC5500 family. Their first destination was Stockholm, Sweden and the second one was Turin, Italy. The last visit was Bologna, Italy which is approximately 500 miles from Turin. On encouragement from the chief engineer in Turin, Richard and Munir agreed to drive from Turin to Bologna by taking a route through the Dolomites. Because of their deep involvement in customer training, a lot of discussion was focused on the MPC5500 microcontroller family during the drive time. Munir suggested to Richard that they write a book to help customers better understand this advanced microcontroller family. Richard concurred. The next morning over breakfast, they started laying out their plans. They shook hands, determined to write and publish the book within six months. In April of 2004 the first edition of "MPC5554/MPC5553 Revealed" was born.

Because their first book was received well by Freescale customers around the world, Munir, Richard and Rudan decided to write the ColdFire® book with encouragement from Freescale management. The authors hope and wish readers will find this book very helpful and useful in understanding the popular ColdFire® architecture.

*Munir Bannoura*

*Richard Soja*

*Rudan Bettelheim*

# *Dedication*

To my mother for her love and dedication, to my brothers, sisters, and to my precious first grandson Christian, I also dedicate this book to those whom I love the most.

~ Munir

To my daughter Joanna, the real author in the family.

~ Rudan

My thanks and love go out to my wife, Linda who has supported me throughout the long nights spent on this book, and to my three children.

~ Richard

# *Preface*

The M68000 Microprocessor family introduced in 1978 became one of the most popular architectures available. The architecture provided solutions to various markets including automotive, consumer and communications. Because of its popularity, Freescale Semiconductor decided to continue the M68000 legacy by offering their customers the same architecture but with greater performance and other functional enhancements.

The M68000 family was based on a Complex Instruction Set Computing (CISC) architecture that was easy to use, but there were needs for higher performance devices based on the same architecture. The solution was to offer customers a Reduced Instruction Set Computing (RISC) version based on the M68000. The ColdFire® core was introduced to address these issues needed by today's demanding complex applications.

The ColdFire® family combines both benefits of CISCs and RISCs worlds by providing a variable-length reduced instruction set (RISC) architecture and keeping the same architecture that most engineers are familiar and comfortable with, but offering a much higher performance than CISC. By utilizing variable-length instruction set, system designers will realize a significant system-level advantages over conventional RISC fixed-length instruction set. The ColdFire® instruction set architecture's (ISA) support of variable-length instructions provides denser code and less memory usage than any fixed-length RISC machines available today. Although many ColdFire® operations involve efficient 16 bit opcodes, the instruction length can be one, two or three 16 bit words depending on the addressing modes used by the instruction.

The ColdFire V1 was optimized for microcontroller applications, while retaining full instruction set, register set, and architectural compatibility with other ColdFire cores (V2, V3, and V4).

The 8-bit M6805 CPU was developed from the popular 8-bit M6800 architecture in 1979, and specifically optimized for microcontroller products. Retaining the von Neumann architecture with a single memory map for both data and program facilitates flexible use system memories, and keeps the system buses relatively simple. Real time control was signigicantly helped by the addition of bit manipulation and test instruction in the M6805 CPU. In addition to the CPU Motorola added a wide variety of peripheral systems for analog, communication, and timing functionality. That, together with a growing selection of memory configurations, quickly lead to one of the most popular microcontroller product families. This family was then developed further to the M68HC05 HCMOS products, followed by the M68HC08, the cost reduced RS08, and eventually the high perfromance HCS08.

As the need for larger memory systems, ever more comples connectivity, and corresponding CPU throughput grew in the 8-bit microcontroller area, and process developments continuoesly reduced the price of 32-bit microcontrollers, Freescale recognized the need for a seamless microcontroller family enabling easy migration in both directions between 8-bit and 32-bit CPU based products. The result is Freescale's Controller Continuum, and the Flexis 8 and 32-bit compatible products.

In this book, our attempt is to cover the Flexis and ColdFire® V1 product systems to give the reader a basic and complete understanding of the core, memoris, and its set of integrated peripherals.

# CHAPTER 1

*Introduction to Flexis*

In this chapter, a brief description is provided for the popular ColdFire V1 and HCS08 microcontroller members along with brief description of the advanced I/O peripheral modules that are shared between the two 8 and 32-bit families.

**ColdFire® V1 and HCS08 Cores**

The ColdFire V1 Core combines both benefits of CISCs and RISCs worlds by keeping the same architecture that most customers are familiar and comfortable with, but offering a much higher performance than CISC.

This family is based on the concept of variable-length reduced instruction set (RISC) architecture. By utilizing variable-length Instruction Set Architecture (ISA), system designers will realize a significant system-level advantages over conventional RISC fixed-length instruction set. The support of variable-length instructions provides a much denser code and less valuable memory than any fixed-length RISC machines available today. Although many of the ColdFire V1 operations involve efficient one 16 bit opcodes, but the instruction length can be one, two or three 16 bit words depending on the addressing modes used by the instruction.

Freescale Semiconductor offers customers with a numerous derivatives in this family that integrates various number of peripherals and different memory sizes allowing each customer to chose the best device that best meets his/her application's requirements.

By integrating a large number of flexible peripherals and memory around the core, the ColdFire V1 became ideal solution for many applications.

These devices are suitable for cost-sensitive applications requiring significant control processing for file management, connectivity, data buffering, and user interface, as well as signal processing in a variety of key markets such as security, imaging, networking, gaming, and medical. This leading package of integration and high performance allows fast time to market through easy code reuse and extensive third party tool support.

Freescale Semiconductor innovation with the ColdFire V1 brought a new concept by allowing their customers to easily upgrade their applications from 8 bit architecture to 32 bit architecture. This concept enables the customer to upgrade their applications with a much higher performance with little or no change.

Today, there are many applications based on 8 bit architectures, in particular the popular Freescale HCS08 microcontroller family. To provide customers with an easy migration path from 8 bit to 32 bit architecture (HCS08 to ColdFire V1), Freescale put together a new cost reduced version of two microcontroller families, one is based on the ColdFire Version 1 core (CFV1) and the other is based on the HCS08 CPU. These two architecture are often referred to as the Flexis or the QE128 family.

The Flexis family is integrated with many I/O Peripheral devices, these I/O devices are also integrated on the HCS08 core. Since these I/O Peripherals are identical on both architectures, the two families (HCS08 and CFV1) are designed and manufactured pin-to-pin compatible. This provides an easy migration path since most system developers write their application software in high level language such as C or C++. This new concept allow system designers to remove an HCS08 MCU from its socket and replace it directly with a ColdFire V1 without any further modification to the hardware, except the crystal oscillator if there is a need to run at a much higher frequency than the HCS08 devices.

Freescale's Flexis family will debut with about a dozen 8 and 32 bit chips, counting all the variations of features and chip packages. These variations boil down to one basic 8-bit MCU (the MC9S08QE) and one basic 32-bit MCU (the MCF51QE). The 8-bit devices are based on Freescale's HCS08 architecture, which traces its lineage to the Motorola 6800 in the 1970s, long before Motorola's semiconductor division spun off and became Freescale in 2004. Flexis 32-bit MCUs are based on ColdFire Version 1 architecture. It may seem puzzling that Freescale is renewing its devotion to ColdFire at a time when the Power Architecture is a rising force. Freescale explains that ColdFire architecture is still very popular for consumer and industrial applications—the same markets for which the initial Flexis MCUs are intended. Freescale currently offers more than 50 ColdFire based products and continues to develop new ones. In addition, millions of embedded developers are familiar with the ColdFire/68K architectures, so they should feel immediately comfortable with the new ColdFire-based Flexis MCUs.

Although pin compatibility is the major selling point of the Flexis family, it's not universal across the whole product line. The only pin-compatible 8- and 32-bit Flexis chips are those with identical packages and higher pin counts. For example, the first 32-bit Flexis MCUs will be available in low-profile quad flat packs (LQFP) with either 64 or 80 pins, depending on the number of general-purpose I/O (GPIO) ports they have. These chips are pin compatible with the 8- bit Flexis MCUs in identical LQFP-64 and LQFP-80 packages.

The 8-bit Flexis MCUs with fewer than 64 pins don't have 32-bit counterparts. Freescale says it may extend pin compatibility to Flexis devices with smaller pin counts in the future, if customers want it.

Even when 8- and 32-bit Flexis MCUs aren't pin compatible, they still share many things in common. Their integrated peripherals are the same, and so are their peripheral drivers. Developers can reuse the software when upgrading a design.

All Flexis MCUs have a one-pin port for background debugging mode (BDM). Their on-chip real-time trace buffers are large enough to capture a few hundred instructions.

In addition, programmers can write software for all Flexis MCUs using the same development tools, although the compiled code will be different, of course. A new version of Freescale's CodeWarrior for Microcontrollers supports both CPU architectures. Freescale offers a free Special Edition, which limits the size of compiled code to 32KB on 8-bit MCUs and 64KB on 32-bit MCUs.

In summary, the 8-bit and 32-bit MCUs are the same design with different processor cores. By keeping on-chip peripherals the same across both sides of the product line, Freescale can maintain a high degree of pin compatibility. Both architectures share the same peripheral drivers and one-bit debug port pin.

In this book, our attempt is to cover some of the most common and popular Flexis derivatives to give the reader a basic and complete understanding of HCS08 and ColdFire V1 cores and their integrated peripherals.

### Interrupt Controller (INTC)

Interrupts provide a mechanism for the processor core to react to real-time events that may be generated internally from the on-chip I/O peripherals or from an external source such as an interrupt request pin or port. Once acknowledged, the CPU saves the current machine state onto the system stack, executes an interrupt service routine (ISR) to service the event, and then returns to the interrupted program by restoring the machine state information to allow resumption of the interrupted program.

Due to a large number of I/O peripherals integrated into each member of the Cold-Fire V1 family derivatives, an interrupt controller (INTC) is integrated on the ColdFire V1 derivatives and designed to handle up to 192 interrupt sources. Currently, only 32 interrupt sources are implemented on both ColdFire V1 and the HCS08 MCUs to provide a complete hardware compatibilities between the 8-bit and 32-bit MCU families.

It is important to mention here that the HCS08 architecture does not integrate an interrupt controller such as the one found in the ColdFire V1 because historically Freescale 8-bit MCUs were designed to handle incoming interrupt request automatically by the core hardware.

### Cryptographic Acceleration Unit (CAU)

The CAU is a coprocessor that executes certain ColdFire coprocessor instructions. The CAU supports acceleration of the DES, 3DES, MD5, SHA-1 and AES cryptographic algorithms, that can provide the security needed when the ColdFire microcontroller is embedded in a network connected device such as a networked home appliance or a router. Other cryptographic applications for the CAU include vending machines, fare collection, and employee or personal identification.

### Random Number Generator Accelerator

The Random Number Generator Accelerator (RNGA) module is a digital circuit capable of generating 32 bit random numbers. It is designed to comply with FIPS-140 standards for randomness and non-determinism. The random bits are generated by clocking shift registers with clocks derived from ring oscillators integrated in the silicon. The configuration of the shift registers ensures statistically good data. Because the oscillators operate at unknown frequencies, they provide the required entropy needed to create random data.

### Embedded Local Memories

This chapter describes the different types and sizes of on-chip memories that are available for all devices in the Flexis family.

### Mini-FlexBus

The external bus interface implementation on this device is called Mini-FlexBus. This bus contains the interface signals to access external memory mapped devices. It provides address, data and control signals to allow the application software access external asynchronous and synchronous peripherals such as RAM, ROM, EPROM, and other logic devices that require parallel address and data buses.

The additional external memory is often required even when a large amount of on-chip memory is integrated. Software development may also benefit from temporarily using external memory prior to programming code into the on-chip flash. The Mini_FlexBus can be configured in a multiplexed and non-multiplexed mode and allows memory expansion by as much as 2 one Mega byte memory banks using a 20 bit address bus with two chip-selects. For applications that do not require external memory, the external bus interface pins may be used as general purpose I/O (GPI/O) with each pin individually configured as general purpose input or output.

### Debug and the BDM

As microcontrollers become more highly integrated, the debug process gets tougher. Software is never perfect, especially during the development phases. Its interaction with multiple hardware modules adds to the burden facing the developer. Hardware interfaces, especially to devices that are external to the MCU may also require debugging. As can be seen, developing any kind of microcontroller system always involves a combination of software and hardware debugging techniques. Tools that can monitor and control the execution of software and also provide an easy method to exercise hardware interfaces reduce the time it takes to determine the cause of software and hardware problems. With these tools, the programmer can rapidly apply software patches as well as dynamically modify the run time environment to verify fixes before proceeding with a more formal method of eliminating errors. Many of these tools also provide methods to gather statistics on the real time performance of the system, which may in turn help designers understand the limitations and sensitivities of the microcontroller system and adapt the design to improve performance and/or cost.

### System Integration

Both Coldfire and HCS08 based devices contain a number of modules that are required for correct system operation, or provide some form of system-level protection. Systems Integration includes the Multipurpose Clock Generator (MCG) and Time Of Day Module (TOD) features that affect the entire system. In addition there are status and control registers that affect pin characteristics, clock gating and selection choices, power management controls plus the Low Voltage Warning (LVW) and Low Voltage Detect (LVD) features.

### I/O Ports and KBI

Input and Output signal capability is described in this chapter along with methods to generate interrupts from external signals.

### Serial Communications Interface (SCI)

Most Flexis (QE128) Microcontrollers contain two or three independent SCI modules. This interface implements the industry standard serial communication protocol that has been available on a wide range of microcontrollers since the 1980's. This protocol is compatible with the RS232 interface used on most desktop computers and also used in many industrial, automotive and commercial applications such as point-of-sale equipment. The SCI communicates asynchronously and supports both half and full duplex serial data transfers and may be programmed to operate at both standard and non-standard baud rates. The interface uses standard non-return-to-zero (NRZ) format and includes an on-chip baud rate generator and implements only two signals, a transmit data (TxD) and receive data (RxD) pins. If needed, other modem control signals, such as request-to-send (RTS) and clear-to-send (CTS) may be controlled by software with some of the available general purpose I/O pins.

**Serial Peripheral Interface**

This Interface provides a method of exchanging serial data between ColdFire V1/HCS08 devices and many Freescale MCUs such as the MPC500, MPC5500/MPC5600, MC68300, M68HC11/12, M68HC08, M68HC05 microcontroller families, and peripheral devices such as serial EEPROMs, A/D converters, Liquid Crystal Displays (LCD), and many other devices that have the popular SPI bus charactristics. The SPI is a full-duplex, high speed synchronous bus targeted for exchanging data with on-board SPI peripherals. The SPI bus is basically a four-wire interface signals, Master Data Out, Slave Data In (MOSI) to transmit/receive serial data, Master Data In, Slave Data out (MISO) to receive/ transmit data, Clock (SPSCK) to clock data in and out, and a Chip Select (CS) to select one of SPI peripherals. To allow the SPI to communicate with a number of SPI devices, any available number of GPI/O pins can be used as chip-select signals.

The SPI block consists of transmitter and receiver sections, a baud rate generator, and a control unit. The transmitter and receiver sections use the same clock, which is derived from the SPI baud rate generator when in master mode. During an SPI transfer, data is exchanged between the master and slave simultaneously.

**Inter-Integrated Circuit Bus ($I^2C$)**

The $I^2C$ is a simple 2 wire serial bus which was developed in the early 1980's by Phillips Semiconductors. Its original purpose was to provide an easy way to connect a microcontroller to peripheral chips in a television.

Today, the $I^2C$ bus is used in many more application fields than just consumer audio and video equipment. The bus is generally accepted in the industry as a de-facto standard. The $I^2C$ bus has been adopted by several leading chip manufacturers to provide an interface to a wide range of peripheral components such as analog-to-digital converters, temperature sensors, real-time clock chips, LCD displays and I/O port expanders. Moreover, $I^2C$ has been used as the basis for a number of other industry standard serial communication protocols. For example, $I^2C$ is the protocol adopted in the VESA DDC standard for communicating setup information and allowing user control of PC monitors. It is also the basis of the System Management Bus (SMBus) used in PC motherboards and smart chargers to provide status and management of power supply systems. Serial Presence-Detect (SPD) operation on SDRAM modules uses the $I^2C$ protocol to interrogate the embedded EPROM containing performance attributes of SDRAM memories used in personal computers.

The $I^2C$ is a multi-master bus allowing interconnection between multiple devices that are all capable of initiating a data transfer. The $I^2C$ specification states that the device that successfully initiates a data transfer on the bus is considered the Bus Master and for the duration of data transfer, all other ICs are Bus Slaves. The $I^2C$ specification incorporates arbitration and clock stretching protocols to deal with situations where multiple devices request simultaneous bus access.

### Controller Area Network (msCAN)

The controller area network (CAN) is a serial communications protocol targeted for automotive and industrial applications that require a high level of data integrity. Some application examples include heavy duty trucks, industrial controls and automotive electronics. The CAN bus is a multi-master protocol which uses non-destructive collision resolution to ensure the highest priority message is transmitted on to the bus first. The CAN bus supports bit rates of up to 1Mbits/second. Not all ColdFire and HCS08 microcontrollers have the MSCAN module, but on some derivatives there may be one or two, supporting CAN specification version 2.0A and 2.0B. Each module has 5 entry receive and 3 entry transmit message buffers.

The bus protocol has built in error detection and error signaling features, along with automatic retransmission of corrupted messages. The CAN bus also distinguishes between temporary errors and permanent node failures and the ability to prevents a faulty node from causing long term disruptions of network traffic. In this chapter we will first cover the CAN bus protocol and later explain the MSCAN architectural details.

### Universal Serial Bus (USB)

The Universal Serial Bus (USB) is an industry standard that extends the architecture of the personal computer to external devices such as keyboards, cameras, printers and disk drives. The USB protocol allows a wide range of peripherals (called devices) to be attached at the same time to a PC (called the host).

USB On-The-Go (OTG) allows two USB devices to talk to each other without requiring the services of a personal computer thus providing dual-role device capable of functioning as either host or peripheral. Part of the magic of OTG is that a host and peripheral can exchange roles if necessary.

## Fast Ethernet Controller (FEC)

The FECs supports both 10 and 100 Mbps ethernet/IEEE® 802.3 networks. An external transceiver interface and transceiver function are required to complete the interface to the media. The FEC supports three different standard MAC-PHY (physical) interfaces for connection to an external ethernet transceiver. The descriptor controller is a RISC-based engine that provides the following functions in the FEC:

- Initialization (those internal registers not initialized by the user or hardware)
- High level control of the DMA channels (initiating DMA transfers)
- Interpreting buffer descriptors
- Address recognition for receive frames
- Random number generation for transmit collision backoff timer

## Timer Systems

The Timer Systems chapter describes the Timer/PWM Module (TPM), the Flex Timer Module (FTM), the Real Time Counter (RTC) and the Independent Real Time Clock (IRTC). A description of the architecture, implementation and usage of the timers is also provided.

## Analog to Digital Converter

In embedded control applications there is a need to monitor analog values because the real world is primarily analog in nature. The real world value, such as temperature, is converted to an electrical value, either voltage or current, by a sensor. The electrical value then needs to be converted to a digital value usable by the embedded processor, and this is done using an Analog to Digital Converter (ADC). Current ColdFire® products offer two main ADC modules, a 12-bit resolution Analog-to-Digital Converter (ADC) and a 10-bit resolution *Queued* Analog to Digital Converter (QADC).

The 10-bit QADC has been used by Freescale on several generations of microcontroller product families, namely the MC68HC16, MC68300, MPC500, MCM2000, and Cold-Fire®. Starting in 2006 Freescale started using the 12-bit ADC module in ColdFire® products.

Since an analog signal may represent an infinite number of values, and may change very quickly, the conversion process invariably leads to some loss of information referred to as the conversion error. The conversion error has a number of components, quantization error, offset, gain, linearity, and noise. These all add up to an overall accuracy of the converter. The resolution of a converter (the quantization error) depends primarily on the number binary bits used to represent the analog value and is determined by the fundamental design of the converter.

**LCD Controller**

The LCD driver module is a CMOS charge pump voltage inverter that is designed for low-voltage, low-power operation. The LCD driver module is designed to generate the appropriate waveforms to drive multiplexed numeric, alpha-numeric, or custom LCD panels. Depending on LCD module hardware and software configuration, the LCD panels can be either 3 V or 5 V. The LCD module also has several timing and control settings that can be software configured depending on the applications requirements. Timing and control consists of registers and control logic for:

- LCD frame frequency
- Duty cycle select
- Frontplane/backplane select and enable
- Blink modes and frequency
- Operation in low-power modes

In a 64-pin package, the LCD module can be configured to drive 8 backplanes/24 frontplanes (192 segments). In a 48-pin package, the LCD module can be configured to drive 8 backplanes/16 frontplanes (128 segments).

**Tools and Software**

Any embedded control system requires a number of supporting tools, such as hardware and software development and debug tools, system calibration tools, and production programming and test tools. For the debug tools to function effectively support systems need to be built directly into the processor, that provide real time trace and other data, and allow access to internal processor resources. Standard software such as Real Time Operating System (RTOS) or communication protocol stacks may also be required. ColdFire® products are supported by an extensive tools and software, with a wide range of both proprietary and Open Source tools and software available.

The ColdFire® processor core debug interface is provided to support system debugging in conjunction with low-cost debug and emulator development tools. Through a standard debug interface, users can access real-time trace and debug information. This allows the processor and system to be debugged at full speed without the need for costly in-circuit emulators. The debug interface is a superset of the BDM interface provided on the 683xx family of parts.

The on-chip breakpoint resources include a total of 8 programmable registers, a set of address registers (with two 32-bit registers), a set of data registers (with a 32-bit data register plus a 32-bit data mask register), an address attribute register, a trigger definition register, and one 32-bit PC register plus a 32-bit PC mask register. These registers can be accessed through the dedicated debug serial communication channel or from the processor's supervisor mode programming model. The breakpoint registers can be configured to generate triggers by combining the address, data, and PC conditions in a variety of single or dual-level definitions. The trigger event can be programmed to generate a processor halt or initiate a debug interrupt exception.

To support program trace, the Version 2 debug module provides processor status ($PST_{[3:0]}$) and debug data ($DDATA_{[3:0]}$) ports. These buses and the PSTCLK output provide execution status, captured operand data, and branch target addresses defining processor activity at the CPU's clock rate.

ColdFire® Devices and I/O Modules Integration Availability is shown in Appendix A. Device Matrix.

# CHAPTER 2

# *The Controller Continuum*

## *Controller Continuum Concept*

The microcontroller landscape is changing, due to multiple pressures and influences. Control applications are becoming more complex, with more sophisticated control algorithms, more user friendly interfaces, built in diagnostics, and increasing use of networking. At the same time there is increasing pressure on development and product update cycles, system costs, expanding product ranges, and the quest for differentiation. This is leading to the need to re-use as much of existing designs as possible, and evolutionary development. Hardware, tools, and particularly software, re-use needs to be maximized.

In many traditionally 8-bit applications, there is a growing need for more program and data memory, and more processing throughput to accommodate new functionality, with wired and wireless networking posing particularly strong demands. Networking is also leading to a growing adoption of third party software in the form of protocol stacks, which opens the door to real time operating systems (RTOS), and other common function software. Due to 16-bit program counters, 8 and 16-bit architectures generally have an architectural address range limit of 64K bytes. This can be extended with memory paging schemes of various sophistication, but these solutions tend to be limiting and unpopular with users.

In applications already using 32-bit microcontrollers it is sometimes highly desirable to be able to offer a very low end and low cost version of a product, without having to redesign the hardware or rewrite the software. 32-bit microcontrollers also tend to have far more sophisticated and complex peripheral systems, leading to longer learn and development cycles for applications that do not require that level of functionality. Many 8-bit microcontroller users are reluctant to use 32-bit microcontrollers due to the higher level of complexity of peripheral and memory systems, despite the ever lower cost of, and growing availability of, 32-bit microcontroller products.

Freescale developed the Controller Continuum to address these requirements. The Controller Continuum seamlessly links the 8 and 32-bit microcontroller worlds, and allows easy and fast migration between the two. Freescale also seriously considered 16-bit products, but concluded that with decreasing 32-bit product costs, the growing 8-bit product performance, there is no longer a need for 16-bit products. The ColdFire V1 core and associated products are specifically developed to bridge the gap between the 8 and 32-bit worlds, forming the Controller Continuum, based on the following concepts:

- Pin compatible (Flexis) microcontrollers
- Common peripheral systems
- Common tools
- Common software components

Within the Controller Continuum the Flexis microcontroller product families offer pin

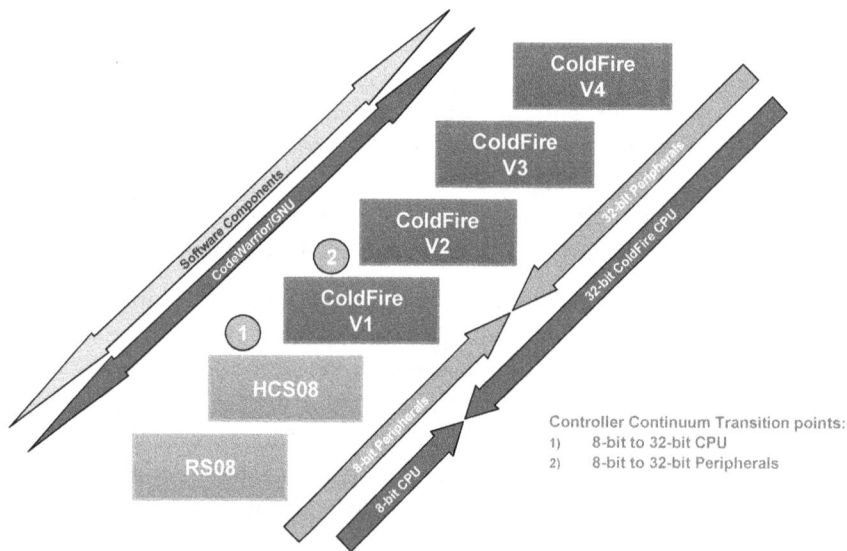

FIGURE 2.1. Freescale's Controller Continuum

compatible products with a choice of either the S08 8-bit core or the ColdFire V1 32-bit core, and the same peripherals.

Figure 2.1, "Freescale's Controller Continuum" shows the Controller Continuum and it's two transition points:

1. S08 core to ColdFire V1 core, retaining the same peripherals and tools
2. 8-bit to 32-bit peripheral systems, retaining tools and a ColdFire core

## Software Compatibility

There are a number of aspects to software compatibility, with instruction set compatibility being just one component. As more and more software development is in C, it is generally quite simple to recompile the source code for a different target processor. In the case of the Freescale Controller Continuum, and Flexis products in particular, the same CodeWarrior development suite supports both the S08 and the ColdFire V1 cores. The required target core is simply a compiler option. Currently there is no assembler translator available to facilitate direct migration between S08 and ColdFire V1 products, but converting S08 assembler code to ColdFire assembler code could be done relatively easily, due to the much larger number of registers in the ColdFire core than there are accumulators in the S08, and due to the existence of functionality equivalent instructions in the ColdFire instruction set. Converting ColdFire assembler code to S08 assembler code would be a much more challenging proposition, and it would most likely be more efficient to recode in C.

As stated earlier, there is much more to software compatibility than re-compilation or instruction translation. The microcontroller peripheral systems, interrupt handling, memory architecture, and system integration control require closely targeted and coupled software. It is not unusual for the entire control software to be architected specifically around key peripherals such as the timer system, or the memory architecture. This is where the Controller Continuum Flexis products are specifically designed to maintain maximum compatibility. In most cases the identical, 8-bit peripherals, from the S08 product family are also used in ColdFire V1 products, thus retaining compiler level software compatibility for peripheral functions. There are some exceptions to this rule, in order to support additional functionality, and utilize the additional processing and memory addressing capabilities of the ColdFire V1 32-bit CPU. Namely, on products with USB connectivity, the S08 versions implement a USB Device controller for the lowest possible cost, and the ColdFire V1 versions implement a USB OTG controller that may be used as either a USB Device or Host, for greater flexibility. Similar differences may apply to future products with complex communication peripherals, such as an Ethernet controller. Interrupt handling and system integration control are similarly kept as close as possible.

The CodeWarrior development suite from Freescale supports easy migration between S08 and ColdFire V1 based versions of Flexis products, by making the target CPU architecture a compiler option. This makes it literally possible to develop a program, compile it for the S08 version of the product and run it in the application, then recompile for the ColdFire V1 CPU, swap the processors in the application socket, and run the application on that. This makes it very easy to compare attributes such as performance and power consumption.

## Hardware Compatibility

Hardware compatibility is much easier to define, it is the ability to insert the compatible processor into the application hardware and have it be fully functional. This means that signals must be on the same physical pins, with the same electrical characteristics, and that oscillator components remain the same. Figure 2.2, "Flexis Hardware Compatibility" shows an example of how either an S08 of ColdFire V1 based processor from the same Flexis product family may be used in the same target hardware, the Freescale evaluation board, in this case.

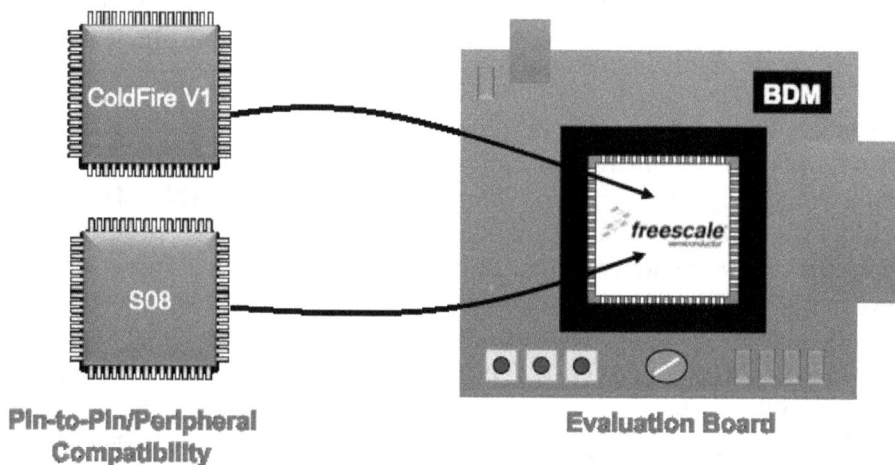

FIGURE 2.2. Flexis Hardware Compatibility

The one area of hardware compatibility where more attention is needed is that of the power supply design. Devices with larger memories are likely to have slightly higher power consumption in all power modes, and the ColdFire V1 typically have about 10% higher run mode power consumption with the same size memory system, and at the same system clock as the S08 based device. However, the ColdFire V1 based devices typically have larger memories, and are capable of operating at signinficantly higher system clocks, so may draw corespodingly more power. When designing for hardware compatibility, the power supply must be designed for the largest device at the highest system clock likely to be used. In many cases it will more economical to limit compatibily by optimizing the power supply design for the actual device that will be used, but allowing the bill of materials to be adjusted according to the processor used.

Although the power consumption in particular operating modes is higher for the ColdFire V1 products, than fro the S08 version, there are many examples where the overall system power consumption is actually less for the ColdFire V1 based system. This typically occurs in applications where the processor is normally in a low power standby mode, only enters run mode to process specific tasks, and then returns to the low power standby mode. Due to the much faster processing capabilities of the ColdFire V1 32-bit CPU the task processing is completed much faster then in the S08 device, resulting in a much lower duty cycle of being in the run mode, and consequently lower over system power consumption.

## Product Overview

Freescale has already launched a number of Flexis and ColdFire V1 product families, with more in the pipeline. Although the following product overview is current at the time of printing, it is recommended to check Freescale's web site at "www.freescale.com/coldfire" for additional details and the latest product information.

| Product Family | Flexis Product | Key Functionality/ Application | Core | MAC | CRC | CAU | Flash K bytes | SRAM K bytes | MiniFlexBus | TPM | FTM | RTC | IRTC | TOD | PDB | CMT | MTIM | ADC (12-bit) | ADC16 | VREF | DAC | ACMP | PRACMP | FRIAMP | UART /SCI | SPI | I2C | MSCAN | USB full speed | 10/100 Ethernet | RGPIO | Segment LCD |
|---|---|---|---|---|---|---|---|---|---|---|---|---|---|---|---|---|---|---|---|---|---|---|---|---|---|---|---|---|---|---|---|---|
| MC9S08QE | √ | Low Power | HCS08 | | | | 4 to 128 | 0.256 to 8 | | √ | √ | | | | | | | √ | | | | √ | | | √ | √ | √ | | | | | |
| MCF51QE | √ | Low Power | ColdFire V1 | | | | 32 to 128 | 4 to 8 | | √ | √ | | | | | | | √ | | | | √ | | | √ | √ | √ | | | | √ | |
| MC9S08JM | √ | USB device | HCS08 | | | | 8 to 60 | 1 to 4 | | √ | √ | | | | | | | √ | | | | | | | √ | √ | √ | | Device | | | |
| MCF51JM | √ | USB otg | ColdFire V1 | | | √ | 64 to 128 | 8 to 16 | | √ | √ | | | √ | | | | √ | | | | | | | √ | √ | √ | √ | Host/ Device (otg) | | √ | |
| MC9S08AC | √ | Motor Control | HCS08 | | √ | | 8 to 128 | 0.7 to 8 | | √ | | | | | | | | √ | | | | | | | √ | √ | √ | | | | | |
| MCF51AC | √ | Motor Control | ColdFire V1 | | √ | | 128 to 256 | 16 to 32 | | √ | √ | | | | | | | √ | | | | √ | | | √ | √ | √ | √ | | | √ | |
| MC9S08JE | √ | Low Power USB | HCS08 | | | | 32 to 128 | 4 to 12 | | √ | | | √ | √ | √ | | √ | √ | √ | √ | | √ | | √ | √ | √ | | Device | | | |
| MCF51JE | √ | Low Power USB | ColdFire V1 | √ | √ | | 128 to 256 | 32 | √ | √ | | | √ | √ | √ | | √ | √ | √ | √ | | √ | | √ | √ | √ | | Host/ Device (otg) | | √ | |
| MC9S08MM | √ | Medical | HCS08 | | | | 32 to 128 | 4 to 12 | | √ | | | √ | √ | √ | | √ | √ | √ | √ | | √ | √ | √ | √ | √ | | Device | | | |
| MCF51MM | √ | Medical | ColdFire V1 | √ | √ | | 128 to 256 | 32 | √ | √ | | | √ | √ | √ | | √ | √ | √ | √ | √ | √ | √ | √ | √ | √ | | Host/ Device (otg) | | √ | |
| MCF51CN | | Ethernet | ColdFire V1 | | | | 128 | 24 | √ | √ | | √ | | | | √ | √ | | | | | | | | √ | √ | √ | | | | √ | √ |
| MCF51EM | | Electricity Meters | ColdFire V1 | √ | √ | | 128 to 256 | 8 to 16 | | √ | | | √ | | √ | | √ | √ | | | | √ | | √ | √ | √ | | | | √ | √ |

FIGURE 2.3. **Product Summary Table**

## Flexis QE Family

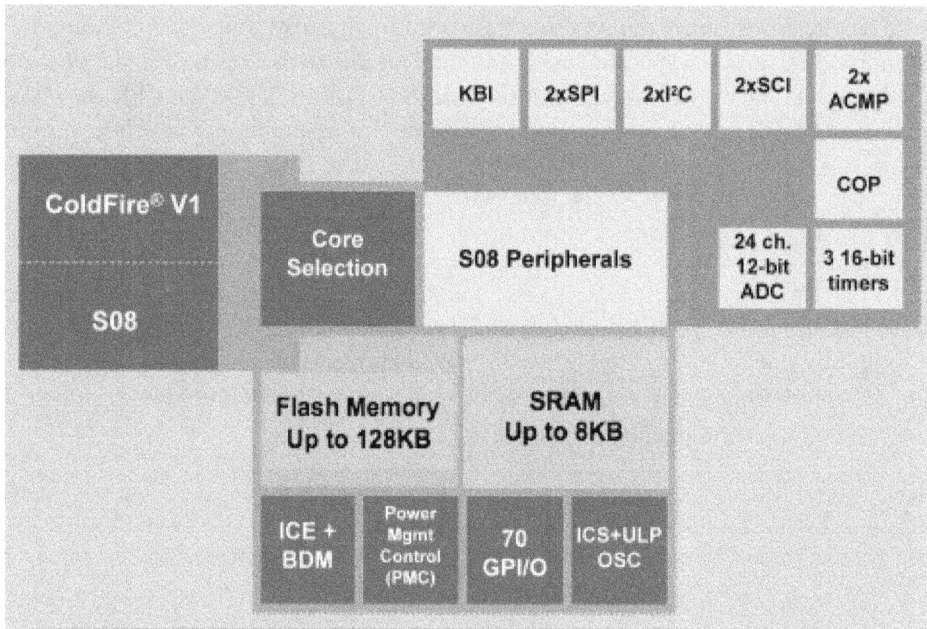

FIGURE 2.4. MC9S08QE and MCF51QE

The Flexis QE family is an ultra low power general purpose microcontroller family with both S08 MC9S08QE and ColdFire V1 MCF51QE products. The family offers from 4K bytes to 128K bytes of integrated Flash memory, and from 256 bytes to 8K bytes of integrated SRAM. The pripheral functions include all the microcontroller classics, with a 12-bit analog to digital converter, synchronous and asynchronous serial interfaces, multiple channels of 16-bit timers, general purpose I/O with interrupt capability, and several low power modes. Packages vary from a small 32 pin LQFP package up to 80 pins, with directly pin compatible versions of MC9S08QE and MCF51QE available in 64 and 80 pin packages. A demo kit available from Freescale includes useful functions, such as the support for a plug-in RF daughter card for the 2.4 GHz 802.15.4 wireless communication protocols that include ZigBee.

## Flexis JM Family

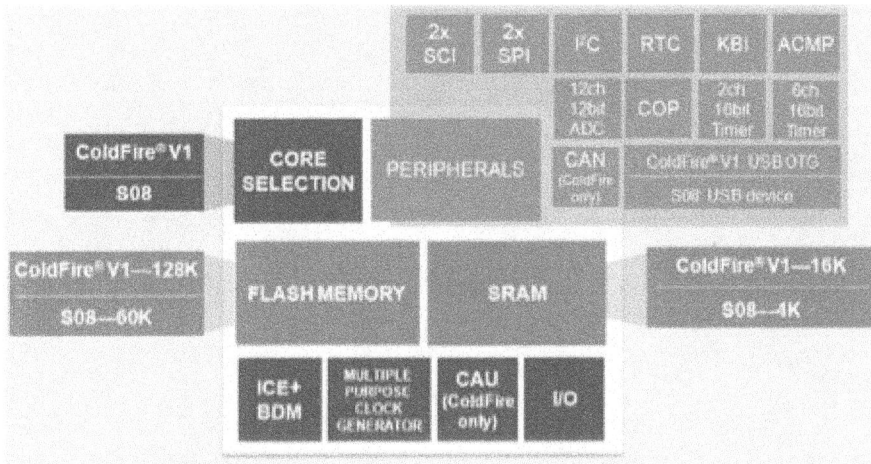

FIGURE 2.5. **MC9S08JM and MCF51JM**

The Flexis JM family adds USB connectivity to low power general purpose microcontroller functionality, with both S08 MC9S08JM and ColdFire V1 MCF51JM products. The family offers from 8K bytes to 128K bytes of integrated Flash memory, and from 1K bytes to 8K bytes of integrated SRAM. The key function of this family is USB connectivity, with the MC9S08JM including a Full Speed USB 2.0 Device controller with integrated physical layer transciever, and complimentary USB protocol stack. The MCF51JM including a Full Speed 2.0 OTG controller capable of operating either as a USB Device or Host controller. The integrated physical layer transiever supports both Device and Host modes, but for on the fly switching additonal external components are required. Complimentary USB Host and Device stacks are also included.

In addition to the USB, standard peripheral microcontroller functions are also included with a 12-bit analog to digital converter, synchronous and asynchronous serial interfaces, multiple channels of 16-bit timers, general purpose I/O with interrupt capability, and several low power modes. Packages vary from a small 44 pin LQFP package up to 80 pins, with directly pin compatible versions of MC9S08JM and MCF51JM available in 44, 64 and 80 pin packages.

The ColdFire MCF51JM products also offer the option of a hardware Cryptography Acceleration Unit (CAU) with a random number generator (RNG), and a Controller Area Network (CAN) communication controller.

## Flexis AC Family

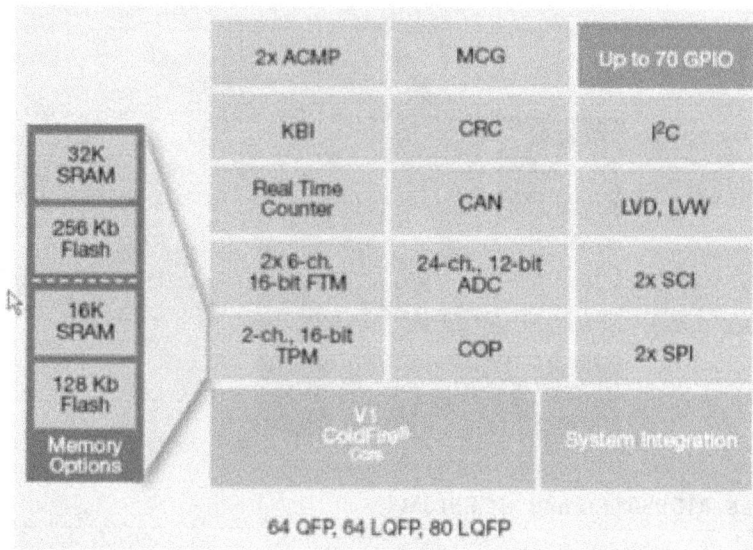

FIGURE 2.6. MCF51AC

The Flexis AC family, S08 MC9S08AC and ColdFire V1 MCF51AC, is designed for a range of electric motor applications, with a 16-bit flexible timer with two independent timebases and center aligned Pulse Width Modulation (PWM). Memory options range from 8K bytes to 256K bytes of integrated Flash memory, and 768 bytes to 32K bytes of integrated SRAM. In addition to the motor control timer, standard pripheral microcontroller functions are also included, with a 12-bit analog to digital converter, synchronous and asynchronous serial interfaces, multiple channels of 16-bit timers, general purpose I/O with interrupt capability, and several low power modes. Packages vary from a small 44 pin LQFP package up to 80 pins, with directly pin compatible versions of MC9S08AC and MCF51AC available in 64 and 80 pin packages.

The ColdFire MCF51AC products also include a Controller Area Network (CAN) communication controller, and hardware Cyclic Redundancy Check (CRC) for fast verification of memory contents. All products in the AC family have been designed for robust EMC/EMI performance in the electrically harsh environment of electric motors.

## MCF51CN128

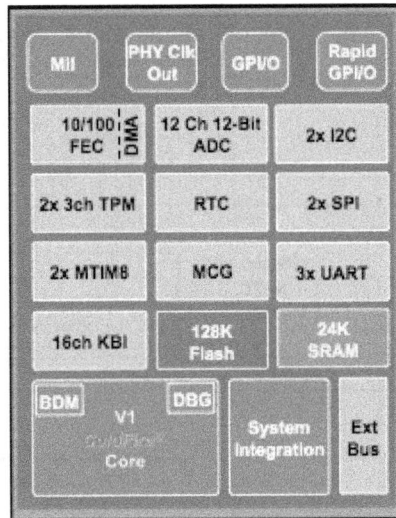

FIGURE 2.7. MCF51CN

The MCF51CN128 adds 10/100 Ethernet connectivity to a general purpose low power microcontroller. Although there are 8-bit microcontrollers with Ethernet in the market, these are proving to have inadequate resources for practical applications. The MCF51CN128 provides the smallest practical solution for Ethernet networking. It is ideally suited for networked control applications that only require a selection of the TCP/IP protocols, such as remote sensors and actuators. Freescale has made adding network capability easy by including for free the MQX operating system with full and configurable networking protocol stacks.

The MCF51CN128 also includes a 12-bit analog to digital converter (ADC), 16-bit timer system, and legacy serial interfaces I2C, SPI, and UARTs. The 128K bytes of integrated Flash memory, and 24K bytes of integrated SRAM may be extended with external memory over the optional external bus. Packages vary from a small 48 pin QFN package up to 80 pins LQFP.

## MCF51EM Family

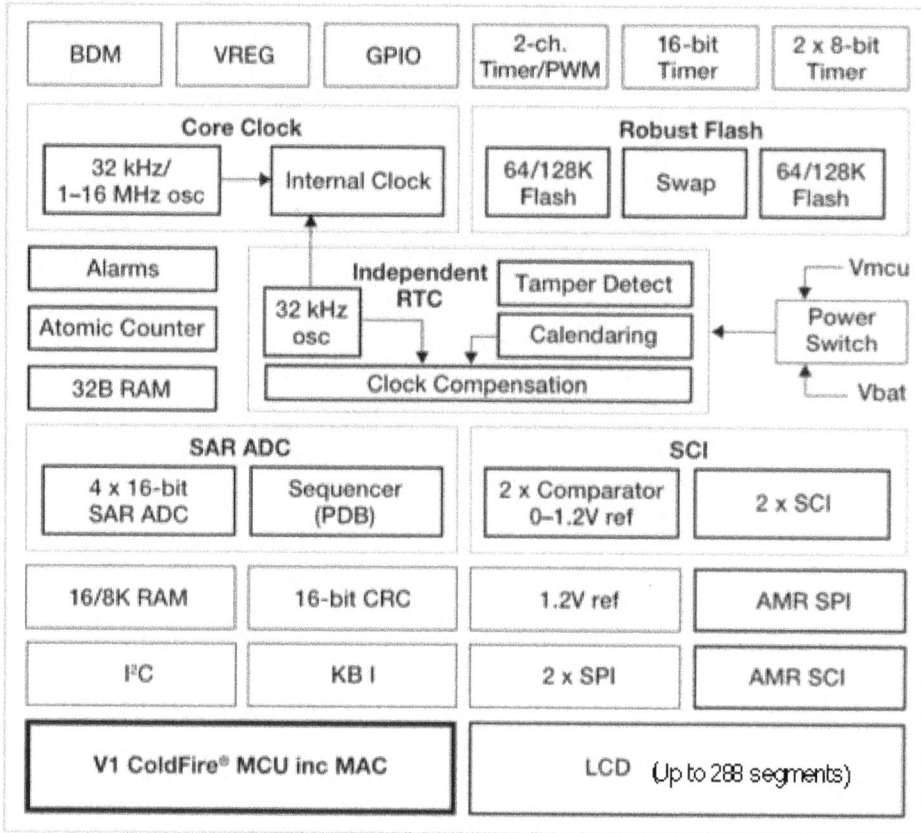

FIGURE 2.8. MCF51EM

The MCF51EM family is designed for electricity meter applications, both 1-phase and 3-phase, and for both 110V and 230V. It combines a high accuracy 16-bit analog to digital converter (ADC) system, with a segment LCD controller, robust firmware update system, and system security including tamper detection Memory options range from 128K bytes to 256K bytes of integrated Flash memory, and 16K bytes to 32K bytes of integrated SRAM. Packages include 80 and 100 pin LQFP packages.

## Flexis MM Family

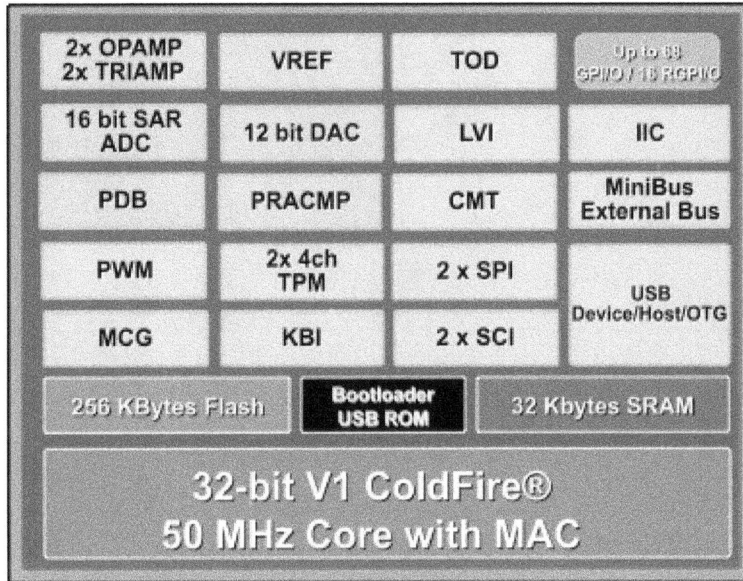

FIGURE 2.9. MCF51MM

The Flexis MM family, S08 MC9S08MM and ColdFire V1 MCF51MM, is designed for a range of medical applications, with a high accuracy analog system, USB connectivity, a segment LDC controller, and the option of an external expansion bus. Memory options range from 128K bytes to 256K bytes of integrated Flash memory, and 12K bytes to 24K bytes of integrated SRAM. The analog system includes two operations amplifiers, two trans-impedence amplifiers, a 16-bit analog to digital converter (ADC), and a 12-bit digital to analog converter (DAC). Packages vary from 80 pin LQFP package up to 104 ball BGA, with directly pin compatible versions of MC9S08MM and MCF51MM available in the 80 pin LQFP package.

The ColdFire MCF51AC products also include a Controller Area Network (CAN) communication controller, and hardware Cyclic Redundancy Check (CRC) for fast verification of memory contents. All products in the AC family have been designed for robust EMC/EMI performance in the electrically harsh environment of electric motors.

# *ColdFire V1® Core*

## *Introduction to the ColdFire® Family*

The ColdFire® family combines both benefits of CISC and RISC worlds by keeping the same architecture that most customers are familiar and comfortable with, but offering a much higher performance than CISC.

The ColdFire® core is based on the concept of a variable-length reduced instruction set (RISC) architecture. By utilizing a variable-length instruction set, system designers will realize significant system-level advantages over conventional RISC fixed-length instruction set. The support for variable-length instructions provides higher density code with less memory usage than any fixed-length RISC machine. While many of the ColdFire® core's operations involve efficient 16 bit opcodes, the instruction length may consist of one, two or three 16 bit opcodes depending on the addressing modes used by the instruction.

As previously stated, the ColdFire® cores are based on the popular 68K architecture with enhanced instruction and operand pipelines to maximize execution speed, and to minimize core stall by preventing bubbles in the pipeline. The core consists of two independent, decoupled pipelines architected to increase performance while minimizing core size.

Within the ColdFire® family of embedded cores, there are multiple versions of processor microarchitectures, supporting varying points of the price/performance curve. Having different processor microarchitectures allows Freescale to select the most appropriate core for a given application space.

The different versions of ColdFire® cores are primarily distinguished by their pipeline organizations. All ColdFire® cores include two independent and decoupled pipelines: an instruction fetch pipeline is used to prefetch instructions and an operand execution pipeline that performs the actual instruction execution. The two pipelines are connected via an instruction buffer, which serves as a FIFO buffer.

Each generation of processor microarchitecture is optimized for a different point along the price/performance curve. Examples include cores optimized for good performance in a minimal size, better performance in a slightly-larger size, maximum performance/price or simply maximum performance. For these different core implementations, the main differentiator is the pipeline organization.

In the simplest view, processor performance is loosely related to the number of pipeline stages since the operating frequency of the core is directly related to the number of stages. When the core is architected with more stages, the higher the operating frequency will be. In reality however, the increased pipeline depth also introduces a number of degradation factors which somewhat diminish processor performance. For example, recovery time from mispredicted conditional branches increases with pipeline length as do certain pipeline stalls related to register dependencies. The result is the careful selection of the processor's pipeline organization is needed to satisfy the price/performance criteria for a given device.

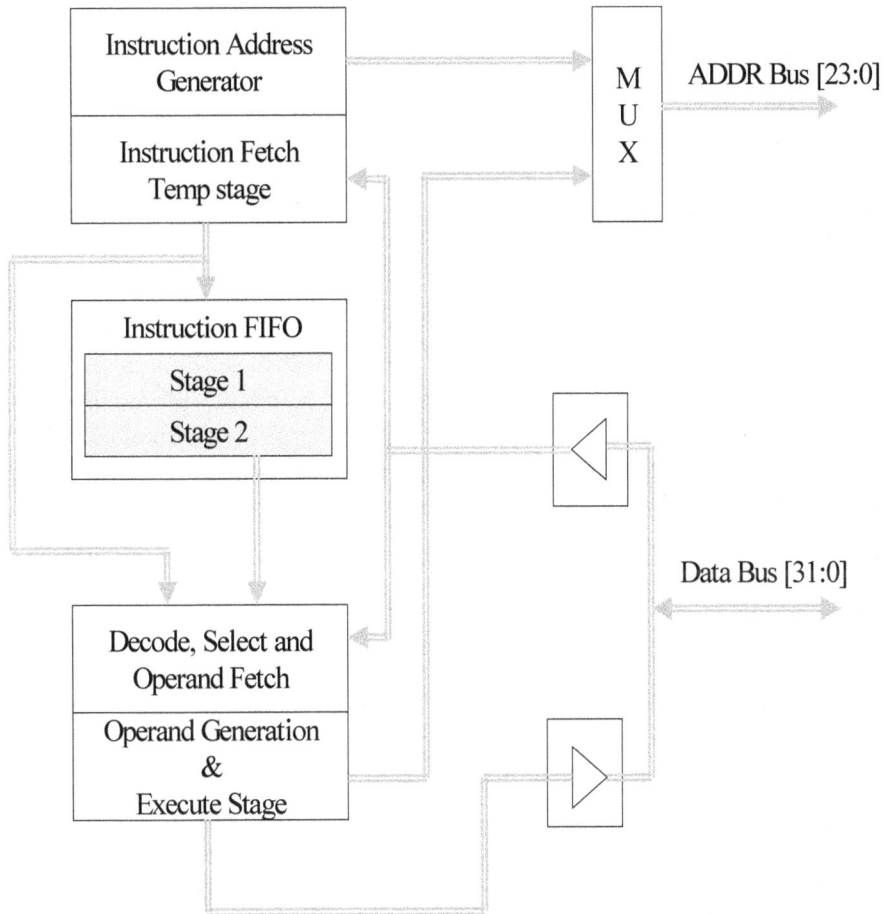

FIGURE 3.1. **ColdFire® V1 Core Pipeline**

After briefly describing the ColdFire family in general, we will from this point on concentrate only on the ColdFire Version 1 (CFV1) core.

Figure 3.1, shows the organization and the structure of the ColdFire V1® pipeline. The prefetched instruction stream is gated into a two-stage operand execution pipeline (OEP), to decode the instruction and fetch the operands needed to carry out the required operation. The instruction fetch and the operand execution pipelines are decoupled by an instruction buffer that serves as first-in-first-out (FIFO), queue allowing for instruction prefetch in advance to their actual use by the OEP, thus minimizing core stall.

As shown in the figure above, the ColdFire® V1 core instruction pipe consists of five main blocks. A description of each follows:

### Instruction Address Generator

In this stage the address of the next instruction to be prefetched is calculated and placed on the address bus to access instruction memory.

### Instruction Fetch

The prefetched instruction received from memory via the data bus is placed into this stage awaiting an empty slot in the FIFO instruction buffer. If the FIFO instruction buffer is empty and the execution stage have retired all previously dispatched instructions, the FIFO instruction buffer is by-passed allowing the instruction to stream into the operand execution pipeline.

### FIFO Instruction Buffer

The FIFO is a two stage deep buffer that receives its prefetched instruction from the instruction fetch stage.

### Decode Stage

In this stage the instruction is decoded and the required operands are selected. The required operands by the currently decoded instruction may be register operand (s) or an operand that needs to be fetch from system memory.

### Address Generation Execute stage

In this stage the actual operation is performed and the operand destination is determined. The destination result can either be a core register or a memory location. Address calculation is initiated by this stage if the operand destination is a memory location. Once the operand is written into the destination, the instruction retires and the decode stage is ready to receive the next instruction either directly from the instruction fetch stage or from the instruction FIFO.

## *User Programming Model*

ColdFire® V1 processor executes instructions in one of two modes: user or supervisor. The user mode provides the execution environment for most application programs, while the supervisor mode allows some additional instructions and privileges intended for use by the operating system and other system software.

Shown in Figure 3.2, is the ColdFire® V1 core user programming model offering sixteen 32-bit, general-purpose registers (D7–D0, A7–A0), a 32-bit program counter (PC), and an 8-bit condition code register (CCR). The first eight registers (D7-D0) are used as data registers for byte (8-bit), word (16-bit), and longword (32-bit) operations. The second set are the seven address registers (A0–A6) and the user stack pointer A7 or SP. All 8 registers may be used as address pointers for memory accesses and can also be used as software stack pointers. In addition, address register A7 is the user hardware stack pointer. All 16 registers may be used as index registers.

### Data Registers operation

Each of data registers D7-D0 is 32 bits wide. Byte operands occupy the low-order 8 bits; word operands occupy the low-order 16 bits; and longword operands, occupy the entire 32 bits.

When a data register is used as either a source or destination operand, only the appropriate low-order byte or word (in byte or word operations, respectively) is used or changed; the remaining high-order portion is unaffected. The ColdFire® instruction set provides two instructions to extend the register from byte to a longword and to extend a word to a long word to allow for algebraic operations when needed.

As shown in the figure below, the least significant bit (LSB) of a long-word integer is addressed as bit zero, and the most significant bit (MSB) is addressed as bit 31. The bit numbering is applicable to register operands and to memory operands when the memory is organized as 32 bit in width.

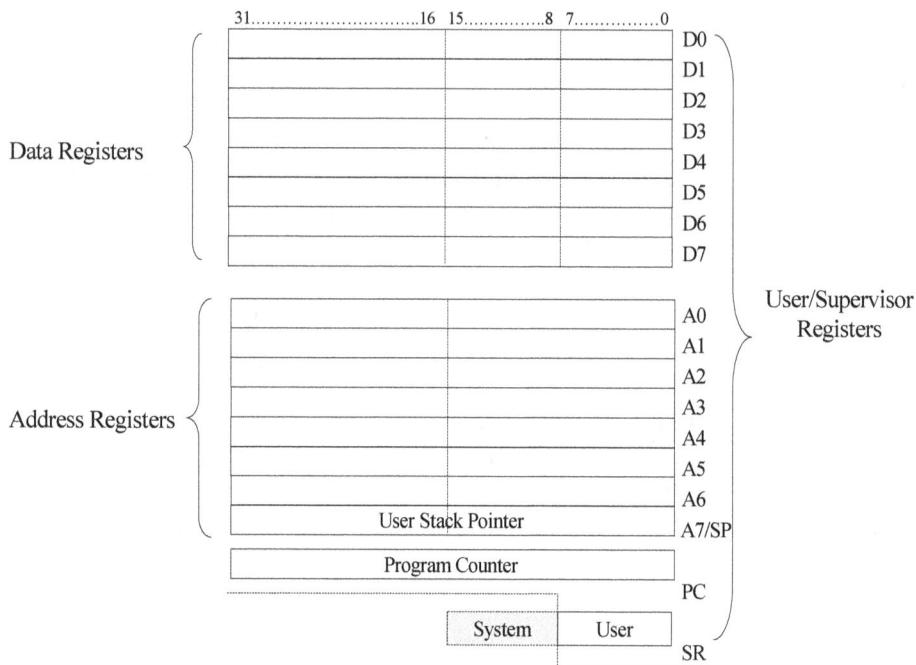

FIGURE 3.2. **User mode programming model**

### Address Registers operation (A0-A7)

The address register A0 -A7 are also general purpose address registers used mainly for addressing memory with a number of various addressing modes available on ColdFire® processors, more on addressing modes later in this chapter. These registers are accessible as 16 bit words and as 32 bit longwords. When an address register is accessed as 16 bit word, its automatically sign extended to a 32 bit longword since these registers are meant to be used as pointers to address memory. Although, A7 is the user stack pointer but still behaves as any address register

### Stack Pointer (A7) or (SP)

The stack is used to save system context during subroutine calls and for temporary data storage onto the user's stack. The stack area can be located anywhere in the MCU memory map RAM address space and can grow to any size up to the total amount of memory available in the system.

Address register A7 is the user hardware stack pointer and may be referenced as SP or A7. Normally, the SP is initialized by one of the first instructions in an application program. The stack grows downward from the address pointed to by SP. Each time a byte of data is pushed onto the stack, the stack pointer is automatically decremented, and each time a byte is pulled from the stack, the stack pointer is automatically incremented. When a subroutine is called, the address of the instruction following the calling instruction is automatically calculated and pushed onto the stack.

Normally, a return-from-subroutine (RTS) instruction is executed at the end of a subroutine. The return instruction loads the program counter with the previously stacked return address and execution continues at that address.

Unlike conventional reduced instruction set computing engines or RISC machines, the stack pointer does not need to be manually maintained in software.

In the supervisor mode, the user stack pointer can be accessed by the processor privileged instructions, MOVE to and MOVE from USP. More information is provided later on these instructions in this chapter.

The instructions below show the stack pointer basic push operation using the MOVE instructions with a predecrement addressing mode.

- MOVE.L  D4, -(A7)  // push D4 on the stack
- MOVE.L  D5, -(A7)  // push D5 on the stack
- MOVE.L  D6, -(A7)  // push D6 on the stack

Note that the stack pointer always points to the last item pushed. In order to avoid losing stacked data, the stack pointer must be decremented for push and incremented for pop operation since the stack grows towards lower memory addresses. Figure 3.3, depict the stack operation by showing SP value before and after the push sequence.

A more efficient method of pushing multiple registers on the stack is to use the load effective address (LEA) and move multiple (MOVEM) instructions instead of pushing each register individually.

**Example 3.1. Move multiple registers**

```
LEA     -12(SP), SP      // allocate 12 bytes of stack space
MOVEM.L  D4-D6, (A7)     // push register D4 thru D6 onto the stack.
```

```
The LEA instruction with a negative displacement allocates the
required stack space while the MOVEM pushes the registers
specified in the instruction.
```

In fact the MOVEM instruction allows pushing and popping all core registers in one instruction. Pushing or popping all registers in one instruction is typically useful and very efficient when performing a system context switch.

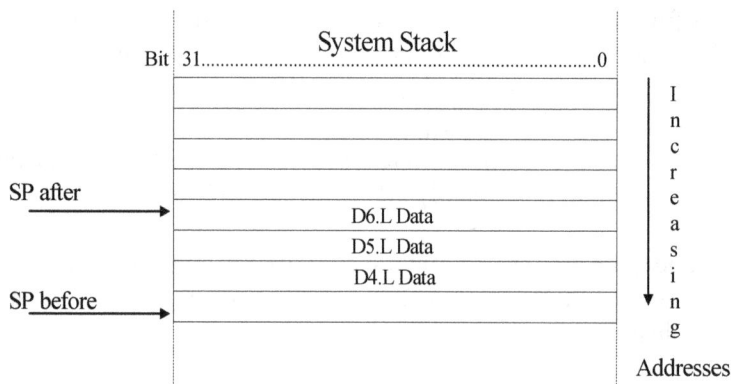

**FIGURE 3.3. Stack Pointer Operation**

**Example 3.2. Push multiple core registers onto the stack**

```
LEA   -60(SP), SP          // allocate 60 bytes of stack space
MOVEM.L  D0-D7/A0-A6, (A7)// push D0-D7 & A0 -A6 registers to the stack
```

The MOVEM instruction can also be used to pop core registers when attempting to restore them during a task dispatching sequence.

**Example 3.3. Restore multiple core registers from the stack**

```
MOVEM.L  (A7), D0-D7/A0-A6  // pop all core registers
LEA 60(A7), A7              // adjust stack pointer to original value
```

Note: the M68K allows postincrement and predecrement addressing modes with MOVEM instruction, but ColdFire® processors do not.

### Program Counter (PC)

The program counter is mainly used as an instruction pointer into the instruction steam. The PC contains the address of the current instruction being executed by the core. It automatically increments to the next instruction address when the previous one retires from the execution stage. The PC is loaded when a subroutine is called and restored upon return.

Upon reset, the PC is automatically loaded from the exception vector table, address 0x4 or vector number 0x1. The value loaded into the PC is a pointer to the first instruction in the reset or system initialization handler.

When an exception or an interrupt is recognized by the core, the program counter is automatically pushed on the supervisor stack and then loaded from the appropriate exception vector based on the exception or the interrupt taken by the core.

The PC can also be used to reference operands when using PC relative addressing modes. In other words, its possible to have the PC point to a data structure with a relative offset and meanwhile, point to the instruction stream.

### Condition Code Register (CCR)

The CCR occupies the least significant byte of the status register (SR). These bits represent indicator flags based on the results of the core operation. The CCR may be tested by a conditional branch instruction when evaluating a condition. The CCR register format is shown in Figure 3.4. The first 4 condition code bits indicate if the result has a carry/borrow, signed overflow, zero or equal, and negative. The X bit is used for multi-precision arithmetic computation. A brief description of each bit follows:

Bit: 15 ... ... ... ... ... ... ... ... ... 8    7    6    5    4    3    2    1    0

| Status Byte | 0 | 0 | 0 | X | N | Z | V | C | CCR |

Condition Code Register (User Byte)
X = Extend
N = Negative
Z = Zero
V = 2's Complement Overflow
C = Carry/Borrow

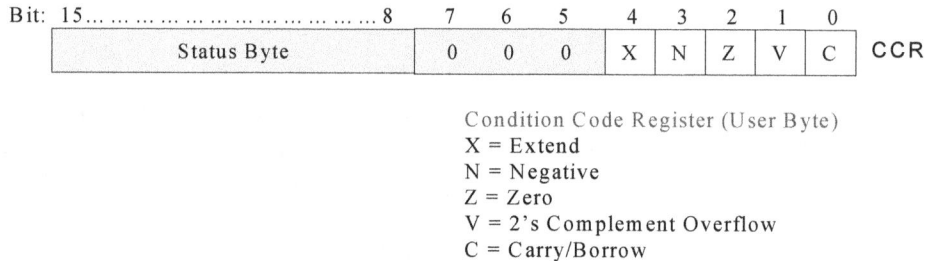

**FIGURE 3.4. Condition Code Register**

- X- Extend condition code bit: This bit is used when multi-precision arithmetic computation is needed. For example, the ADDX and SUBX instructions use this bit as the carry/borrow depending on whether the operation is addition or subtraction, respectively. Note when there is a carry/borrow, the C bit is also set, but when a condition is evaluated the C bit is updated but the X bit is not. Having the X and C bit provides the application software an easy method to add/subtract extended by using the X bit as the carry/borrow bit, while the C bit is used for conditional testing inside a multiprecision addition/subtraction routine.

- N- Negative condition code bit: This bit sets anytime the result of arithmetic operation ends up with the most significant bit (MSB), is set. For algebraic operations, the MSB of the result represent the sign bit. This bit is cleared when the result is positive and set when the result is negative.

- Z- Zero condition code bit: The Z flag is set when the result of an arithmetic or logical operation is zero. This flag may be used to evaluate conditions for equality after a compare instruction is executed, for instance.

- V- Overflow condition code bit: This bit is set when an overflow condition is detected after the execution of arithmetic or a logical operation. For example, if addition or subtraction of two operands having the same sign bit and the result ends up with opposite sign, this bit is set to indicate an overflow condition. This condition implies that the result cannot be represented in the result operand size.

- C- Carry condition code bit: This bit is set due to a carry from the most significant bit of the result during addition, or if a borrow occurs during subtraction. Note that the X bit also gets set if a carry or borrow occurs during addition or subtraction, respectively. As stated earlier, the C bit may change during an evaluation of a condition but the X bit is only updated when a carry or borrow occurs.

**Chapter 3**

## *Supervisor programming Model*

The supervisor mode is entered out of system reset allowing the ColdFire V1 core to access all resources without any restrictions. The supervisor versus user mode is determined by the 'S' bit in the core status register (SR). There are 4 registers in the ColdFire® V1 core supervisor space which cannot be accessed when the core is running in the user mode. This provides a method to change system environment while executing only in the supervisor mode and protects system from being corrupted when running user tasks. Any attempt by a user task to access supervisor resources will result in a privilege violation exception. The following section describes all supervisor registers shown in Figure 3.5.

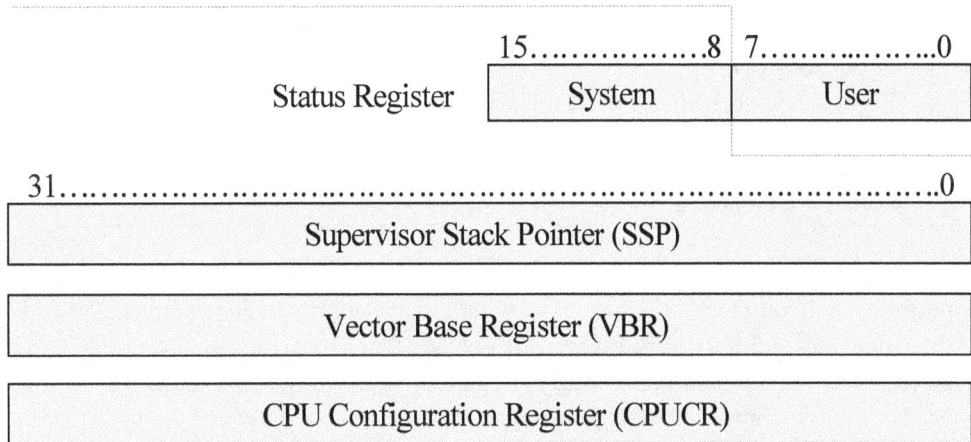

FIGURE 3.5. **CFV1 Core Supervisor Registers**

### Status Register (SR)

The status register most significant 8 bits are referred to as the system byte, while the least significant is the condition code byte discussed earlier in this section. Figure 3.6 depicts the status register. A brief description of each bit follows:

- T- Trace bit: To allow for development and debug, the trace bit provides a mechanism to single step through the program being debugged. After each instruction executed, the core takes a trace exception, if the T bit is set.
- Supervisor/user bit: This bit denotes whether the core is executing in supervisor or user mode. When set, the core is executing in the supervisor mode, otherwise the core is

executing in the user mode. By default, this bit is set out of system reset allowing start-up code to execute in the supervisor mode. Applications that require to run the in the supervisor mode need not modify this bit.

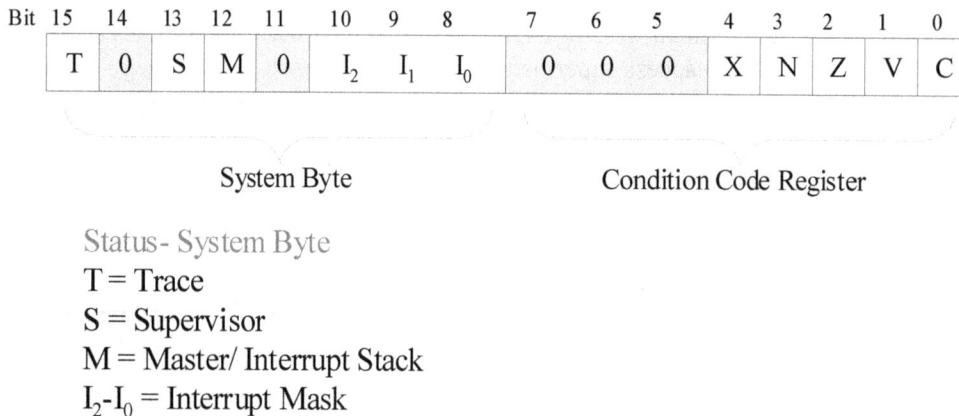

| Bit 15 | 14 | 13 | 12 | 11 | 10 | 9 | 8 | 7 | 6 | 5 | 4 | 3 | 2 | 1 | 0 |
|---|---|---|---|---|---|---|---|---|---|---|---|---|---|---|---|
| T | 0 | S | M | 0 | $I_2$ | $I_1$ | $I_0$ | 0 | 0 | 0 | X | N | Z | V | C |

System Byte                              Condition Code Register

Status- System Byte

T = Trace

S = Supervisor

M = Master/ Interrupt Stack

$I_2$-$I_0$ = Interrupt Mask

FIGURE 3.6. **Status Register (System Byte)**

- M- Master/interrupt state: The main purpose of this bit is to provide downward compatibility with other M68K processors which distinguishes between exception and interrupt stack pointers. Since the ColdFire® V1 cores have only one supervisor stack pointer, this bit can be used to emulate the other stack pointer found on some M68K processors. This bit is cleared by an interrupt exception and may be set by software.

- $I_{[2:0]}$ - Interrupt level mask: This 3 bit field defines the current interrupt priority. Out of reset, $I_{[2:0]}$ are set to '$111_2$' preventing any interrupt from being acknowledge except level seven. Interrupt level seven is the highest priority and is non-maskable. Typically these bits should be set '000' after system initialization is complete to enable all interrupt levels to be acknowledge by the CPU core. When an interrupt occurs, these bits are automatically set to the interrupt level being acknowledged to prevent interrupt nesting at the same or lower levels than the one being serviced. The only exception to this rule is level 7 since its *edge* triggered and non-maskable. If interrupt level 7 is taken for instance, another level 7 can interrupt while in the service routine.

The status register can be read and written only in the supervisor mode using the following instructions:

Example 3.4. **Read from and Write to SR Instructions**

```
MOVE.W   SR, D1          // read SR into data register D1
MOVE.W   #0x2400, SR     // write SR with immediate data
STLDSR   #0x2700         // save SR on the stack and load it with 0x2700
```

### Supervisor Stack (SSP)

The supervisor stack pointer (SSP) behaves identical to the user stack pointer (USP), but only accessible in the supervisor mode. Upon system reset, the SSP is automatically loaded from the exception vector table address 0x0, allowing it to immediately point to the supervisor stack area from start-up.

ColdFire® processors save information on the stack whenever the flow of control in a program is altered. The supervisor stack is an area in memory addressed by the SSP which is used to save context during subroutine calls, and when an exception or an interrupt event is taken. When an exception or interrupt is recognized by the core, the program counter and the status register (machine state information), are automatically pushed onto the supervisor stack allowing a return back to interrupted program. The return from exception (RTE), instruction restores the PC and the SR back to their original values.

### Vector Base Register (VBR)

The vector base register contains the base address of the ColdFire® vector table. Out of reset, this register is loaded with all 0's which locates the exception vector table at 0x0. Bits 19-0 are not used and always read as '0's since the ColdFire® V1 vector table is relocatable only on 1 mega byte boundary. Furthermore, the upper 8 bits are not used either, and always read as '0's because the core can only address maximum 16 mega bytes of address space. In other words, the CFV1 core uses only bits 23-20 of the vector base register to allow relocation of the vector table to any of the 16 one mega byte boundary. Figure 3.7 below depict the vector base register and shows its default reset value.

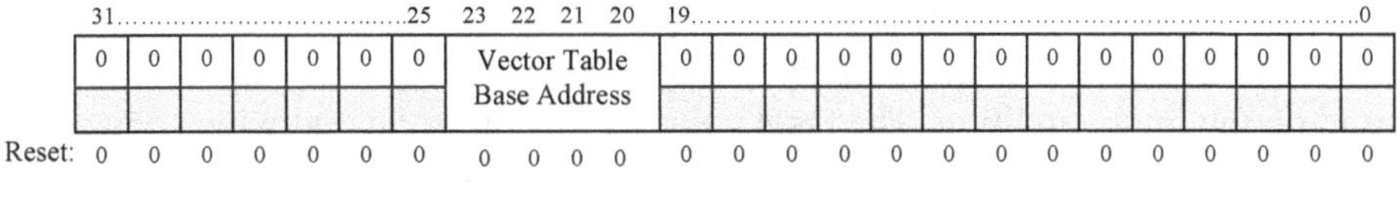

| 31 ............................... 25 | 23 22 21 20 | 19 ............................................................... 0 |
|---|---|---|
| 0 0 0 0 0 0 0 | Vector Table Base Address | 0 0 0 0 0 0 0 0 0 0 0 0 0 |
| Reset: 0 0 0 0 0 0 0 | 0 0 0 0 | 0 0 0 0 0 0 0 0 0 0 0 0 0 |

FIGURE 3.7. Vector Base Register (VBR)

When an exception or an interrupt is acknowledge, the ColdFire® core determines the vector number for the event causing the exception or the interrupt and uses it as an offset from the exception vector base pointed to by VBR.

Since the VBR is a control register that can not be loaded with immediate value as applicable to data and address registers, a read from or write to the VBR is accomplished by the move control (MOVEC) instruction. The MOVEC is supervisor only instruction with data size of 32 bits even though the control register may be implemented with fewer bits.

**Example 3.5. Reading from or writing to VBR**

```
MOVEC  VBR, D5    // read VBR into data register D5
MOVEC  D1, VBR    // write VBR with the value in data register D1
```

### CPU Configuration Register (CPUCR)

The CPU configuration is another supervisor only register that provides system level control and may be used to enable or disable hardware implementation specifics to the CFV1 core. The CPUCR register format is shown in Figure 3.8. A brief explanation of these control bits follows:

| | 31 | 30 | 29 | 28 | 27 | 26 | 25 | 24…0 |
|---|---|---|---|---|---|---|---|---|
| | ARD | IRD | IAE | IME | BWD | 0 | FSD | 0…0 |
| Reset: | 0 | 0 | 0 | 0 | 0 | 0 | 0 | 0…0 |

**FIGURE 3.8. CPU Configuration Register (CPUCR)**

- Address-Related Reset Disable (ARD)

This control bit provides a mechanism to enable or disable system reset upon the occurrence of an address error, bus error, RTE format error or a fault-on-fault conditions. Typically the application designer sets this bit during application debug and development to aide in determining the cause of the fault by having the CPU take the appropriate exception based on the fault condition. The reset value of this bit is cleared which causes the CPU to automatically reset on any of the above fault conditions.

- Instruction-Related Reset Disable (IRD)

This control bit provides a mechanism to enable or disable system reset upon an attempt to execute illegal instruction, unimplemented line-A and line-F instructions or privilege violation. Typically the application designer sets this bit during application debug and development to aide in determining the cause of the fault by having the CPU take the appropriate exception based on the fault condition. The reset value of this bit is cleared which forces the CPU to automatically reset on any of the above fault conditions.

- Interrupt Acknowledge Enable (IAE)

When set, this bit forces the CPU to execute an interrupt acknowledge cycle from the interrupt controller during exception processing to obtain the vector number of the interrupt request. The reset value of this bit is cleared, causing the CPU to use the vector number provided by the interrupt controller without executing an interrupt acknowledge (IACK) cycle.

- Interrupt Mask Enable (IME)

The function of this control bit is useful when porting an HCS08 code to ColdFire® V1. ColdFire® processors are able to nest up to 7 interrupt levels, whereas HCS08 processors are not able to nest interrupts. Setting this bit to a logic '1', forces the ColdFire core to raise its mask in the status register ($SR_{[12:10]}$) to level 7 to inhibit nesting of level 1 -6 during interrupt processing. Since level 7 is non-maskable, the ColdFire core will only be able to acknowledge level 7. The default reset value of this bit is '0' which allows nested interrupts. In other words, this bit should only be set to mimic an HCS08 interrupt processing.

- Buffered Peripheral Bus Write Disable (BWD)

To improve performance, the ColdFire CPU peripheral write cycles are buffered and terminated immediately to prevent the CPU from waiting for a bus termination status. The default reset value of the BWD bit is cleared to force write buffered operation. The caveat having buffered write enabled, if a bus error occurs during write operation the CPU will not be signaled with a fault if the peripheral is not responding. When set, this bit disables buffering and forces the processor to wait until the peripheral acknowledge reception of the data. In this way, any write operation terminating with a bus error will immediately be signaled to the CPU with the fault condition.

- Flash Speculation Disable (FSD)

When cleared, this bit provides additional Flash access performance by enabling speculation read accesses. More information on the use of this bit is provided in Chapter 6 on memories.

## *Addressing Modes*

The ColdFire® ISA has numerous addressing modes that provide flexible and efficient accesses to operands in registers and memory. With contrast to RISC machines, ColdFire® processors are able to manipulate operands in memory and capable of performing read-modify-write operations without having to bring the operands to the core registers. Below is a brief description of the addressing modes along with an examples on each.

### Register Direct Addressing Mode

This mode is used to access any of the core general purpose data registers. An example of this mode is shown below:

**Example 3.6. Register Direct addressing**

```
MOVE.S    D2, D5      // copy data in register D2 into register D5.
 The .S is an option used to specify the operand size. Operand sizes of
 8, 16, and 32 bit are supported.
```

The following example combines 2 addressing modes in one instruction. The two modes are data register direct and address register direct. Since the destination is an address register, the data size indicated by the .S option must be a word or a longword. No byte size operand is allowed with this operation. When the operand size is a 16 bit word, the address register is automatically signed extended to a 32 bit longword.

**Example 3.7. Address/Data Register Direct addressing**

```
MOVE.S D2, A4           // copies word or long word in D2 into A4
 depending on the operand size specified by the .S option.
```

### Address Register Indirect Addressing

Register indirect addressing is useful in addressing memory quickly and efficiently. Consider the examples below:

**Example 3.8. ARI addressing**

```
MOVE.B     D2, (A5)   // this instruction moves the least significant
 byte in D2 to memory location pointed to by address register A5.

MOVE.W  (A3), D1      // this instruction moves a word from memory
 location pointed by A3 to D1.
```

```
MOVE.L   D5, (A6)     // this instruction moves a longword from D5 into
memory location pointed to by A6.
```

## Address Register Indirect with Post Increment Addressing

This addressing mode is a similar to the previous one but allows the address register to increment to the next entry in memory. The address register increments by a value determined by the operand size. For a byte size access, the address register increments by 1, for a word size access, its incremented by 2 and for a longword access, its incremented by 4. Incrementing the address register by the operand size to point to the next entry into memory does not increase instruction execution time, and provides a convenient and efficient method to update the register to the next entry so it does need to be incremented manually. An example of indirect addressing with post increment mode is shown below:

**Example 3.9. ARI with Post Increment addressing**

```
MOVE.S   D2, (A5)+      // moves the data in register D2 to the memory
location pointed by address register A5. A5 is then incremented by the
size of the operand determined by the .S option. Where .S can be a
byte, word or longword.
```

The short program below copies source memory to destination memory to perform a data block move to demonstrates usage of address register indirect with post increment addressing.

**Example 3.10. Block Move using ARI with Post Increment addressing**

```
        MOVEQ.L  #count, D1      // init loop counter
        MOVEA.L  #SRC, A2        // get data source address
        MOVEA.L  #DEST, A5       // get destination address
Loop:   MOVE.L   (A2)+, (A5)+    // copy data ARI with post increment
        SUB.L    #1, D1          // decrease loop count by 1 and
        BNE      Loop            // keep looping until done
```

## Address Register Indirect with Predecrement Addressing

Predecrement addressing is similar to the previous mode discussed, but the address register is first decremented before the operand is fetched. Consider the instruction below:

**Example 3.11. ARI with Predecrement**

```
MOVE.S    D2, -(A5)        // moves the contents of data register D2
to memory location addressed by A5 predecremented first by 1, 2 or 4
depending on whether the operand size is byte, word or a long word,
respectively.
```

Address register indirect with predecrement mode is very useful when there is a need to access data table in memory organized in descending order. A short example may be helpful to demonstrate the use of this addressing mode.

Figure 3.9 depicts two source operands and a result operand in memory.

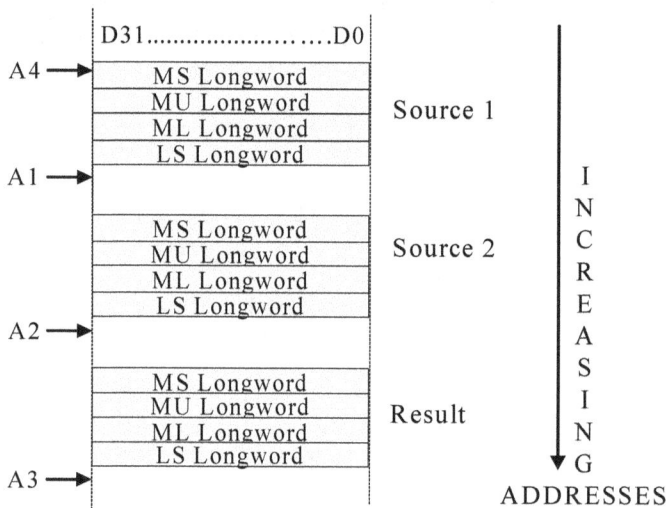

FIGURE 3.9. Operand Organization in Memory for Multiprecision Add Operation

The code snippet below is a multi-precision addition example. The two operands to be added are located in memory and are accessed in descending order. The result operand is stored into memory locations addressed by register A3. The size of the operands are 4 longwords each.

Example 3.12. Multiprecision addition using ARI with Pre-decrement addressing

```
       MOVE.W    CCR, D0       // get value of CCR in D0
       ANDI.W    #0xEF, D0     // clear extend bit before starting addition
       MOVE.W    D0, CCR       // write CCR with X bit = 0
Loop:  MOVE.L    -(A1), D1     // get 1st. operand
       MOVE.L    -(A2), D2     // get 2nd. operand
       ADDX.L    D1, D2        // add extended
       MOVE.L    D2,-(A3)      // store result
```

```
CMPA.L  A4, A1          // all numbers added?
BHS     Loop            // loop if A1 is higher or same as A4
```

Here address register A4, points to the last memory location used in the addition. The CMPA.L A1, A4 instruction checks when A1 becomes lower than A4, and the loop is exited when the condition is met.

### Address Register Indirect with 16 bit Displacement

To enable programs to access a particular element in a data table or array, the address register indirect with displacement addressing mode is used. The 16 bit displacement in this addressing mode is sign extended to allow an element to be accessed with + or - 32K bytes from the current address pointed to by the address register. The instruction example below demonstrates this addressing mode:

**Example 3.13. ARI with 16-bit Displacement**

```
MOVE.B   D4, d16(A3)  // moves least significant byte in D4 to the
memory location pointed by register A3 plus the 16 bit signed dis-
placement.
```

For example, if A3 contains the value 0x1000 and the 16 bit signed-displacement equal to 0x100, then D4.B will be stored in memory location 0x1100. Likewise, if the displacement is -0x100, then D4.B will be stored at location 0x0F00. Figure 3.10 shows an example of this addressing mode.

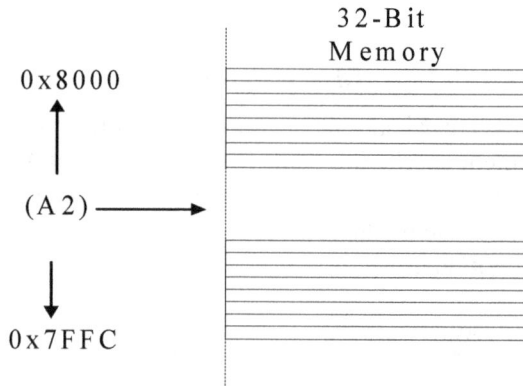

FIGURE 3.10. **ARI with 16 bit Signed Displacement**

### Address Register Indirect with Index and 8 bit displacement

Probably this is the most interesting addressing mode available on ColdFire® processors. It uses 3 elements to calculate the effective address. These 3 elements are an address register, an index register that can be scaled by a value of 1, 2 or 4, and an 8 bit displacement.

This addressing mode requires one extension word that contains an index register indicator and an 8-bit displacement. The index register indicator includes size and scale information. In ColdFire® processors, the size of the scaled indexed register is always a longword.

The operand's effective address is the sum of the address register contents; the sign-extended displacement value in the extension word low-order eight bits; and the index register contents (possibly scaled by 2 or 4). The user specify the address register, the displacement, and the index register in this mode.

The example below in Figure 3.11 shows how this addressing mode could be used to access a particular element in one of four I/O peripheral devices. The I/O peripherals are located at starting address 0x00028000 -0x0002800F and are pointed to by address regis-

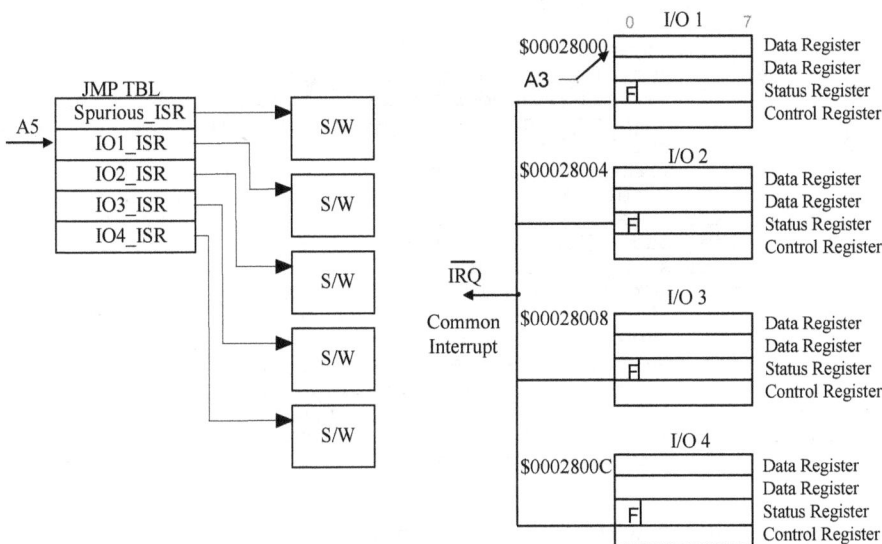

FIGURE 3.11. ARI with Indexed and 8-bit Signed Displacement

ter A3. In this example, the four I/O devices are identical and each consists of 4 memory mapped locations: 2 data registers, a status register and a control register. Furthermore, they share one common interrupt request signal. The IRQ signal will assert when any of the four I/O flag (F) bits get set. Since the IRQ signal is shared, a polling routine is required by the application software inside the interrupt handler to determine which flag is asserting the request. The polling routine below uses address register indirect with index scaled by 4, and 8 bit displacement to test the flag bit in each I/O device.

Example 3.14. Using ARI with Indexed and 8-bit Signed displacement

```
        MOVEQ.L  #3, D1              // set loopcount to 3 to loop 4 times
Next:   BTST.B¹  #7, (2,A3,D1*4)  // test I/O flag bit number 7
        BNE      Done               // if set, go to done to calculate ISR addr
        SUBQ.L   #1, D1             // else lets poll the next I/O status flag
        BPL      Next               // continue until loop counter goes negative
Done:   LEA      (A5,D1*4), A3   // calculates ISR address in jump table
        JMP      (A3)               // jump to the appropriate I/O service routine

        The first MOVEQ.L instruction loads register D1 with a loop count
        of 3 to allow iteration of up to 4 times. The next instruction
```

---

1. ColdFire processors are able to execute a read-modify-write operation in one instruction cycle.

calculates the effective address to point to the last I/O device
flag, bit number 7. Note that A3 points to the base address of the
four I/O devices, and the data register is used as an offset from
that address to point to 1 of 4. Since D1 is scaled by 4, the
effective address will be the last I/O address which is location
0x0002800C.
The base displacement in this instruction is 0x2 which will be used
to offset the sum of register A3, and register D1 scaled by 4. The
effective address calculated by this instruction is address of the
last I/O device status register. The bit test (BTST.B) instruction
checks bit number 7 if set.
The polling routine continues until it finds one of the I/O devices
flag 'F' bit is set, at which time a branch instruction will be
taken to calculate the appropriate I/O device interrupt service
address. The load effective address (LEA) instruction calculates
the ISR address into register A3, and the last instruction
jumps to the address shown in the jump table.
The loop count in register D1 is decremented by one with each loop
iteration. If the loop count decrements to minus one, it indicates
that *none* of the flag bits are set and the spurious interrupt
address is calculated and taken. Spurious interrupt indicates to
the application software that an interrupt was requested but with
drawn for some unusual condition. For more information on spurious
interrupts, refer to exceptions and interrupt controller chapter.

## Program counter Indirect with 16 bit displacement

This addressing mode is similar to address register indirect (ARI) with 16 bit displacement but the PC is used as the indirect address register to allow generation of position-independent code. Programs can be relocated anywhere in memory when using PC relative addressing because no absolute addresses are generated. In other words, if all memory references instructions use PC relative addressing, the program will be dynamically position independent since the effective addresses will be calculated as each instruction executes. Moving a program and restarting it at the point where it was temporarily suspended will not cause any problem as long as the program and its data are moved together as a block to other areas in memory.

## Program Counter with Index and 8 bit Displacement

Again, this mode is similar to address register with index and 8 bit displacement. To provide for a flexible addressing to relocating programs in memory, this mode may be used.

## Absolute Short

In this modes, the operand address is in the extension word to allow accessing the first and the last 32K bytes of the memory space. An extension word value from 0x0 - 0x7FFF will access the first 32K bytes and a value from 0x8000 - 0xFFFF will access the last 32K bytes. In absolute short addressing the extension word is signed extended to 32 bit address.

**Example 3.15. Absolute Short addressing**

```
MOVE.W   D5, 0x2000     // store D5 to address 0x2000
JMP      0x8000         // jump to address 0xFFFF8000 (EA is signed
                        //extended to 32 bits)
```

## Absolute Long addressing Modes

This addressing mode allows the application software to access any memory location in the 16 mega byte address space since two extension words are specified in the opcode to address the memory operand.

For absolute long addressing mode, two extension words are used in the instruction allowing accesses to any memory location in the 4G byte address space.

**Example 3.16. Absolute Long addressing**

```
MOVE.L   0x12340000, D4     // load D4 from address 0x12340000
```

## Immediate Addressing Mode[2]

To provide loading of constants into registers, immediate addressing mode is used. The immediate data is in the instruction stream can be one or two extension words depending the data size used in the instruction.

**Example 3.17. Immediate Addressing**

```
MOVE.W   #0x4000, D1      // load D1 with immediate value of 0x4000
MOVE.W   #0x12345678, D7 // load D7 with immediate value of 0x12345678
```

---

2. This addressing mode is not applicable to use with control/configuration registers such as VBR and CPUCR, etc.

## Special instructions

A number of instructions were covered in the addressing mode section of this chapter. The next section briefly covers the ColdFire® special instructions and later will provide a complete list of the ColdFire® instruction set. The reader may refer to the ColdFire® programming reference manual for more information. All ColdFire® reference and user's manuals can be downloaded from Freescale website by visiting: www.freescale.com

### LINK and UNLK Instructions

The LINK instruction provide a convenient method to manage parameters accesses on the stack inside a subroutine. When the main program calls a subroutine, it is expected that the subroutine should have no visible effect except for linkage as defined by the calling program.

Typically, the calling program passes a set of parameters on the stack by using the MOVEM.L instruction before calling a subroutine. The branch (BSR) or jump-to-subroutine (JSR) instruction pushes the program return address (PC), onto the stack and control is transferred to the subroutine. The subroutine can use the LINK instruction to allocate a stack space for local variables.

**Example 3.18. LINK A5, - #40    // allocate 40 bytes and create a stack frame**

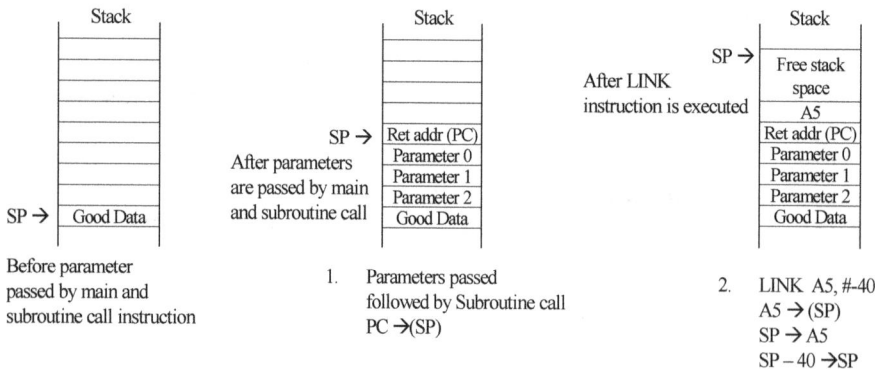

FIGURE 3.12. Using the LINK Instruction

This instruction saves the value of A5 on the stack and replaces it with the SP value. Address register A5 becomes the stack frame pointer since it remembers the SP location before allocating $40_{10}$ bytes in memory stack. The frame pointer can be used to access parameters passed-in by the main program by using a positive offset from the location pointed by the frame pointer (A5). Figure 3.12 above, shows the steps executed by the calling program and the subroutine as follows:

1. In this example, the calling program passes 3 parameters, executes a Jump to a subroutine which causes the return address in the PC to be pushed onto the stack.

2. The first instruction in the subroutine allocates 40 bytes on the stack by executing a LINK instruction using A5 as a frame pointer, the contents of address register A5 is automatically pushed onto the stack, and then A5 is loaded with the SP pointer value.

3. Finally, the 16 bit displacement is added to the stack pointer to open up 40 bytes of work space for the subroutine.

The subroutine may use the stack pointer to access local variables and the frame pointer A5 in this example, may be used to access global variables passed in by the main program.

Before returning to the main program, the subroutine executes UNLK A5. The UNLK instruction first loads the SP with the current value in A5, then the old value of A5 is restored from the stack. This causes the stack pointer to increment to the return address, and control is transferred to the calling program by executing a return-from-subroutine (RTS) instruction.

Its important to note that when using the LINK and UNLK instructions, the subroutine becomes reentrant. This means if the subroutine is interrupted and the interrupt handler calls the same subroutine, a new stack frame is built on the stack and previous process information loss is prevented.

## Stop Instruction

This instruction loads a 16 bit value into the core status register (SR) to mask certain interrupt levels prior to entering the stopped state.

**Example 3.19.** **STOP #0x2400 // set supervisory (S) bit and interrupt mask to level 4**

Note: If the S bit is cleared prior to the execution of STOP, the processor will take privilege violation exception. A wake up event is required to exit the low-power stop mode and return to run mode. Wake up events can be of any of these conditions:

- Any type of reset
- Any valid, unmasked interrupt request

Exiting from low power mode via an interrupt request requires:

- An interrupt request whose priority higher than the value programmed in the interrupt priority mask ($I_2$-$I_0$) field of the core's status register (SR).

- An interrupt request from a source which is not masked in the interrupt controller's interrupt mask register.

- An interrupt request which has been enabled at the module of the interrupt's origin.

## FF1- Find First One

The FF1 instruction scans a data register from the most-significant down to least-significant bit searching for the first set bit. This instruction is useful when polling a peripheral device status register having multiple interrupt requests. The interrupt handler may want to find out if other interrupts are requesting inside the interrupt service routine.

**Example 3.20. Find First Bit Set in a Register**

```
FF1 Dn    // where Dn can be any one of the 8 data registers

If the most-significant bit (bit 31) is set, a value of '0' is
returned, if all bits are cleared except the least-significant a
value of '31' is returned, but if none of the bits are set in the
register then a value of '32' is returned in Dn.
The value returned into the register may be used as an offset to
point to a particular entry into a branch or jump table, for
instance.
To accurately position the calculated offset to a 32 bit longword
boundary, the result in the data register should be shifted left
twice which effectively multiplies it by 4.
```

## BITREV - Bit Reverse

The contents of the destination data register are bit-reversed; that is, new Dn[31] = old Dn[0], new Dn[30] = old Dn[1], ..., new Dn[0] = old Dn[31].

**Example 3.21. Reversing Bits in a Register**

```
BITREV  Dn    // reverse all bits in data register Dn

The instruction may be used to change the priority order of group of
bits before executing the FF1 instruction, for instance. In other
words, executing a BITREV before FF1 instruction reverses the priority
order, giving bit '0' the highest priority and bit '31' the lowest.
```

## BYTEREV- Byte Reverse

The contents of the destination data register are byte-reversed; that is, new Dn[31:24] = old Dn[7:0], ..., new Dn[7:0] = old Dn[31:24].

**Example 3.22. Re versing Bytes in a Register**

```
BYTEREV  Dn  // reverse all bytes in data register Dn

Data is always right-justified when loaded or manipulated into a reg-
ister. For example, if the application program needs to process byte
or bytes residing in the upper portion of the register, it must shift
the register the appropriate number of times first. Shifting the reg-
ister 8 or 16 bits could be a significant latency instead, the
BYTEREV instruction may be used to quickly position the byte into the
desired in the data register.
```

## STLDSR - Store-Load Core Status Register

The store-load status register pushes the contents of the status register onto the stack and then reloads the status register with the immediate data value specified in the instruction.

**Example 3.23. STLDSR #0x2500 // push SR raise the core mask interrupt to level 5**

The STLDSR may be executed as the first instruction in the interrupt handler to raise the mask level in the core status register. Since the first instruction in the interrupt handler is part of exception processing sequence, raising the mask value in the core status register prevents nesting of higher priority interrupts during the execution of a critical code inside an interrupt service routine.

## Move to and from USP

ColdFire® V1 core supports two unique stack pointers: the supervisor stack pointer (SSP) and the user stack pointer (USP). Both of these registers are referenced as SP or A7. In other words, if the supervisor reads from or writes to SP the access will always default to the supervisor stack pointer. In order for the supervisor to access both of these registers, the instruction move from and move to USP differentiate between the user stack pointer and the supervisor stack pointer.

**Example 3.24. Accessing user stack pointer from supervisor mode**

```
MOVE.L  D5, A7  // copy D5 into A7

This instruction is valid for both user and supervisor states. In
```

user mode, the USP is loaded from D5, but in the supervisor mode, the
supervisor stack pointer (SSP) is loaded.

```
MOVE.L   D5, USP // copy D5 into the user stack pointer (USP)
```

This instruction is privileged and if attempted in the user mode the
core will transition into a privileged violation exception, and of
course when executed in the supervisor mode the user stack pointer
(USP) is accessed.

The hardware implementation of these two programmable-visible, 32-bit registers does
not uniquely identify one as the SSP and the other as the USP. Rather, the hardware uses
one 32-bit register as the currently-active A7, and the other register is named simply the
'other_A7.' Thus, the contents of the two hardware registers is a function of the operating
mode of the processor as determined by the supervisory bit of the core status register (SR).

### LEA - Load Effective Address

The LEA instruction is used to calculate an effective address based on the source addressing mode and transfer a 32 bit value into an address register.

**Example 3.25. Calculate <EA>**

```
LEA  <EA>, An         // calculate the source effective address and //
                      //loaded into An
Consider the effective address when register indirect with
scaled index and 8 bit displacement is used in accessing memory in a
program loop shown below.

MOVE.L   (4, A4, D6*4), D3  // load D3 from memory
```

Since the address has 3 elements in the source effective address, it
will take extra clock cycles to calculate the operand effective
address. The extra clock cycles become significant if this mode is
used in a loop with multiple iterations. It is much more efficient to
calculate the effective address and then use the MOVE instruction
with indirect addressing mode to reduce the time required to access
the operand as shown in the following example:

```
   LEA (4, A4, D6*4), A4    // calculate effective address & load it
                            // into A4
Loop: MOVE.L   (A4)+ D3        // load D3 from the calculated address
```

The MOVE.L instruction is a one 16 bit opcode and executes much
faster than the previous example.

## PEA - Push Effective Address

The PEA instruction also calculates an effective address based on the source addressing mode and pushes a 32 bit address onto active stack.

**Example 3.26.  Calculate and Push <EA> onto Stack**

```
PEA    (8, A2, D4*2)
```

```
The stack pointer is first decremented by 4 and the calculated address
is pushed to that location.
```

## *Multiply-Accumulate Unit*

The MAC design provides a set of DSP operations which can be used to improve the performance of embedded code while supporting the integer multiply instructions of the base-line ColdFire® architecture.

The MAC provides functionality in three related areas:

- Signed and unsigned integer multiplies
- Multiply-accumulate operations supporting signed and unsigned integer operands as well as signed, fixed-point, fractional operands
- Miscellaneous register operations

The ColdFire® family supports two MAC implementations with different performance levels and capabilities different silicon costs.

The original MAC uses a three-stage execution pipeline optimized for 16-bit operands and featuring a 16x16 multiply array with a single 32-bit accumulator. The EMAC features a four-stage pipeline optimized for 32 bit operands, with fully pipelined 32x32 multiply array and four 48 bit accumulators. Either can be attached to any of the ColdFire® version (V2, V3, or V4) as dedicated by application requirements.

The first ColdFire® MAC supported signed and unsigned integer operands and was optimized for 16x16 operations, such as those found in a variety of applications including servo control and image compression. As ColdFire® based systems proliferated, the desire for more precision on input operands increased. The result was an improved ColdFire® MAC with user-programmable control to optionally enable use of fractional input operands.

EMAC improvements target three primary areas:

- Improved performance of 32x32 multiply operations.

- Addition of three more accumulators to minimize MAC pipeline stalls caused by exchanges between the accumulator and the pipeline's general-purpose registers.

- A 48-bit accumulation data path to allow the use of a 40-bit product plus the addition of 8 extension bits to increase the dynamic number range when implementing signal processing algorithms.

The three areas of functionality are addressed in detail in following sections. The logic required to support this functionality is contained in a MAC module, as shown Figure 3.13

.

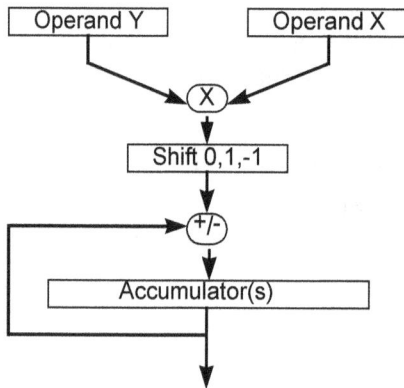

FIGURE 3.13. Multiply-Accumulate Operation

The MAC is an extension of the basic multiplier found in most microprocessors. It is typically implemented in hardware within an architecture and supports rapid execution of signal processing algorithms in fewer cycles than comparable non-MAC architectures. For example, small digital filters can tolerate some variance in an algorithm's execution time, but larger, more complicated algorithms such as orthogonal transforms may have more demanding speed requirements beyond the scope of any processor architecture and may require full DSP implementation.

Rather than having an expensive full digital signal processing engine that typically not needed for embedded applications, the ColdFire® architecture was intended mainly for low-speed signal processing.

To strike a balance between speed, size, and functionality, the ColdFire® MAC is optimized for a small set of operations that involve multiplication and cumulative additions. Specifically, the multiplier array is optimized for single-cycle pipelined operations with a possible accumulation after product generation. This functionality is common in many signal processing applications. The ColdFire® core architecture also has been modified to allow an operand to be fetched in parallel with a multiply, increasing overall performance for certain DSP operations. Consider a typical filtering operation where the filter is defined as in Figure 3.14.

Here, the output y(i) is determined by past output values and past input values. This is the

$$y(i) = \sum_{k=1}^{N-1} a(k)y(i-k) + \sum_{k=0}^{N-1} b(k)x(i-k)$$

FIGURE 3.14. **Infinite Impulse Response (IIR) Filter**

general form of an infinite impulse response (IIR) filter. A finite impulse response (FIR) filter can be obtained by setting coefficients a(k) to zero. In either case, the operations involved in computing such a filter are multiplies and product summing. To show this point, reduce the above equation to a simple, four-tap FIR filter, shown in Figure 3.15 in which the accumulated sum is a sum of past data values and coefficients.

$$y(i) = \sum_{k=0}^{3} b(k)x(i-k) = b(0)x(i) + b(1)x(i-1) + b(2)x(i-2) + b(3)x(i-3)$$

FIGURE 3.15. **Four-Tap FIR Filter**

**Multiply-Accumulate General Description**

The MAC speeds execution of ColdFire® integer multiply instructions (MULS and MULU) which also provides additional functionality for multiply-accumulate operations. By executing MULS and MULU in the MAC, execution times are minimized and deterministic compared to the 2-bit/cycle algorithm with early termination that the operand execution pipeline (OEP) normally uses if no MAC hardware is present.

The added MAC instructions to the ColdFire® ISA provide for the multiplication of two numbers, followed by the addition or subtraction of the product to or from the value in the accumulator. Optionally, the product may be shifted left or right by 1 bit before addition or subtraction. Hardware support for saturation arithmetic can be enabled to minimize software overhead when dealing with potential overflow conditions. Multiply-accumulate operations support 16- or 32-bit input operands of the following formats:

- Signed integers
- Unsigned integers
- Signed, fixed-point, fractional numbers

All arithmetic operations use register-based input operands, and summed values are stored internally in the accumulator. Thus, an additional move instruction is needed to store data in a general-purpose register. One new feature found in EMAC instructions is the ability to choose the upper or lower word of a register as a 16-bit input operand. This is useful in filtering operations if one data register is loaded with the input data and another is loaded with the coefficient. Two 16-bit multiply accumulates can be performed without fetching additional operands between instructions by alternating the word choice during the calculations.

## EMAC Programming Model

The EMAC has four accumulator registers versus the MAC's single accumulator. The additional registers improve the performance of some algorithms by minimizing pipeline stalls needed to store an accumulator value back to general-purpose registers. Many algorithms require multiple calculations on a given data set. By applying different accumulators to these calculations, it is often possible to store one accumulator without any stalls while performing operations involving a different destination accumulator.

The need to move large amounts of data presents an obstacle to obtaining high throughput rates in DSP engines. New and existing ColdFire® instructions can accommodate these requirements. A MOVEM instruction can move large blocks of data efficiently by generating line-sized burst references. The ability to simultaneously load an operand from memory into a register and execute a MAC instruction makes some DSP operations such as filtering and convolution more manageable.

The programming model includes a 16-bit mask register (MASK), which can optionally be used to generate an operand address during MAC + MOVE instructions. The application of this register with auto-increment addressing mode supports efficient implementation of circular data queues for memory operands.

The additional MAC status register (MACSR) contains a 4 bit operational mode field and condition flags. Operational mode bits control whether operands are signed or unsigned and whether they are treated as integers or fractions. These bits also control the overflow/ saturation mode and the way in which rounding is performed. Negative, zero, and multiple overflow condition flags are also provided.

The following is a list of MAC and EMAC instructions shown in Table 3.1 below:

TABLE 3.1 MAC Instructions

| *Command* | *Mnemonics* | *Description* |
|---|---|---|
| Multiply Signed | MULS <ea>y,Dx | Multiplies two signed operands yielding a signed result |
| Multiply Unsigned | MULU <ea>y,Dx | Multiplies two unsigned operands yielding an unsigned result |
| Multiply Accumulate | MAC Ry,RxSF,ACCx<br><br>MSAC Ry,RxSF,ACCx | Multiplies two operands and adds/ subtracts the product to/from an accumulator |
| Load Accumulator | MOV.L {Ry,#imm},ACCx | Loads an accumulator with a 32-bit operand |
| Store Accumulator | MOV.L ACCx,Rx | Writes the contents of an accumulator to a CPU register |
| Copy Accumulator | MOV.L ACCy,ACCx | Copies a 48-bit accumulator |
| Load Accumulator Extensions01 | MOV.L {Ry,#imm},ACCext01 | Loads the accumulator 0,1 extension bytes with a 32-bit operand |
| Load Accumulator Extensions23 | MOV.L {Ry,#imm},ACCext23 | Loads the accumulator 2,3 extension bytes with a 32-bit operand |
| Store AccExtensions01 | MOV.L ACCext01,Rx | Writes the contents of accumulator 0,1 extension bytes into a CPU register |
| Store AccExtensions23 | MOV.L ACCext23,Rx | Writes the contents of accumulator 2,3 extension bytes into a CPU register |
| Load MACSR | MOV.L {Ry,#imm},MACSR | Writes a value to MACSR |
| Store MACSR | MOV.L MACSR,R | Write the contents of MACSR to a CPU register |
| Store MACSR to CCR | MOV.L MACSR,CCR | Write the contents of MACSR to the CCR |

**TABLE 3.1 MAC Instructions**

| Load MAC Mask Register | MOV.L {Ry,#imm},MASK | Writes a value to the MASK register |
|---|---|---|
| Store MAC Mask Register | MOV.L MASK,Rx | Writes the contents of the MASK to a CPU register |

Thus, the 48-bit accumulator definition is a function of the EMAC operating mode. Given that each 48-bit accumulator is the concatenation of 16-bit accumulator extension register (ACCext$n$) contents and 32-bit ACC$n$ contents, the specific definitions are as follows:

if MACSR[6:5] = 00 /* signed integer mode */

Complete Accumulator[47:0] = {ACCext$n$[15:0], ACC$n$[31:0]}

if MACSR[6:5] = 01 or 11 /* signed fractional mode */

Complete Accumulator [47:0] = {ACCext$n$[15:8], ACC$n$[31:0], ACCext$n$[7:0]}

if MACSR[6:5] = 10 /* unsigned integer mode */

Complete Accumulator[47:0] = {ACCext$n$[15:0], ACC$n$[31:0]}

The four accumulators are represented as an array, ACC$n$, where $n$ selects the register. Although the multiplier array is implemented in a four-stage pipeline, all arithmetic MAC instructions have an effective issue rate of 1 cycle, regardless of input operand size or type. All arithmetic operations use register-based input operands, and summed values are stored internally in the accumulator. Thus, an additional move instruction is needed to store data in a general-purpose register. One new feature found in EMAC instructions is the ability to choose the upper or lower word of a register as a 16-bit input operand. This is useful in filtering operations if one data register is loaded with the input data and another is loaded with the coefficient. Two 16-bit multiply accumulates can be performed without fetching additional operands between instructions by alternating the word choice during the calculations.

The EMAC has four accumulator registers versus the MAC's single accumulator. The additional registers improve the performance of some algorithms by minimizing pipeline stalls needed to store an accumulator value back to general-purpose registers. Many algorithms require multiple calculations on a given data set. By applying different accumulators to these calculations, it is often possible to store accumulator without any stalls while performing operations involving a different destination accumulator.

The need to move large amounts of data presents an obstacle to obtaining high throughput rates in DSP engines. New and existing ColdFire® instructions can accommodate these requirements. A MOVEM instruction can move large blocks of data efficiently by generating line-sized burst references. The ability to simultaneously load an operand from memory into a register and execute a MAC instruction makes some DSP operations such as filtering and convolution more manageable.

The programming model includes a mask register (MASK), which can optionally be used to generate an operand address during MAC + MOVE instructions. The application of this register with auto-increment addressing mode supports efficient implementation of circular data queues for memory operands.

## Differences between the 680x0 and ColdFire® families

Although the ColdFire® architecture is closely related to the 680x0, there are many simplifications to the instruction set. Nearly all of the differences are omissions from the 680x0 instruction set and addressing modes. This means that (with a few important exceptions detailed below), a 680x0 instruction which is implemented in a ColdFire® core behaves in exactly the same way under the two architectures. In fact, almost all user-level (and much supervisor-level) ColdFire® code can be run unchanged on a 68020 or later 680x0 processor (apart from new instructions introduced in the Version 4 ColdFire® core). The converse, however, is not the case.

In outline, the main omissions fall into five categories:

- Missing addressing modes
- Missing instructions
- Non-availability of word- and byte-forms of nearly all arithmetic and logical instructions
- Many instructions act only on registers, not on memory
- Restrictions on available addressing modes for particular instructions
- Simplification of the supervisor-level programming model

### Principles behind the differences

In order to understand the ColdFire® instruction set in relation to that of the 680x0, it helps to have an appreciation of why the simplifications have been made. The philosophy behind the ColdFire® architecture is influenced by the success of RISC processors in providing high performance - for a given degree of chip complexity - by eliminating seldom-used instructions and complex addressing modes, and by regularizing the instruction set to make it easier for the hardware to optimize dispatch of the instruction stream.

However, standard RISC processors achieve high performance at the expense of low code density, in part because all instructions are the same width (generally 4 bytes) and also because only very simple addressing modes are available. In addition, RISC processors do not allow direct modification of memory locations; all memory reads and writes have to go via registers. This all means that programs compiled for RISC processors tend to be substantially larger than those compiled for CISC architectures such as the 680x0. This penalty does not greatly matter for desktop systems or servers with 32MB or more of RAM, but for embedded applications it can be a significant disadvantage, both in terms of system cost and power consumption.

The ColdFire® architecture - which Freescale Semiconductor characterizes as "Variable-Length RISC" - aims to share many of the speed advantages of RISC, without losing too much of the code density advantages of the 680x0 family. Like most modern processor architectures, it is optimized for code written in C or C++, and instructions which are not frequently generated by compilers are amongst those removed from the instruction set. Some of the complex addressing modes - again not important for compilers - are eliminated, and the additional hardware complexities involved in supporting arithmetic operations on bytes and words also disappear. In order to regularize the instruction stream, all ColdFire® instructions are either 1, 2 or 3 sixteen bit opcodes; this is why certain combinations of source and destination operands are not available.

### Missing addressing modes

The ColdFire® addressing modes are quite similar to those of the original 68000, i.e. without the extensions introduced in the 68020 and later processors, but with some differences in indexed addressing. Compared with a 68020 or later processor, the comparison is as follows:

### Fully supported:
- Data Register Direct
- Address Register Direct
- Address Register Indirect
- Post-increment
- Pre-decrement
- Displacement (16-bit displacement)
- PC Displacement (16-bit displacement)
- Absolute Short
- Absolute Long

- Immediate

**Partially supported:**

- Indexed
- PC Indexed

The restrictions on these two modes are:

- The displacement constant is 8-bit only;
- "Zero-suppressed" registers are not supported;
- The Index register can only be handled as a Long. Word-length index registers are not supported.
- The scale factor must be 1, 2, or 4. Scale factors of 8 are not supported.

**Not implemented at all:**

- Memory-indirect post-indexed
- Memory-indirect pre-indexed
- PC-indirect post-indexed
- PC-indirect pre-indexed

Note that further restrictions may be imposed on the addressing modes supported by particular instructions, even if a particular addressing mode is itself available on ColdFire® cores.

**Missing instructions**

A number of instructions are not implemented at all in the ColdFire® ISA. These include:

- DBcc, EXG, RTR, RTD, CMPM, ROL, ROR, ROXL, ROXR, MOVE16, ABCD, SBCD, NBCD, BFCHG, BFCLR, BFEXTS, BFEXTU, BFFFO, BFINS, BFSET, BFTST
- CALLM, RTM, PACK, UNPK
- CHK, CHK2, CMP2, CAS, CAS2, TAS *(restored in V4 core)*
- BKPT, BGND, LPSTOP, TBLU, TBLS, TBLUN, TBLSN
- TRAPV, TRAPcc, MOVEP, MOVES, RESET
- ORI to CCR, EORI to CCR, ANDI to CCR

In addition, DIVS and DIVU (with some differences from the 680x0 equivalents) are available on some ColdFire® processors but not others. MULU and MULS producing a 64-bit result are not implemented, but 16 x 16 producing 32-bit, and 32 x 32 producing (truncated) 32-bit, are available.

## Long-word forms only

Most arithmetic and logical instructions act on Long words only. This applies to:

- ADD, ADDA, ADDI, ADDQ, ADDX, AND, ANDI, ASL, ASR
- CMP, CMPI *(word/byte forms re-introduced in version 4 core)*
- CMPA, EOR, EORI, LSL, LSR,
- NEG, NEGX, NOT, OR, ORI
- SUB, SUBA, SUBI, SUBQ, SUBX

MOVEM.W has also been removed from the instruction set.

In fact, the only instructions which do act on the full set of byte, word and long operands are CLR, MOVE and TST (and CMP and CMPI in the version 4 core). EXT.W, EXTB.L and EXT.L survive, as do MULx.W and MULx.L

## Instructions which act only on registers, not on memory

Some arithmetic instructions cannot act directly on memory - the destination must be a register. This applies to:

- ADDI, ADDX, ANDI, CMPI, ASL, ASR, LSL, LSR,

*Note: ADDQ and SUBQ can act directly on memory.*

- NEG, NEGX, NOT, EORI, ORI, SUBI, SUBX, Scc

## Restrictions on addressing modes for particular instructions

Even where a particular memory addressing mode does exist in a ColdFire® instruction, some instructions are subject to further restrictions. Often, this is because of the limit of six bytes as the maximum length of a single instruction. Specific restrictions include:

- Some combinations of addressing modes for MOVE are disallowed. If the source addressing mode is Displacement or PC Displacement, the destination addressing mode cannot be Indexed or Absolute. If the source addressing mode is Indexed, PC-Indexed, Absolute or Immediate, the destination addressing mode cannot be Indexed, Displacement, or Absolute.

- The addressing modes for MOVEM are restricted to only Displacement and Indexed - no Pre-decrement or Post-increment!

- For BTST, BSET, BCLR and BCHG, if the source operand is a static bit number, the destination cannot be Indexed or Absolute memory.

## Miscellaneous Omissions

There are a few miscellaneous omissions for specific instructions:

- LINK.L is not supported
- MOVE to CCR/SR: Source must be Immediate or Data Register
- MOVE from CCR/SR: Destination must be data register
- BSR and Bcc accept only an 8- or 16-bit displacement in version 2 and version 3 cores (32-bit displacements are reintroduced in version 4)

## Instructions which behave differently from the 680x0 equivalent

In most cases, an instruction/addressing mode which does exist in the ColdFire® ISA behaves exactly like its 680x0 equivalent, which makes it easy for experienced 680x0 programmers to understand ColdFire® code. It also means that user-mode code written for a ColdFire® core can generally run unchanged on a 680x0 processor, provided the new ColdFire® only instructions are not used.

However, there are a few subtle cases where the ColdFire® instruction is not exactly the same as its 680x0 counterpart. The most important of these is that multiply instructions (MULU and MULS) do not set the overflow bit. This means that a 680x0 code sequence which checks for overflow on multiply may assemble and run on a ColdFire® device, but give incorrect results.

ASL and ASR also differ in that they do not set the overflow bit - but this is less likely to cause problems for real programs!

## Simplification of the supervisor programming model

Various members of the 68000 family have different register sets available at the supervisor level. The most important simplification in the ColdFire® supervisor-level model is that on some of the early ColdFire® V2 processors have only one stack pointer, shared for all code including interrupts, supervisor-level services, and user code. It follows from this that, on a ColdFire® architecture, it is never safe to write below the stack, since any interrupt which occurs would overwrite the stored data. (Writing below the stack, though not recommended, is possible in some 680x0 systems in user mode, because interrupts cause a switch to the Interrupt or Supervisor Stack Pointer). A further issue is that ColdFire® processors automatically align the stack to a four-byte boundary when an exception occurs, which can cause problems if code is reading or writing at a fixed offset from the stack pointer. In fact, it is strongly recommended (for performance reasons) that the ColdFire® stack should be kept long-word aligned at all times.

## *ColdFire® Instruction Set Architecture Enhancements*

This section describes the new opcodes implemented as part of the Revision A+ enhancements to the basic ColdFire® ISA.

**BITREV**                                              **Reverse Register**

(Supported Starting with ISA A+)

**Operation:**               Bit Reversed Dx Æ Dx

**Assembler Syntax:**      BITREV.L Dx

**Attributes:**            Size = longword

The contents of the destination data register are bit-reversed; that is, new Dx[31] = old Dx[0], new Dx[30] = old Dx[1], ..., new Dx[0] = old Dx[31].

**Condition Codes:**      Not affected

**BYTEREV**                                **Byte Reverse Register**

(Supported Starting with ISA A+)

**Operation:**               Byte Reversed  Dx

**Assembler Syntax:**      BYTEREV.L Dx

**Attributes:**            Size = longword

The contents of the destination data register are byte-reversed as defined below:

| | |
|---|---|
| new Dx[31:24] | = old Dx[7:0] |
| new Dx[23:16] | = old Dx[15:8] |
| new Dx[15:8] | = old Dx[23:16] |
| new Dx[7:0] | = old Dx[31:24] |

**Condition Codes:**      Not affected

**FF1**                                    **Find First One in Register**
(Supported Starting with ISA A+)

**Operation:** Bit Offset of the First Logical One in Register Æ Destination

**Assembler Syntax:**          FF1.L Dx

**Attributes:**                Size = longword

The data register, Dx, is scanned, beginning from the most-significant bit (Dx[31]) and ending with the least-significant bit (Dx[0]), searching for the first set bit. The data register is then loaded with the offset count from bit 31 where the first set bit appears, as shown below. If the source data is zero, then an offset of 32 is returned.

| Old Dx[31:0] | New Dx[31:0] |
|:---:|:---:|
| 0b1---- . . . ---- | 0x0000 0000 |
| 0b01--- . . . ---- | 0x0000 0001 |
| 0b001-- . . . ---- | 0x0000 0002 |
| ... | ... |
| 0b00000 . . . 0010 | 0x0000 001E |
| 0b00000 . . . 0001 | 0x0000 001F |
| 0b00000 . . . 0000 | 0x0000 0020 |

Condition Codes:

| X | N | Z | V | C |
|:---:|:---:|:---:|:---:|:---:|
| — | * | * | 0 | 0 |

X   Not affected
N   Set if the msb of the source operand is set; cleared otherwise
Z   Set if the source operand is zero; cleared otherwise
V   Always cleared
C   Always cleared

**STLDSR**                                    **Store/Load Status Register**
(Supported Starting with ISA A+)

**Operation:** If Supervisor State

Then SP - 4 Æ SP; zero-filled SR Æ (SP); immediate data Æ SR
Else TRAP

**Assembler Syntax:** STLDSR #<data>

**Attributes:**              Size = word

**Description:** Pushes the contents of the Status Register onto the stack and then loads the Status Register with the immediate data value. This instruction is intended for use as the first instruction of an interrupt service routine to permit raising the current interrupt to a higher priority to prevent preemption. It allows the level of the just-taken interrupt request to be stored in memory (using the core status register interrupt mask $[I_2:I_0]$ field), and then masks interrupts by loading the SR[IML] field with 0x7 (if desired).

If execution is attempted with the supervisor (S) bit of the immediate data cleared (attempting to place the processor in user mode), a privilege violation exception is generated. The opcode for STLDSR is 0x40E746FC.

Condition Code Register

| X | N | Z | V | C |
|---|---|---|---|---|
| * | * | * | * | * |

X  Set to the value of bit 4 of the immediate operand
N  Set to the value of bit 3 of the immediate operand
Z  Set to the value of bit 2 of the immediate operand
V  Set to the value of bit 1 of the immediate operand
C  Set to the value of bit 0 of the immediate operand

## Introduction to ColdFire V1 Exceptions

A system exception can be thought of as a deviation from normal processing which can occur when there is a system error, an interrupt request from an I/O device or when a trap instruction is executed. The CPU reacts to these conditions by suspending main program execution, saving the machine state information on the supervisor stack and then transitions to the appropriate service routine to handle the exceptional condition. Once the CPU handles the event causing the exception or the interrupt, the machine state is restored and the application program resumes.

The ColdFire processor have numerous events that cause system exceptions. Each of these events will be covered in details later in this chapter. But first let's cover the activity performed by the core once an exception condition is detected and acknowledged.

### Exception Processing Sequence

When an exception or an interrupt is detected and taken, the ColdFire core transitions from normal execution state to exception processing state. There are four main steps performed by the ColdFire core. These steps are referred to as exception processing state which do not include the execution of exception or the interrupt service routine. Each of the 4 steps is explained below:

4. The ColdFire core copies the current value of the 16 bit core status register (SR) into a temporary location within the core so it can be saved on the system stack later without losing any machine state information. It then sets the S-bit and clears the T-bit in the status register. Setting the S-bit and clearing the T-bit transitions the core into the supervisor or privileged mode and disables tracing, respectively. If the exception condition is an interrupt request, the master/interrupt M-bit will also be cleared and the interrupt mask in the core status register will be set to the current interrupt level being acknowledged.

5. Next, the ColdFire determines the exception vector number based on the exception or the interrupt and then calculates the vector address by multiplying the vector number by 4 and adding it to the vector base register (VBR). Recall that by default the VBR is cleared at reset, thus the vector table is located at address 0x0. The vector table can be relocated to any of 16 one mega byte boundaries since only bits 20-23 of the VBR are used to determine the vector table location. The vector number generated by the processor is based on the exception type and will be used as an offset into the exception vector table. For interrupt exceptions, the ColdFire performs an interrupt acknowledge (IACK) bus cycle to obtain the vector number from the I/O peripheral device requesting service. The IACK cycle is mapped into a special acknowledge address space with the interrupt level encoded in address bits 4:2. More on interrupts later in this chapter.

6. In this step, the ColdFire saves the state of the machine by creating an exception stack frame on the supervisor stack and pushes the PC and the saved value of the SR. The exception type determines whether the program counter saved in the exception stack frame defines the location of the faulty instruction or the address of the next instruction to be executed. As a result of an exception or an interrupt condition, a stack frame is created at a 0-modulo-4 address on the top of the current supervisor stack area.

7. Once the machine state information is saved onto the stack, the CPU fetches the exception vector from the table located at the address defined in the vector base register. The vector fetched is loaded into the PC which points to the address of the first instruction in the interrupt handler. After the first instruction is executed, exception processing terminates and normal instruction execution commences in the exception or interrupt service routine.

## Exception Vector Table

The exception vector table (VCT) contains 256 entries, entry number 0x0 - 0x3F or the first 64 vectors are reserved for ColdFire processors as shown in Table 4. Entry number 0x40 - 0xFF, or the last 192 vectors are reserved for the user and can be assigned[3] on as needed basis to I/O peripherals. Once a vector is assigned to an I/O device, the device will supply its vector number when the core acknowledges its interrupt request. The core executes an interrupt acknowledge bus (IACK) cycle to obtain the 8 bit vector number to identify the appropriate I/O device interrupt service routine (ISR). The vector number is either generated by the core when an exception event is detected or supplied by the I/O device when an interrupt request is acknowledged. Since the vector number is in the range from 0x0 - 0xFF (0 - $255_{10}$), ColdFire processors multiply this field by 4 to generate a vector offset aligned on a long word boundary. The vector offset is then used as index value from the base to select the vector table entry needed by the core to process the exception or interrupt event.

TABLE 4. Exception Vector Assignments

| Vector Number(s) | Vector Offset (Hex) | Stacked PC | Assignment |
|---|---|---|---|
| 0 | 0x000 | - | Initial Supervisor SP |
| 1 | 0x004 | - | Initial value of PC |
| 2 | 0x008 | Fault | Access Error |

3. On V1 cores, interrupt vectors are pre-assigned at the chip level and can not be re-assigned. However, the interrupt controller provides a mechanism to elevate 2 requests to the highest maskable priority. Priority elevation will be discussed in more details later in this chapter.

**Chapter 3**

**TABLE 4. Exception Vector Assignments**

| Vector Number(s) | Vector Offset (Hex) | Stacked PC | Assignment |
|---|---|---|---|
| 3 | 0x00C | Fault | Access Error |
| 4 | 0x010 | Fault | Illegal Instruction |
| 5-7 | 0x014-0x01C | - | Reserved |
| 8 | 0x020 | Fault | Privilege Violation |
| 9 | 0x024 | Next | Trace |
| 10 | 0x028 | Fault | Unimplemented Line-A Opcode |
| 11 | 0x02C | Fault | Unimplemented Line-F Opcode |
| 12 | 0x030 | Next | Debug Interrupt |
| 13 | 0x034 | - | Reserved |
| 14 | 0x038 | Fault | Format Error |
| 15-23 | 0x03C-0x05C | - | Reserved |
| 24 | 0x060 | Next | Spurious Interrupt |
| 25-31 | 0x064-0x07C | - | Reserved |
| 32-47 | 0x080-0x0BC | Next | Trap # 0-15 Instructions |
| 48-60 | 0xC0-0x0F0 | - | Reserved |
| 61 | 0x0F4 | Fault | Unsupported Instructions |
| 62-63 | 0x0F8-0xFC | - | Reserved |
| 64-95 | 0x100-0x17C | Next | Device Specific Interrupts |
| 96-102 | 0x180-ox198 | Next | Level 1-7 Software Interrupts |
| 103-255 | 0x19C-0x3FC | - | Reserved, Unused for V1 |

In general, vectors 0x40 - 0xFF or (64 - $255_{10}$) of the exception vector table are reserved for the user and in particular, vector 64 - 102 are reserved for I/O peripheral requests and seven software interrupts. Vector number 103 - 255 are currently reserved and not used. As noted earlier, the first 64 exception vectors are reserved for the processor to handle reset and error conditions such as access errors, address errors, system calls (Trap Instructions), etc. When an interrupt request is acknowledged by the core, the vector number of the interrupt source is retrieved. Once retrieved, the processor creates a stack frame in memory pointed by the SSP before it begins execution of the exception or the interrupt handler.

For ColdFire processors, all exception stack frames are two longwords in length containing the original values of the SR and PC. After the status register is saved onto the supervisor stack, the $SR_{[I2, I1, I0]}$ mask field is set to the level of the interrupt request being acknowledged, effectively masking that level and all lower levels while in the service routine. The PC value saved onto the stack is either the address of the instruction that caused the exception or the address of the next instruction that would have executed if the program was not interrupted, depending on whether the exception is an error or interrupt condition, respectively. After the exception stack frame is created in memory, the processor obtains the exception vector number and uses as an offset into the vector table, it then branches to that address and begins execution of the service routine.

For some peripheral devices, the processing of the IACK cycle directly negates the interrupt request, while other devices require that request be explicitly negated by the core during service routine processing.

The processing of the interrupt acknowledge cycle on the V1 core is fundamentally different than previous ColdFire processors. In the new approach, all IACK cycles are directly handled by the interrupt controller (INTC), so the requesting peripheral device is not accessed during the IACK cycle. As a result, the interrupt request must explicitly be cleared in the peripheral by software during interrupt servicing.

The exception vector table (VCT) is relocatable only on 1 mega byte boundary because the lower 20 bits of the VBR are hardwired to zero. Also, since the ColdFire V1 cores support only 16 MByte address space, the top two hexidecimal digits (bits 31-24) are not used and assumed to be zero. This leaves bits 23-20 of the VBR as the vector table base address.

At start up, the VBR is initialized out of reset to 0x0000_0000 which effectively relocates the VCT to address 0x0. This means that the reset vector will always be fetched from address 0x0 and address 0x4. Reset uses 2 vectors, the first initializes the supervisor stack (SSP), and the second loads the PC with the reset routine start address. Since the VBR register can be written anytime while in the supervisor mode, the application software can relocate the vector table from the default position in Flash memory at address 0x0 to system RAM at address 0x80_0000. The VBR provides a convenient mechanism that enables a ColdFire based system to have multiple vector tables by simply changing the VBR value. Having multiple vector tables may be useful during debug and development stages to support different set of routines according to system requirements.

## *ColdFire Exception Events*

There are many exceptions defined by the ColdFire architecture, these exceptions will be detailed in the following paragraphs in the order shown in the vector table above. Prior to handling the event causing the exception, the ColdFire transitions into exception processing. The sequence executed by the core upon entry into exception processing consists of the 4 steps outlined above. We begin our coverage of each exception starting from the highest priority down to the lowest.

### Reset Exception (Vector # [0:1]

ColdFire V1 processor has numerous conditions that causes it to enter the reset state. Reset is entered due to one or more of the following events: power-on (POR), assertion of the external reset pin, computer operating properly (COP) time-out, illegal address (ILAD), illegal opcode (ILOP), and low voltage detect. Furthermore, an illegal address reset may also be caused by an access error, format error or fault-on-fault condition. Whereas, the ILOP reset may occur due to an attempt to execute illegal instruction, line-A opcode, line-F opcode or privilege violation.

The ILAD and ILOP events cause the ColdFire V1 core to reset only if the appropriate CPU configuration register address-related reset disable (ARD) and instruction-related disable (IRD) control bits are cleared. More information is provided on exception errors later in this chapter.

This exception is highest priority and is entered whenever the reset signal is asserted. The recognition of the reset signal causes the core to enter the supervisor mode by asserting the S- bit in the core status register (SR). As previously stated, two exception vectors are used for reset, the first is loaded into the stack pointer (SSP), and the second is loaded into the program counter (PC). These two vector are fetched from address 0x0 and 0x4 because the VBR is always cleared on a reset. Once the SSP and PC are loaded from the vector table, the reset handler is started and executed entirely in the supervisor mode. After determining the cause of reset by examining the reset status register (RSR), the core initializes the MCU as required by the target application.

Typically, the last few steps in the reset handler are used to dispatch the application task for execution. Since the SSP has been loaded at reset automatically by the hardware, and is pointing to supervisor stack area in RAM, the core can push the task entry (PC value) and the task state (SR) onto the stack and then execute return-from-exception (RTE) instruction. The RTE instruction is normally the last instruction in the reset handler which pops the task state into the SR and the entry point into the PC, causing normal program execution at the location pointed by the PC.

Reset is the only exception that does not save any state information or pushes anything on the stack. In other words, there is no recovery back to the main program when a reset is initiated since the machine state information is completely lost.

A few things should be kept in mind on the initial state of the core when a reset occurs. From the previous section of this chapter, we learned that the supervisory (S) bit and the master stack pointer (M) bit of the SR, are respectively set and cleared upon system reset. It is also important to note that the interrupt mask level $[I_2, I_1, I_0]$ bit field is set to the highest level, level 7. This means that while in the reset handler all interrupts are masked except level 7 because it is non-maskable.

Power-on and external resets are most common and considered normal. The other reset sources are generated internally due to some error or fault condition. The ColdFire core records type of the reset condition in the reset status register (RSR) permitting the applica-

**TABLE 5. Reset Status Register (Reset Sources)**

| Reset Source | Reset Condition Description |
|---|---|
| (POR) Bit 7 | Power-On reset was caused by a transition on VDD pin. The low-voltage reset (LVD) status bit is also set to indicate that the reset occurred while the internal supply was below the LVD threshold. |
| PIN-Bit 6 | Reset was caused by an active-low level on the external reset Pin. |
| COP-Bit 5 | Computer Operating Properly (COP) Watchdog. Reset was caused by the COP watchdog timer timing out. This reset source is blocked if COPE is cleared in the system options (SOPT) register. |
| ILOP-Bit 4<br><br>Illegal Opcode[a] | Reset was caused by an attempt to execute an unimplemented line-A opcode, line-F opcode, illegal[b] instruction or a privilege violation (execution of a supervisor instruction in user mode).<br><br>All illegal opcode resets are enabled when CPUCR[IRD] is cleared. If CPUCR[IRD] is set, then the appropriate processor exception is generated instead of a reset. |
| ILAD Bit 3 | Reset was caused by the processor attempted access of an illegal address, an address error, an RTE format error, or a fault-on-fault condition. All the illegal address resets are enabled when CPUCR[ARD] is cleared. When CPUCR[ARD] is set, the appropriate processor exception is generated instead of a reset, or if a fault-on-fault condition is reached, the processor simply halts. |
| LVD Bit 1 | If LVDRE is set and the supply drops below the LVD trip voltage, an LVD reset occurs. This bit is also set by POR. |
| Bit 2 and 0 | Reserved |

a. The STOP instruction is considered illegal if [STOPE or WAITE] control bits are cleared in the system option (SOPT) register.

b. This includes any illegal instruction (except the ILLEGAL (0x4AFC) opcode).

tion program to figure out the exact nature of the error or the fault, if desired. Table 5 depicts the RSR bit fields and provides a brief description of each reset event. More information is provided in the next paragraphs that explain the details of all internally generated resets.

**Access Error (vector # 2)**

This fault forces the ColdFire V1 processor to generate a system reset since the CPU configuration (CPUCR) register address-related reset disable (ARD) bit is cleared by default out of reset. An access error reset condition sets the illegal address status (ILAD) bit in the reset status (RSR). The setting of the ILAD status bit may be used to aide system designers during debug and development to determine the cause of reset. However, if CPUCR[ARD] control bit is set, the reset is disabled and a processor exception is taken instead, as detailed below:

An access error is sometimes referred as a bus error exception that may occur due to accessing a memory region illegally. An example of access error is when attempting to access a non-resident memory location, or when attempting to write to a block of memory region that has write protection attribute. The ColdFire behavior to an access error depends on the type of memory access being executed. For an instruction fetch, the core defers transition into the error routine until the instruction causing the exception is attempted for execution. Thus, faults which occur during instruction prefetches which are followed by a change of instruction flow will not generate an access error. If the CPU attempts to execute an instruction with a faulty address, an access error will be signaled, the instruction is aborted immediately and an access error exception is taken. For this type of exception, the programming model has not been altered by the instruction generating the access error.

If the access error occurs on an operand read, the ColdFire immediately aborts the current instruction's execution and initiates exception processing. In this situation, any address register updates due to the auto-addressing modes, {e.g., (An)+, -(An)}, will already have been performed and the address register will contain the updated value when the access error exception is taken. If the access error is generated during load multiple register operation with the MOVEM instruction, any of the registers that have been loaded prior to detecting the access error will have been updated with the operands fetched from memory.

For access errors on writes, ColdFire processors use an imprecise reporting mechanism. Since the actual write cycle may be decoupled from the execution of the instruction requesting the write, the signaling of an access error appears to be decoupled from the instruction which generated the write cycle. Accordingly, the PC value pushed onto the exception stack frame merely represents the location in the program when the access error was signaled, not when the offending instruction was executed. This may be an issue during development and debugging stages since its hard to figure out the exact operand faulty address. To resolve this problem, the NOP instruction can be used for purposes of collecting access errors for writes. The NOP is a memory synchronization instruction that delays the execution of the next instruction until all previous operations, including all pending writes to memory resources are completed.

To further aid system designers during debug and development, the ColdFire creates a stack frame with the access error fault status to indicates the exact nature of the fault. Table 6 shows the faulty access status encoding pushed onto the stack.

**TABLE 6. Fault Status Bit Definition** [a]

| FS [3:0] | Definition |
|---|---|
| 0100 | Error on instruction fetch |
| 1000 | Error on operand write |
| 1001 | Attempt to write to write protected memory space |
| 1100 | Error on operand read |
| All others | Reserved |

   a. Refer to Figure 1 for more information on exception stack frame.

When an access error is generated or detected, the core branches to access error handler via vector # 2 at address VBR + 8.

**Address Error (Vector #3)**

By default, an address error forces the ColdFire V1 processor to generate a system reset since the CPU configuration (CPUCR) register address-related reset disable (ARD) bit is cleared at reset. An access error sets the illegal address status (ILAD) bit in the reset status (RSR). The setting of the ILAD status bit in the RSR may be used to aide system designers during debug and development to determine the cause of the fault condition. If CPUCR[ARD] control bit is set, the reset is disabled and a processor exception is taken instead, as detailed below:

Address error exception occurs when an attempt is made to access program space at misaligned address. For example, a branch or a jump that attempts to fetch instructions at odd addresses will result in an access error, (i.e., if bit 0 of the target address is set to 1). For conditional branch instructions, the exception is generated regardless of the taken/not-taken resolution of the branch condition.

                                                                    **Chapter 3**

Also, any attempted use of a word-sized index register (Xn.w) or an invalid scale factor on an indexed effective addressing mode generates an address error. Accessing operands on misaligned addresses are permitted in the ColdFire architecture and will not generate an address error.

An address error event causes ColdFire processors branch to the exception handler via vector number 3 at address VBR+0x00C.

### Illegal Instruction Exception (Vector # 4)

By default, an illegal instruction forces the ColdFire V1 processor to generate a system reset since the CPU configuration (CPUCR) register instruction-related reset disable (IRD) bit is cleared at reset. An illegal instruction reset sets the illegal opcode (ILOP) bit in the reset status (RSR). The setting of the ILOP status bit in the RSR may be used to aide system designers during development and debug to determine the cause of the fault condition. If CPUCR[IRD] control bit is set, the reset is disabled and forces the processor to take illegal instruction instead, as detailed below:

ColdFire processors do not decode every illegal instruction and the core behavior will be unpredictable if an attempt was made to execute an instruction that is not defined by instruction set architecture (ISA). Only opcode 0x4AFC is recognized as illegal instructions and always generates an illegal instruction exception, regardless of the CPUCR[IRD]. This instruction may be used during the development stage for the purpose of software breakpoints or instruction emulation.

An Illegal instruction causes the CPU to vector to the exception handler via vector # 4 at address VBR+0x010.

### Privilege Violation (Vector #8)

By default, a privilege violation forces the ColdFire V1 processor to generate a system reset since the CPU configuration (CPUCR) register instruction-related reset disable (IRD) bit is cleared at reset. A privilege violation reset condition sets the illegal opcode (ILOP) bit in the reset status (RSR). The setting of the ILOP status bit in the RSR may be used to aide system designers during debug and development to determine the cause of the fault. If CPUCR[IRD] control bit is set, the reset is disabled and a privilege violation exception is taken instead, as detailed below:

ColdFire processors execute code in either supervisor or user modes as determined by the S- bit of the core SR. For the sake of system security, only the supervisor can access critical resources and executes privileged instructions. Any attempt by a user level task to access these resources or execute any of the privileged instructions will signal a privilege violation exception to the core. Instructions that cause privilege violation are listed below:

- STOP - Load SR with immediate data and stop the processor core.
- RTE - Restore processor state from the system stack and return from exception.
- MOVE from and MOVE to SR - Read or write the core status register.
- STLDSR - Push core SR on the stack and loads it with a new value.
- MOVEC - Move the contents of a general purpose register to a control register.
- MOVE USP - Read or write USP. User level tasks should use MOVE.L A7 instead.
- HALT - Halt the processor core.
- WDEBUG - Write debug control register
- CPUSHL - Cache push line and possibly invalidate (This instruction is currently not implemented on CFV1 core).

The privilege violation exception handler is invoked via vector #8 at address VBR + 20.

### Trace Exception (Vector #9)

To aid in program development, the ColdFire cores provide an instruction-by-instruction tracing capability when the trace (T) bit is set in the core SR. While in trace mode, the completion of each instruction execution triggers a trace exception to allow debuggers to monitor program execution.

The core vectors to the trace exception handler via vector # 9 at address VBR+0x024. For additional information, refer to Chapter 8.

### Unimplemented Line-A and Line-F Opcodes (Vector # 0xA & 0xB)

By default, an attempted execution of line-A or line-F opcodes forces the ColdFire V1 processor to generate a system reset since the CPU configuration (CPUCR) register Instruction-related reset disable (IRD) bit is cleared at reset. A line-A or line-F reset condition sets the illegal opcode (ILOP) bit in the reset status register (RSR). The setting of the ILOP status bit in the RSR may be used to aide system designers during debug and development to determine the cause of the fault. If CPUCR[IRD] control bit is set, the reset is disabled causing the processor to take either line-A or line-F exception instead, as detailed below:

When attempting to execute line-A opcodes not used by the MAC unit, and line-F opcodes not used by debug unit, an unimplemented opcode exception is taken. The unimplemented line-A and line-F exception handlers are invoked via vector number 0xA and 0xB, at address VBR+0x028 and VBR+0x02C, respectively.

### Debug Interrupt (Vector # 0xC)

This exception is another mechanism implemented in ColdFire processors as an aid during system debug and development. It is initiated by a hardware breakpoint trigger when detected by the breakpoint module. More information is available in Chapter 8.

### Format Errors (Vector #0xE)

By default, a format error forces the ColdFire V1 processor to generate a system reset since the CPU configuration (CPUCR) register instruction-related reset disable (IRD) bit is cleared at reset. A format error condition sets the illegal opcode (ILOP) bit in the reset status (RSR). The setting of the ILOP status bit may be used to aide system designers during debug and development to determine the cause of the fault. If CPUCR[IRD] control bit is set, the reset is disabled and a processor format error exception is taken instead, as detailed below:

When a return-from-exception (RTE) instruction is executed to exit the handler, the Cold-Fire core examines the 4 bit format field in the exception stack frame to validate the frame type. Any attempted execution of an RTE instruction when the format is not equal to 0x4, 0x5, 0x6, or 0x7 will generate a format error. depicts the stack frame format for all exceptions. The exception stack frame for the format error is created without disturbing the original stack frame. The PC field in the new exception stack frame will point to the RTE instruction. Table 7 shows the stack format codes to indicate the SSP value before the exception was taken. Upon a return from exception, the SSP is adjusted according to the stored format value on the stack.

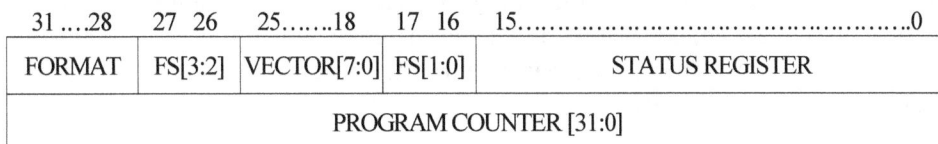

| 31 ....28 | 27  26 | 25.......18 | 17  16 | 15.................................................................0 |
|-----------|--------|-------------|--------|------------------------------------------------------------------|
| FORMAT | FS[3:2] | VECTOR[7:0] | FS[1:0] | STATUS REGISTER |
| PROGRAM COUNTER [31:0] | | | | |

FIGURE 3.16. **Exception Stack Frame**

If the format field on the stack defines a valid type, the ColdFire will:

1. restore SR
2. fetch the return address into PC
3. adjust the stack pointer by adding the format value to the auto-incremented address after the fetch of the first long word, and finally,
4. transfers control to the instruction address defined by the PC (fetched in step 2).

RTE format error exception handler is invoked with vector #0xE at address VBR+0x038.

**TABLE 7.  Exception Stack Format Encoding**

| Original SSP @ Time of Exception. SSP Bits [1:0] | SSP Value @ Start of Exception Handler | Format Field Based on Original Value of SSP |
|:---:|:---:|:---:|
| 00 | Original SP - 8 | 4 |
| 01 | Original SP - 9 | 5 |
| 10 | Original SP - 10 | 6 |
| 11 | Original SP - 11 | 7 |

### Trap #0 - #15 Instructions (Vectors #0x20 -0x2F)

Trap instructions can be thought of as software interrupts and may be used to invoke system calls to the operating system (OS). System calls are special routines defined by the OS and are used to provide services to application tasks. ColdFire processors have 16 trap instructions numbered #0 - #15. Each trap instruction has its own unique vector to allow multiple calls to different OS functions.

### Spurious Interrupt (Vector #0x018)

When an interrupt request is acknowledged, the core executes an IACK bus cycle to retrieve the vector number from the requester. If the IACK cycle terminates with an access error, the CPU will vector to the spurious interrupt handler via vector number 0x18. A spurious interrupt event implies that the I/O device asserted an interrupt request and during the acknowledge cycle the request was withdrawn. This condition may indicate some sort of error in the I/O device behavior.

### Unsupported Instructions (Vector #0x03D)

Some operations in the ColdFire instructions set architecture (ISA) are defined as legal but currently not supported if the required hardware is not present in the core. If unsupported instruction is attempted, the core will transition into unsupported exception via vector # 0x3D. If desired, the handler may emulate the unsupported instructions in software.

**Fault-on-Fault**

If for any reason the ColdFire processor encounters any type of a fault during the exception processing of another fault, it immediately halts execution and signals a fault-on-fault condition. This condition is indicated by continuously driving the processor status (PST[3:0]) signals with an encoded value of 0xF. The fault-on-fault is entered if while the processor is trying to handle an access error during exception processing of another access error or address error occurs, for instance.

By default, a fault-on-fault condition forces the ColdFire V1 processor to generate a system reset since the CPU configuration (CPUCR) register address-related reset disable (ARD) bit is cleared at reset. If CPUCR[ARD] control bit is set, the reset is disabled and a processor is halted. The only way to exit the halted condition is to assert a system reset.

**Hardware Configuration**

Some ColdFire processors may integrate additional hardware such as a multiply-accumulate (MAC) and/or EMAC unit, a memory management unit (MMU), and possibly a floating point unit (FPU), etc. To enable the application software to determine the exact configuration of the MCU, any of the reset events discussed above causes the processor to load two longwords of hardware configuration information into general purpose registers D0 and D1.

The following hardware configuration fields are loaded into general purpose data register D0. The configuration shown in Table 8 are MCF51QE128 device specifics.

TABLE 8. Hardware Configuration Loaded into D0

| Field in D0 | Bit [s] | Value | Meaning |
|---|---|---|---|
| PF | [31-24] | 0xCF | ColdFire Core is present |
| VER | [23-20] | 0001 | ColdFire Version 1 (CFV1) |
| MAC | [15] | 0 | MAC unit is not present on this core |
| DIV | [14] | 0 | Divide Engine is not present on this core |
| EMAC | [13] | 0 | Enhanced MAC engine is not present on this core |
| FPU | [12] | 0 | Floating Point Unit in present on this core |
| MMU | [11] | 0 | No Memory Management Unit present on this core |

**TABLE 8. Hardware Configuration Loaded into D0**

| Field in D0 | Bit [s] | Value | Meaning |
|---|---|---|---|
| ISA | [7-4] | 0x2 | Instruction Set Architecture Revision C on this core |
| Debug | [3-0] | 0x1 | Debug B+ Revision implemented on this core |
| Reserved | [10-8] | 0 | Not used |

The second longword containing the other hardware configuration information is loaded into general purpose register D1. This information is used to indicate the amount of memory available on this device as shown in Table 9 below:

**TABLE 9. Hardware Configuration loaded into D1**

| Fields | Bits | Field Description |
|---|---|---|
| Flash Size | [23:20] | 128K integrated on this device |
| Ram Size | [7:4] | 8K Bytes integrated on this device |
| Other fields | Not used | Reserved |

## Interrupts

Interrupts are signals generated from on-chip I/O peripherals, external I/O devices and or input signals. These signals are used to request services from the core when an I/O peripheral devices have events requiring CPU attention. An input device may need the core to read its input data, and an output device may need additional data, for instance.

When a request is signaled by an I/O device, the CPU will suspend program execution upon the completion of the current instruction, assuming the request is not masked. Masked interrupts will not be serviced until they become unmasked. When an interrupt source requests service and currently is masked, it simply becomes pending until it is unmasked, and only then will it be serviced by the CPU. Typically, each interrupt request must eventually be serviced by the core in order not to lose any I/O events or data.

Often, the CPU may be servicing a higher priority interrupt when a lower priority interrupt is signaled. In this case, the lower priority interrupt simply become pending and will be serviced as soon as all higher priority requests are serviced. The priority scheme provides a method to service events based on their importance, so that critical events will be processed before non-critical ones.

ColdFire V1 processor have 7 interrupt levels and each interrupt source implemented on this architecture is pre-assigned to a level between 1 and 7. Level 1 is the lowest and level 7 is the highest.

As noted earlier, the current interrupt mask level is indicated by the $[I_2\text{-}I_0]$ bit field in the core status register. This field is set to 7 at system start up to mask all interrupt levels except level 7. Level 7 is *edge* triggered and is non-maskable interrupt. This means that anytime level 7 is signaled, the CPU will acknowledge it at the end of the current instruction being executed. Because level 7 is edge triggered, another level 7 request will not be acknowledged as long as the current mask in the SR is equal to 7. Thus in order to acknowledge another level 7 during the servicing of a non-maskable interrupt, the level 7 signal must be negated first or the mask level must be lower than 7.

All other interrupt levels, 1 through 6, are level sensitive and are maskable. The 6 levels can only be acknowledged by the CPU if the requested level is higher than the interrupt mask bit field in the core SR. For example, to acknowledge level 1 interrupt request, the $[I_2\text{-}I_0]$ must equal [000], and to acknowledge level 2, the $[I_2\text{-}I_0]$ must be set [000] or [001], etc. This implies that the interrupt request will only be serviced by the core if the interrupt level signaled is higher than the current mask. Table 10 shows the relationship between interrupt request levels and the mask field in the SR.

As can be seen from Table 10, to recognize level 4 interrupt request for instance, the interrupt mask must be set to 3 or lower. In other words, if the priority of the interrupt request is lower than or equal to the mask bit field in the core SR, execution continues with the next instruction, and the requesting interrupt is deferred until the mask becomes lower than the requested level.

ColdFire processors raise the mask in the SR to the current interrupt request level when acknowledged. For example, if interrupt request 4 is acknowledged, the core automatically raise the interrupt mask to level 4. This ensures only levels higher than 4 will be able to interrupt the current level being serviced.

From our discussion in the above section, it is clear that the core can nest up to 7 interrupt levels. For example, level 2 can interrupt level 1, level 3 can interrupt level 2, and so on.

**Note:** Interrupt level 7 is always acknowledged regardless of the interrupt mask setting in the core status register (SR).

**TABLE 10. Relationship between Interrupt Requests & Mask**

| Interrupt Level Requested | Interrupt mask level required to acknowledge the request |
|---|---|
| 0 | No Request |
| 1 | 0 |
| 2 | 0 - 1 |
| 3 | 0 - 2 |

TABLE 10. Relationship between Interrupt Requests & Mask

| Interrupt Level Requested | Interrupt mask level required to acknowledge the request |
|---|---|
| 4 | 0 - 3 |
| 5 | 0 - 4 |
| 6 | 0 - 5 |
| 7 | xxx[a] |

a. Where xxx may be any value between 0 -7.

Once an interrupt is acknowledged, the core proceeds with the usual exception processing by saving the program counter and the status register onto the stack. The saved value of the program counter is the address of the instruction that would have been executed had the interrupt not been taken. The appropriate interrupt vector contents is then fetched into the program counter, and normal instruction execution commences in the interrupt handler.

It is sometimes undesirable for an interrupt service routine during the execution of critical code to be interrupted even by higher priority ones. To allow the handler to mask higher levels than the one currently being serviced, ColdFire processors inhibit interrupt sampling during the first instruction in the exception or interrupt service routine. This allows any handler to effectively disable interrupts, if necessary, by raising the interrupt mask level contained in the status register as the first instruction in the ISR. In addition, the CFV1 instruction set architecture (ISA_C) includes the ideal instruction to do just that, the store/load SR (STLDSR). This instruction stores the current interrupt mask level and loads SR with a new value specified in the operand field. The STLDSR is specifically intended for use as the first instruction in the interrupt service routine. This way, the ISR can raise the interrupt mask level to prevent the core from acknowledging another request with a higher priority level than the one currently being serviced. Below is an example of store/load SR instruction:

**Example 3.27. STLDSR #0x2700 ; Push SR on the stack and raise the mask to level 7.**

Once the critical code execution is completed, the handler restores the SR from the stack. Normally, the last instruction in the service routine is a return to the previous program by executing the return-from-exception (RTE) instruction. The RTE instruction restores the SR and the PC from the system stack to their original values, and a return is initiated back to the main program.

The above concludes all exception types considered as part the ColdFire core VI system.

# CHAPTER 4

# *Cryptographic Acceleration Unit & RNGA*

## *Introduction to the Cryptographic Acceleration Unit*

The CAU is a coprocessor that executes certain Coldfire® coprocessor instructions. The CAU supports acceleration of the DES, 3DES, MD5, SHA-1 and AES cryptographic algorithms, that can provided the security needed when a ColdFire® microcontroller is embedded in a network connected device, such as a networked home appliance or a router.

Other cryptographic applications for the CAU include vending machines, fare collection, and employee or personal identification.

MD5 (Message-Digest algorithm 5) is a widely-used cryptographic hash function with a 128-bit hash value. MD5 has been used in a wide variety of secure Internet applications, and is also commonly used to check the integrity of files. It was designed by Ronald Rivest in 1991 but has since become questionable due to the discovery of some flaws.

SHA-1 is a member of the SHA (Secure Hash Algorithm) family of related cryptographic hash functions. The most commonly used function in the family, SHA-1, is employed in a large variety of popular security applications and protocols, including TLS, SSL, PGP, SSH, S/MIME, and IPSec. SHA-1 is considered to be the successor to MD5.

DES (Data Encryption Standard) is an encrypting algorithm selected by the United States in 1976, and which has subsequently spread to international use. DES is now considered to be insecure for many applications, chiefly due to the 56-bit key size being too small; DES keys have been broken in less than 24 hours.

3DES (Triple DES) was developed to guard against the brute force attacks on DES. 3DES was chosen as a simple way to enlarge the key space without a need to switch to a new algorithm. 3DES is gradually being replaced by AES.

AES (Advanced Encryption Standard) is a secret key cryptosystem adopted in November of 2001 by the National Institute of Standards and Technology (NIST) as the successor to DES. AES is a form of the block cipher, Rijndael. AES has a fixed block size of 128 bits and a key size of 128, 192 or 256 bits, while Rijndael can be specified with key and block sizes in any multiple of 32 bits, with a minimum of 128 bits and a maximum of 256 bits. The Rijndael cipher was named and developed by two Belgian cryptographers, Joan Daemen and Vincent Rijmen. AES is fast, relatively easy to implement, and requires little memory.

## CAU Architecture and Configurations

The CAU register file consists of eight, 32-bit registers as shown in Table 4.1. Depending on the algorithm executed, each register is assigned to a specific variable of the algorithm, as shown in the DES, AES, SHA-1 and MD5 columns of the table. All registers can be read with the coprocessor store instruction (cp0st.l) and written with the coprocessor load instruction (cp0ld.l). However, only bits 0-1 of the CASR are writable. Bits 2-27 of CASR loads should be 0 for compatibility with future versions of the CAU. The CAU only supports long word (32 bit) accesses and register codes 0x8-0xF are reserved.

**TABLE 4.1 CAU Register File**

| Code | Name | Description | DES | AES | SHA-1 | MD5 |
|------|------|-------------|-----|-----|-------|-----|
| 0 | CASR | status register | -- | -- | -- | -- |
| 1 | CAA | accumulator | -- | -- | T | a |
| 2 | CA0 | general purpose 0 | C | W0 | A | -- |
| 3 | CA1 | general purpose 1 | D | W1 | B | b |
| 4 | CA2 | general purpose 2 | L | W2 | C | c |
| 5 | CA3 | general purpose 3 | R | W3 | D | d |
| 6 | CA4 | general purpose 4 | -- | -- | E | -- |
| 7 | CA5 | general purpose 5 | -- | -- | W | -- |

A cp0ld.l instruction is used to write to CAU registers and specify CAU operations. For instructions that take two operands, the first operand of the instruction is the source address. The CAU destination register is encoded into the second operand, which also contains the "load command" encoding.

Single operand instructions have the command encoded in the operand.

All CAU load instruction commands have an effective execution time of 0 clocks. This means the effect of executing a load command is observable in the next cpu instruction.

A **cp0st.l** instruction is used to read CAU registers. The first operand is the effective address of the destination. The CAU source register is encoded in the second operand with the "store command" encoding. The CAU store instruction command has an effective execution time of 0 clocks. This means the next instruction after the cp0st.l may immediately operate on the result stored in the destination.

The CAU status register (CASR), shown in Figure 4.1, has 3 bit fields defined, but only one of these, the DPE bit, would be used in normal programming. The CASR[IC] bit indicates an illegal command was attempted, and the four bit CASR[VER] value indicates the CAU version number. At the time of writing, only one CAU version has been implemented.

31 30 29 28 27 26 25 24 23 22 21 20 19 18 17 16 15 14 13 12 11 10 9 8 7 6 5 4 3 2 1 0

**FIGURE 4.1. CAU Status Register (CASR)**

The CAA register is used as an accumulator to reduce the execution time of certain cryptographic algorithms. The contents of the CAA register are normally initialized by a cpl0.l instruction prior to executing the appropriate CAU command. When the CAU executes the command the intermediate result is accumulated with the contents of the CAA register.

The remaining six registers, CA0 to CA5, can be used as general purpose registers for the more general CAU commands, (e.g. XOR and ROTL), while the cryptographic commands (e.g. AESR, DESK), require parameters to be loaded into specific CAx registers.

### CAU Commands

The CAU supports 22 commands that are shown in Table 4.2 and described in the following sections. The command name shown in the table is defined as a numeric value that is part or all of the second operand of the co-processor instruction. Values for these command names are given in Table 4.6 later in this chapter. All other encodings are reserved. The IC bit in the CASR will be set if any command is issued that is not defined in the encodings described in this section. A specific illegal command (ILL) is defined to allow for software self checking. Reserved commands should not be issued to ensure compatibility with future implementations.

**TABLE 4.2 CAU Commands**

| Inst Type | Command Name | Description | CMD[8:4] | CMD[3:0] | Operation |
|---|---|---|---|---|---|
| cp0ld.l | CNOP | No Operation | 0x00 | 0x0 | none |
| cp0ld.l | LDR | Load Reg | 0x01 | CAx | Op1 -> CAx |
| cp0st.l | STR | Store Reg | 0x02 | CAx | CAx -> Destination |
| cp0ld.l | ADR | Add | 0x03 | CAx | CAx + Op1 -> CAx |
| cp0ld.l | RADR | Reverse and Add | 0x04 | CAx | CAx + ByteRev(Op1) -> CAx |
| cp0ld.l | ADRA | Add Reg to Acc | 0x05 | CAx | CAx + CAA -> CAA |
| cp0ld.l | XOR | Exclusive Or | 0x06 | CAx | CAx ^ Op1 -> CAx |
| cp0ld.l | ROTL | Rotate Left | 0x07 | CAx | CAx <<< Op1 -> CAx |
| cp0ld.l | MVRA | Move Reg to Acc | 0x08 | CAx | CAx -> CAA |
| cp0ld.l | MVAR | Move Acc to Reg | 0x09 | CAx | CAA -> CAx |
| cp0ld.l | AESS | AES Sub Bytes | 0x0A | CAx | SubBytes(CAx) -> CAx |
| cp0ld.l | AESIS | AES Inv Sub Bytes | 0x0B | CAx | InvSubBytes(CAx) -> CAx |
| cp0ld.l | AESC | AES Column Op | 0x0C | CAx | MixColumns(CAx)^Op1 -> CAx |
| cp0ld.l | AESIC | AES Inv Column Op | 0x0D | CAx | InvMixColumns(CAx^Op1) -> CAx |
| cp0ld.l | AESR | AES Shift Rows | 0x0E | 0x0 | ShiftRows(CA0-CA3) -> CA0-CA3 |
| cp0ld.l | AESIR | AES Inv Shift Rows | 0x0F | 0x0 | InvShiftRows(CA0-CA3) -> CA0-CA3 |
| cp0ld.l | DESR | DES Round | 0x10 | IP FP KS[1:0] | DES Round(CA0-CA3)->CA0-CA3 |
| cp0ld.l | DESK | DES Key Setup | 0x11 | 0 0 CP DC | DES Key Op(CA0-CA1)->CA0-CA1 Key Parity Error & CP -> CASR[1] |
| cp0ld.l | HASH | Hash Function | 0x12 | 0 HF[2:0] | Hash Func(CA1-CA3)+CAA->CAA |
| cp0ld.l | SHS | Secure Hash Shift | 0x13 | 0x0 | CAA <<< 5 -> CAA, CAA->CA0, CA0->CA1, CA1 <<< 30 -> CA2, CA2->CA3, CA3->CA4 |
| cp0ld.l | MDS | Message Digest Shift | 0x14 | 0x0 | CA3-> CAA, CAA->CA1, CA1->CA2, CA2->CA3, |
| cp0ld.l | ILL | Illegal Command | 0x1F | 0x0 | 0x1->CASR[0] |

The following details of each CAU command use the assembler syntax that is supported by the Codewarrior assembler integrated with the Codewarrior Development Studio for ColdFire® Architectures Version 6.3. The value of the symbols used in each command are defined in Table 4.6.

**CNOP**

cp0ld.l#CNOP

This command performs no operation, and is provided for synchronization of other implementations of co-processors.

## LDR

cp0ld.l<ea>,#(LDR+CAx)

Loads one register, specified by x with the data contained at memory location <ea>

Usage examples are given in assembly code segments shown later in this chapter.

## STR

cp0st.l<ea>,#(STR+CAx)

Stores the contents of the CAU register specified by x to the memory location <ea>.

Usage examples are given in assembly code segments shown later in this chapter.

## ADR

cp0ld.l<ea>,#(ADR+CAx)

Adds the contents of memory location <ea> to the contents of the CAU register specified by x and stores the result in CAU register specified by x.

**RADR**

cp0ld.l<ea>,#(RADR+CAx)

The RADR command does a byte reverse on the source operand specified by <ea>, adds that value to CAx and stores the result in CAx. An example is shown in Figure 4.2.

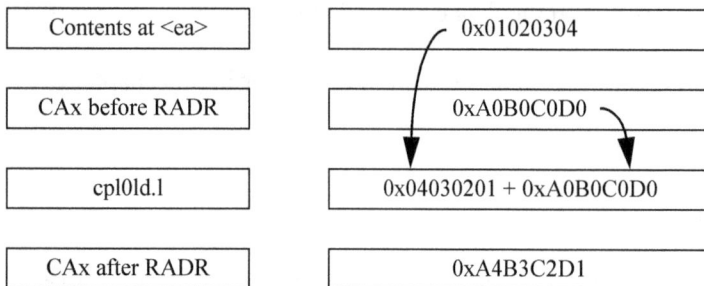

| Contents at <ea> | 0x01020304 |
|---|---|
| CAx before RADR | 0xA0B0C0D0 |
| cpl0ld.l | 0x04030201 + 0xA0B0C0D0 |
| CAx after RADR | 0xA4B3C2D1 |

FIGURE 4.2. **Example register reverse and add**

**ADRA**

cp0ld.l#(ADRA+CAx)

This command adds the contents of register CAx to CAA and stores the result in CAA

**XOR**

cp0ld.l<ea>,#(XOR+CAx)

This command performs an exclusive-or of memory location <ea> with CAx and stores the result in CAx.

**ROTL**

cp0ld.l<ea>,#(ROTL+CAx)

This command rotates left the bits contained in CAx. The number of bit positions rotated is defined by the 5 least significant bits in <ea>, giving a minimum of zero shifts, and a maximum of 32.

### MVRA

cp0ld.l#(MVRA+CAx)

This command copies the contents of CAx to CAA. Note this is a single operand command.

### MVAR

cp0ld.l#(MVAR+CAx)

This command copies the contents of CAA to CAx.

### AESS

cp0ld.l#(AESS+CAx)

This command executes the SubBytes step of the AES algorithm on the contents of CAx, and places the result in CAx.

### AESIS

cp0ld.l#(AESIS+CAx)

This command is the opposite of AESS. It executes the inverse SubBytes step of the AES algorithm on the contents of CAx, and places the result in CAx.

### AESC

cp0ld.l<ea>,#(AESC+CAx)

This command performs an in-place columns then exclusive-or operation on the contents of CAx as required by the AES algorithm.

### AESIC

cp0ld.l<ea>,#(AESC+CAx)

This command performs an in-place exclusive-or then inverse mix columns operation on the contents of CAx as required by the AES algorithm.

## AESR

cp0ld.l#AESR

This command performs the shift-rows operation on the contents of CA0, CA1, CA2 and CA3 as required by the AES algorithm. (To be more precise, the operation is a "rotate rows" rather than a shift rows"). Note this command has only one operand, and the CAU register selection is fixed, corresponding to the 4x4 array of bytes referred to as a *state* in the AES algorithm. Table 4.3 shows the contents of the four affected registers before and after execution of this instruction. Note that the rows appear transposed since each row is actually composed of one byte from each register. Row 0 is unaffected, while the remain-

TABLE 4.3 CAU Registers before and after AESR command execution

| | | Registers | | | | | | | |
|---|---|---|---|---|---|---|---|---|---|
| | | Before | | | | After | | | |
| | | CA0 | CA1 | CA2 | CA3 | CA0 | CA1 | CA2 | CA3 |
| Row Number | 0 | 01 | 05 | 09 | 0D | 01 | 05 | 09 | 0D |
| | 1 | 02 | 06 | 0A | 0E | 06 | 0A | 0E | 02 |
| | 2 | 03 | 07 | 0B | 0F | 0B | 0F | 03 | 07 |
| | 3 | 04 | 08 | 0C | 00 | 00 | 04 | 08 | 0C |

ing three rows have bytes rotated left by 1, 2 and 3 bytes respectively. Example 4.1 lists the assembly code that produces the results in Table 4.3.

```
// First place the constants for CA3,CA2,CA1,CA0 on the stack
move.l   #0x01020304,-(sp)
move.l   #0x05060708,-(sp)
move.l   #0x090a0b0c,-(sp)
move.l   #0x0d0e0f00,-(sp)

// load CA3,CA2,CA1,CA0
cp0ld.l  (sp)+,#(LDR+CA3)
cp0ld.l  (sp)+,#(LDR+CA2)
cp0ld.l  (sp)+,#(LDR+CA1)
cp0ld.l  (sp)+,#(LDR+CA0)

// Execute AES shift rows command
cp0ld.l #AESR
```

Example 4.1. **CAU AES Shift Rows Command**

### AESIR

cp0ld.l#AESIR

This command performs the inverse shift-rows operation on the contents of CA0, CA1, CA2 and CA3 as required by the AES algorithm. Note this command has only one operand, and the CAU register selection is fixed.

### DESR

cp0ld.l#(DESR+[IP]+[FP]+[KSx])

The DESR command performs a round operation as required for the DES algorithm and a key schedule update with the following source and destination designations: CA0=C, CA1=D, CA2=L, CA3=R. If the IP bit is set then the DES initial permutation is performed on CA2 and CA3 before the round operation. If the FP bit is set then the DES final permutation (inverse initial permutation) is performed on CA2 and CA3 after the round operation. The round operation uses the source values from registers CA0 and CA1 for the key addition operation. The KSx field specifies the shift to use for the key schedule operation used to update the values in CA0 and CA1. The specific shift function performed is based on the KSx field as defined in Table 5-6.

**TABLE 4.4 Key Shift Symbol Function**

| KSx Symbol | KSx Code | Shift Function |
|:----------:|:--------:|:--------------:|
| KSL1 | 0 | Left 1 |
| KSL2 | 1 | Left 2 |
| KSR1 | 2 | Right 1 |
| KSR2 | 3 | Right 2 |

### DESK

cp0ld.l<ea>,#(DESK+CP+DC)

The DESK command performs an in-place initial key transformation (permuted choice 1, or pc1) defined by the DES algorithm on the contents of registers CA0 and CA1. CA0 contains bits 1-32 of the key and CA1 contains bits 33-64 of the key.

Decryption is performed when the DC value is present (a value of 1). In this case, the result of the Permuted Choice 1 transformation is stored in CA0 and CA1 without shifting.

Encryption is performed when the DC value is not present. In this case, a left shift by 1 is also performed when the result of the transformation is stored back to CA0 and CA1.

If the CP bit is set and a key parity error is detected then the DPE bit of the CASR is set, otherwise it is cleared.

The code in Example 4.2 executes the key transformation using the DESK command, but its function in this example is to simply return the parity of the key. A pointer to the two 64 bit key is passed on the stack. In the example the key is stored as an array of two 32 bit integers. The DESK command is executed with the CP bit set, so the DPE bit in the CASR register indicates the parity of the key. It is extracted with a cp0st.l command and the parity returned in the register d0.

**uint32 key[2] = {0x12345678,0x23456789};**

**cau_des_chk_parity(key);**

...

**_cau_des_chk_parity:**

```
    move.l   4(sp),a0              //get argument: *key

// load the 64-bit key into the CAU's CA0/CA1 registers
    cp0ld.l  (a0)+,#(LDR+CA0) // load key[i]   -> CA0
    cp0ld.l  (a0),#(LDR+CA1)  // load key[i+1] -> CA1

//  perform the key schedule and check the parity bits
    cp0ld.l  d0,#(DESK+CP)       // key setup + parity check

// the CAUSR[DPE] reflects the DES key parity check
    cp0st.l  d0,#(STR+CASR)      // store CAUSR -> d0
    btst     #1,d0               // test CAUSR[DPE]
    sne      d0                  // if DPE, then d0 = 0xff, else d0 = 0x00
    extb.l   d0                  // return 32-bit int
    rts
```

**Example 4.2.  Parity check using the CAU DESK command**

## HASH

cp0ld.l #(HASH+HFx)

This command performs various hashing operations on CA1, CA2 and CA3 and adds the result to the value in CAA. The hash function selected is based on the HFx field as defined in Table 4.5. C style notation is used for the Hash Operation description. The HASH command is normally executed as part of MD5 and SHA algorithms.

**TABLE 4.5 HASH Command function codes**

| HFx Symbol | HFx Code | Hash Function | Hash Operation |
|---|---|---|---|
| HFF | 0 | MD5 F() | CAA += CA1&CA2 \| ~CA1&CA3) |
| HFG | 1 | MD5 G() | CAA += CA1&CA3 \| CA2&~CA3 |
| HFH | 2 | MD5 H(), SHA Parity() | CAA += CA1^CA2^CA3 |
| HFI | 3 | MD5 I() | CAA += CA2^(CA1\|~CA3) |
| HFC | 4 | SHA Ch() | CAA += CA1&CA2 ^ ~CA1&CA3 |
| HFM | 5 | SHA Maj() | CAA += CA1&CA2 ^ CA1&CA3 ^ CA2&CA3 |

Example 4.3 is part of an MD5 algorithm that takes advantage of the HASH command to reduce execution time.

```
# initialize the md5_state variables, both in memory and the CAU
    move.l      #0x67452301,(a1)+// initialize a
    move.l      #0xefcdab89,(a1)+// initialize b
    move.l      #0x98badcfe,(a1)+// initialize c
    move.l      #0x10325476,(a1)+// initialize d

    cp0ld.l     -(a1),#(LDR+CA3)// load d -> CA3
    cp0ld.l     -(a1),#(LDR+CA2)// load c -> CA2
    cp0ld.l     -(a1),#(LDR+CA1)// load b -> CA1
    cp0ld.l     -(a1),#(LDR+CAA)// load a -> CAA

    cp0ld.l     #(HASH+HFF)    // add F(b,c,d)
```

**Example 4.3. HASH Command initialization and execution**

**SHS**

cp0ld.l#SHS

This command performs register to register move and shift operations in parallel, to accelerate the execution of the SHA-1 algorithm. The SHS command performs the following specific operations:

    CAA=CAA<<<5
    CA0=CAA
    CA1=CA0
    CA2=CA1<<<30
    CA3=CA2
    CA4=CA3

## MDS

cp0ld.l#MDS

The MDS command does a set of register to register move operations in parallel that is useful for implementing MD5. The following source and destination assignments are made:

    CAA=CA3
    CA1=CAA
    CA2=CA1
    CA3=CA2

## ILL

cp0ld.l#ILL

The ILL command is a specific illegal command that sets the IC bit in the CASR.

### CAU Command Equates

The values of the symbols used to specify CAU commands in the co-processor load and store instructions are given in Table 4.6.

The Codewarrior assembler syntax is <**Symbol**> .set <**Value**>

TABLE 4.6 CAU Command Symbol reference

| Symbol | Value | Symbol | Value |
|--------|-------|--------|-------|
| TL | 0 | AESR | 0x0e0 |
| TS | 0 | AESIR | 0x0f0 |
| CASR | 0 | DESR | 0x100 |
| CAA | 1 | DESK | 0x110 |
| CA0 | 2 | HASH | 0x120 |
| CA1 | 3 | SHS | 0x130 |
| CA2 | 4 | MDS | 0x140 |
| CA3 | 5 | ILL | 0x1f0 |
| CA4 | 6 | IP | 8 |
| CA5 | 7 | FP | 4 |
| CNOP | 0x000 | DC | 1 |
| LDR | 0x010 | CP | 2 |
| STR | 0x020 | KSL1 | 0 |
| ADR | 0x030 | KSL2 | 1 |
| RADR | 0x040 | KSR1 | 2 |
| ADRA | 0x050 | KSR2 | 3 |
| XOR | 0x060 | HFF | 0 |
| ROTL | 0x070 | HFG | 1 |
| MVRA | 0x080 | HFH | 2 |
| MVAR | 0x090 | HFI | 3 |
| AESS | 0x0a0 | HFP | 2 |
| AESIS | 0x0b0 | HFC | 4 |
| AESC | 0x0c0 | HFM | 5 |
| AESIC | 0x0d0 | | |

## Random Number Generator Accelerator

The Random Number Generator Accelerator (RNGA) module is a digital circuit capable of generating 32 bit random numbers. It is designed to comply with FIPS-140 standards for randomness and non-determinism. The random bits are generated by clocking shift registers with clocks derived from ring oscillators integrated in the silicon. The configuration of the shift registers ensures statistically good data. Because the oscillators operate at unknown frequencies, they provide the required entropy needed to create random data.

It is highly recommended that the random data produced by this module be used as an input seed to a NIST approved (based on DES or SHA-1) or cryptographically secure (RSA Generator or BBS Generator) random number generation algorithm. It is also recommended that other sources of entropy be used along with the RNGA to generate the seed to the pseudorandom algorithm. The more random sources combined to create the seed the better. The following is a list of sources which could be combined with the output of the RNGA module.

- Current time using highest precision possible.
- Mouse and keyboard motions (or equivalent if being used on a cell phone or PDA).
- Other entropy supplied directly by the user.

The RNGA programmer's model consists of ten 32 bit registers, as shown in Figure 4.3. All registers should be accessed as 32 bits only. Note that the Address column shown in the figure refers to the offset from the base address for the RNGA.

| Address | Use | Access |
|---|---|---|
| 0x00 | Control | R/W |
| 0x04 | Status | R |
| 0x08 | Entropy Register | W |
| 0x0c | Output Register | R |
| 0x10 | Mode | R/W |
| 0x14 | Verification Control | R/W |
| 0x18 | Oscillator Control Counter | R/W |
| 0x1c | Oscillator #1 Counter | R |
| 0x20 | Oscillator #2 Counter | R |
| 0x24 | Oscillator Counter Status | R |

FIGURE 4.3. **RNGA Programmers Model**

In normal operation, software may only access the Control, Status, Entropy and Output Registers. The remaining registers, shown shaded in the programmers model, are intended only for factory test purposes.

After a hardware reset, the RNGA immediately starts operation, but the hardware outputs random data to the Output Register only after software has set the Go bit in the Control Register and 256 system clock cycles have elapsed. Hardware will continue to write a new random 32 bit value every 256 system clocks, unless the Output Register is full, in which case the Output Register is not updated. There is no indication that random data is discarded. Typically, software should poll the Status Register and examine the Output Register Level bit field to determine whether a new random value is available before reading the Output Register.

Note that the Output Register Level bit field in the Status Register is 8 bits wide, allowing the Output Register to be implemented as a FIFO (containing up to 256 values). However, the Output Register Size bit field, which indicates the maximum number of values possible, is set to 0x01, indicating that the Output Register on the RNGA is in fact implemented as a single register. This means the Output Register should not be read in succession without first checking the Output Register Level field

If the Output Register is empty when software reads from it, an underflow error condition occurs, which can optionally generate an Error Interrupt, if enabled by the Control Register Interrupt Mask bit. To clear the Interrupt, sofware should write a 1 to the Control Register Clear Interrupt bit.

If software sets the High Assurance bit in the Control Register, an underflow will cause the Security Violation bit to be set. Both High Assurance and Security Violation bits are sticky and are cleared only by a hardware reset. An underflow will also set the Output Register Underflow and Last Read bits in the Status Register, though these bits are cleared automatically when software reads the register.

The Entropy Register is write only, and allows software to provide a new seed for the RNGA hardware while it is operational. This provides a means for software to increase the randomness of the RNGA output.

For information on the test and verification registers, please refer to the relevant Freescale supplied documentation.

# *HCS08 Core*

## *Introduction to the HCS08 Core*

System developers are often tasked to design a broad portfolio of products from the low end to the high end in order to meet the highest performance required by many applications. Typically, the higher end architecture does not have hardware or software compatibilities with the low end. This forces system designers or developers to learn new product and new tools. In general, this leads to a longer design cycle time and eventually longer time to market. To address these needs, Freescale Semiconductor provides system designers with the ideal scalable architecture by introducing the Flexis products.

The Flexis product consists of two families, the HCS08 and ColdFire. The HCS08 devices are based on 8-bit architecture, while the ColdFire is 32-bit architecture based on version 1 (CFV1) core. The most important objective here is that both architectures are compatible since they integrate the same on-chip hardware with the difference between the two being the CPU core.

An HCS08 device with the suffix "QE128" integrates 128K of flash memory. There are also a number of ColdFire V1 devices that integrate 128K flash memory and have the same suffix "QE128".

QE128 devices (HCS08 and CFV1) are designed with identical I/O peripherals and memory size. Freescale's intention for manufacturing these devices is to allow pin-to-pin compatibility between the two families. The hardware and software compatibility provides the system designer the ability to replace an HCS08 based MCU directly with a ColdFire V1 device without any software changes, assuming the application software was written in a high level language such as C or C++. In another words, programs written in C or C++ will work 100% properly on an HCS08 device when compiled by an HCS08 compiler. Likewise, the very same programs will work on CFV1 device when compiled by a CFV1 compiler.

Besides being pin-to-pin compatible, the Flexis 8- and 32-bit MCUs have a built in debugger consisting of one bidirectional port pin for the background debugging mode (BDM). This gives programmers the ability to write software for all Flexis MCUs using the same development tools, although the compiled code will be different, of course.

In addition, to instantly provide the system designers with a test vehicle, Freescale's CodeWarrior for Microcontrollers supports both architectures and offers a free compiler, limited in size of compiled code to 32KB on 8-bit MCUs and 64KB on 32-bit MCUs

The HCS08 family is based on the original CPU08 core implemented on the previous 8 bit MC68HC08 microcontroller family. However, the HCS08 core is an enhanced version and is 100 percent backward compatible with the HC08 CPU. This improvement is mainly based on a new technology that allows higher frequencies than their predecessors and at a lower cost. However it also includes new instructions and addressing modes added to the HCS08 CPU to further improve the performance and compiler efficiency.

Along with being high performance and low cost, the upward compatibility makes the HCS08 is very attractive since it is no longer limited to 64K address space as was the case in HC08 core. In fact, the HCS08 is capable of addressing up to 128K of address space by integrating an on-chip memory management unit controller capable of addressing multiple memory pages through an address window.

In this chapter we begin our coverage with the HCS08 core architecture and programming model followed by addressing modes with some examples. Later we will cover memory expansion, resets, interrupt processing, and handling along with vector table and stack operation, and finally, low power operating modes will be covered.

## *Programming Model*

The HCS08 CPU programming model consists of a simple register set shown in Chapter 5Figure 5.1. These registers are core resources used by the application program as a scratch pad to accomplish a task and/or a complex mathematical operation. The operation of each of these registers will be briefly explained in the following paragraphs.

### Accumulator (A)

The Accumulator is an 8-bit general purpose register that may be used to hold intermediate, partial and/or final results. This register may be loaded from memory, stored to memory, added to, subtracted from, logically or arithmetically shifted right or left, rotated right or left, incremented, decremented, compared, ANDed, ORed, XOR'd, pushed to the stack and pulled from the stack. It may also be used to perform an 8x8 bit multiply and 16-bit by 8-bit divide. Accumulator A is the most busy register since most operations require its use.

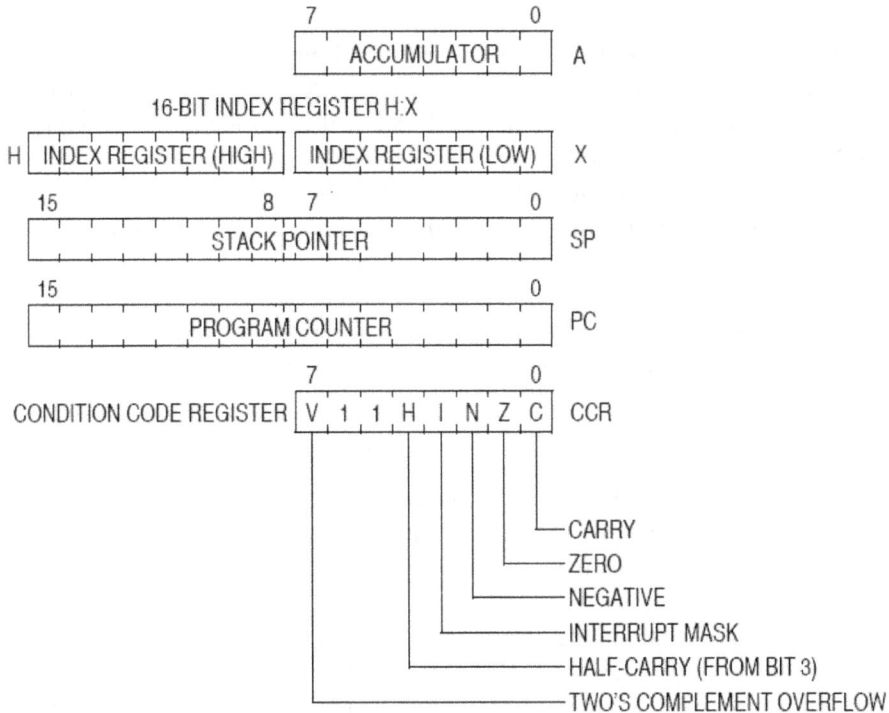

FIGURE 5.1. S08 Programming Model

### Index Register (H:X)

Historically, the X register was only an 8-bit implementation on the HC05 microcontroller family. To improve the efficiency of the compiler and to provide 16-bit operations as well as 16-bit address pointers for indexed addressing modes, the introduction of the CPU08/ HCS08 included the H register. The H register (most significant byte) may be concatenated with X (least significant byte) to provide a 16-bit register. To stay backward compatible with the previous HC05 CPU, the H register default value after reset is $00. This allows backward code compatibility between the CPU05 and the CPU08/S08 cores due to the identical programmer's model.

When used as a 16-bit register, the H:X register may be used for indexing memory with various addressing modes. Some of these addressing modes are: indexing with no offset, indexing with 8-bit offset and indexing with 16-bit offset, which allows easy accesses to the entire 64K memory map. Also, there are other index addressing modes using H:X that allows auto increment after the operand has been address and captured from memory.

The low order byte of the H:X is the X register. As an 8-bit register, it may be used as a general purpose accumulator similar to accumulator A. The X register may be loaded from memory, stored to memory, pushed and pulled, arithmetically and logically shifted left or right, rotated left or right, compared, cleared, complemented, negated, incremented and decremented. Many of the addressing modes that used on the A accumulator will work on the X register. Remember that operation described above only work on the low order part of the H:X register.

As a 16-bit, the H:X register pair can be loaded from memory or stored into memory but with a fewer addressing modes than that of accumulator A or X.

**Stack Pointer (SP)**

The stack pointer is a 16-bit register used to point to a memory area in RAM called the stack. This register and stack memory provide automatic context save and restore upon subroutine calls and returns, respectively.

For example, execution of a subroutine call instruction (JSR or BSR) causes the program counter (PC) containing the return address to the main program to automatically be pushed onto the stack. Likewise, when a return-from-subroutine (RTS) instruction at the end of the subroutine is executed, the return address is automatically restored from the stack.

More importantly, when the CPU acknowledges an interrupt from an external pin or a peripheral device, the contents of the CPU registers are automatically saved (pushed), and when the CPU execute a return-from interrupt (RTI) instruction at the end of the interrupt service routine, the CPU context is automatically restored (pulled) from the stack. A graphical description of the interrupt stack frame is shown in Figure 5.2.

Toward Lower Addresses

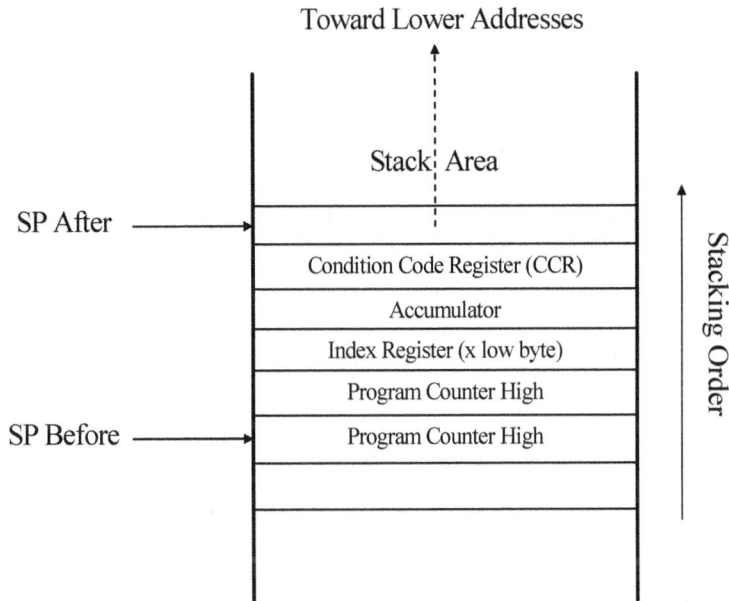

FIGURE 5.2. **Interrupt Stack Frame**[a]

a. For downward compatibility with the HC05 CPU, the H register is not pushed onto the stack.

When interrupted, the automatic context save and restore allows the CPU to gracefully return to the main program as if the CPU was never interrupted.

The stack pointer always points at the next available location on the stack. When a value is pushed onto the stack, it is written to the address pointed to by the SP and then SP is automatically decremented to point to the next available location. When a value is pulled from the stack, SP is first incremented to point at the most recent data that was pushed on the stack, and then the data is read from the address now pointed to by SP. Saved data on is not changed in the process of pulling it from the stack. Old data values can be seen at memory addresses below the current SP address. When new values are pushed onto the stack, they over-write what ever is in those memory locations. If the stack memory area was cleared during startup code or initialization, the maximum depth the stack has grown to can be seen by noticing which memory locations are still clear.

For compatibility with the earlier M68HC05, SP is set to $00FF by reset. Typically the SP is initialized to point to the highest address in the on-chip Ram.Since the RAM space is located at the end of input/output (I/O) and control registers, it is optimal to leave the direct page at address $0000 through $00FF available for more valuable and frequently accessed variables. The direct page can be accessed using the direct addressing mode that requires only one byte address, which saves program space and executes faster than general accesses to other memory locations.

Also, for compatibility with the M68HC05, the reset stack pointer (RSP) instruction forces the low-order half of SP to $FF. In the M68HC05 processors, this instruction forced the SP to the same value ($00FF) it had after reset. RSP is seldom used in the HC08/S08 application codes because it doesn't affect the high-order half of SP, and therefore, it doesn't necessarily restore SP to its reset value.

In new HCS08 programs you would typically initialize SP to point to the highest address in the on-chip RAM.

Again, for compatibility issues with the M68HC05 family, HCS08 core hardware does not push the H register on the stack when interrupted. Therefore, it is recommended that you include a PSHH instruction as the first instruction in the interrupt handler (to save H), and to also include a PULH instruction (to restore H) as the last instruction before the RTI instruction that ends the interrupt service routine.

Often, the application program may need to allocate memory on the stack as a work space during a function call to be able to pass local or global variables. Memory allocation on the stack can be achieved by using the add immediate value to SP (AIS) instruction with a negative offset. The same instruction should be used to deallocate the stack space but with a positive instead of a negative offset. This ensures the stack area is cleaned up before returning from the function. The following code example demonstrates allocation and deallocation of space for local variables on the stack.

**Example 5.1. AIS # -$10 ; Allocate 10 hexidicimal bytes on the stack for local variables**

```
Instruction          ; Start of Function
-                    ;
-                    ; Code
Instruction          ;
AIS  #$10            ; Deallocate local variables
Return               ; End
```

**Chapter 5**

## Program Counter (PC)

The program counter is a 16-bit register which is mainly used as a pointer into the instruction stream. The PC contains the address of the next instruction being fetched from program memory. It automatically increments to the next instruction address when the previous one retires from the execution stage. The PC is also loaded when a subroutine is called and restored to its original value upon return.

When an interrupt event is recognized by the core, the program counter is automatically pushed on the stack along with the other CPU registers. It is then loaded from the appropriate interrupt vector based on the exception or the interrupt taken by the core. The new PC value loaded from the vector table is the address of the first instruction in the interrupt handler.

Upon reset, the program counter is loaded with the reset vector located at address \$FFFE and \$FFFF. The value loaded into the PC contains the address of the first instruction to be executed in the startup routine or reset handler.

The PC can also be used to reference operands when using PC relative addressing modes.

## Condition Code Register (CCR)

This register contains 5 condition code flags plus the interrupt mask bit. The flag bits represent indicator based on the results of the core operation. The CCR is tested by conditional branch instructions to evaluate conditions to determine program flow. The CCR is typically updated after a logical or mathematical operation to indicate if the result has a carry/borrow, zero/equal, negative, signed overflow, and half carry. The CCR register format is shown in Figure 5.3 below. A brief description of each bit follows:

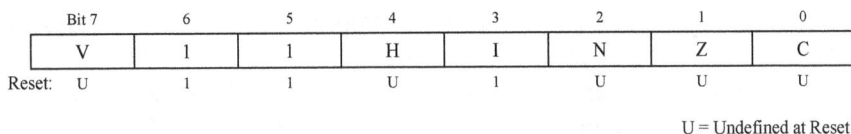

| Bit 7 | 6 | 5 | 4 | 3 | 2 | 1 | 0 |
|-------|---|---|---|---|---|---|---|
| V | 1 | 1 | H | I | N | Z | C |

| Reset: | U | 1 | 1 | U | 1 | U | U | U |
|--------|---|---|---|---|---|---|---|---|

U = Undefined at Reset

FIGURE 5.3. Condition Code Register (CCR)

- C- Carry condition code bit: This bit is set due to a carry from the most-significant bit of the result during addition, or if a borrow occurs during subtraction. The CPU may use this bit when the application needs to perform multi byte addition or subtraction. The add with carry (ADC) and subtract with carry (SBC) instructions are used to support these operations.

- Z- Zero condition code bit: The Z flag is set when the result of an arithmetic or logical operation is zero. This flag may also be used to evaluate conditions for equality after a compare instruction is executed, for instance.

- N- Negative condition code bit: This bit is set when the result of arithmetic operation ends up with the most-significant bit (MSB), is set. For algebraic operations, the MSB represents the sign bit. This bit is cleared when the result is positive and set when the result is negative.

- V- Overflow condition code bit: This bit is set when a 2's complement overflow condition is detected after the execution of arithmetic or a logical operation. For example, if addition or subtraction of two operands having the same sign bit and the result ends up with opposite sign, this bit is set to indicate an overflow condition. This condition implies that the result cannot be represented in the result operand size.

- I - Interrupt mask bit: This bit is the global interrupt mask. Out of reset, the I bit is set to a logic '1' preventing any interrupt from being acknowledged until it is cleared by the application software. Typically the I bit should be cleared after system initialization is complete to enable interrupts.

When an interrupt occurs, this bit is automatically set after the CCR[1] is pushed onto the stack to prevent interrupt nesting during interrupt service routine execution. Upon a return from interrupt, the CCR is automatically restored and thus interrupt are.

Since the CCR is stacked before the 'I' bit is set, the RTI normally clears the 'I' bit, and thus re-enables interrupts. Interrupts can be re-enabled by clearing the 'I' bit within the service routine, but implementing a nested interrupt management scheme requires great care and seldom improves system performance.

If currently set, the WAIT and STOP instructions automatically clear the 'I' bit because interrupts are the normal way to wake up the CPU from stop or wait modes.

- H - Half Carry bit: This flag is intended to support operations involving binary-coded-decimal (BCD) numbers. A BCD number is a decimal number from 0 through 9 which is coded into a single 4-bit binary value. This allows a single 8-bit value to hold exactly two BCD digits. The hexadecimal values $A through $F are considered illegal BCD values. The ALU's normal binary addition function can be used to add BCD numbers but the results need to be checked and corrected so the result is still a valid BCD value.

The HCS08 includes the decimal adjust accumulator (DAA) instruction to simplify the

---

1. HCS08 CPU pushes all core registers when an interrupt is acknowledged and pulls all core register when RTI instruction is executed.

checking and correction operation into a single instruction.

The 'H' bit is affected by a few instructions, but the add instructions (ADD and ADC) are the only two instructions that affect the 'H' bit in a meaningful way. These instructions set the 'H' bit if there was a carry out of bit 3 into bit 4 of the result (from one BCD digit to the next). Execution of the DAA instruction immediately after ADD or ADC, guarantees the result in the Accumulator will be provided in a valid binary coded decimal format.

## Addressing Modes

The HCS08 CPU has numerous and powerful addressing modes that provide flexible and efficient accesses to operands in registers and memory. Every instruction uses an addressing mode to efficiently access the operand needed to complete the operation. Many instructions such as load accumulator (LDA) or load index (LDX) allow several different ways to specify the memory location to be operated on, and each addressing mode variation requires a separate opcode. All of these variations use the same instruction mnemonic, and the assembler determines which opcode to use based on the syntax of the operand field. In some cases, special characters are used to indicate a specific addressing mode (such as the # [pound] symbol which indicates immediate addressing mode). In other cases, the value of the operand tells the assembler which addressing mode to use. For example, the assembler chooses direct addressing mode instead of extended addressing mode if the operand address is between $00 and $FF.

Below is a brief description of the addressing modes along with examples provided on each.

### Inherent Addressing Mode (INH)

This is the simplest addressing mode available in the HCS08 CPU due to the fact that no addressing information is required in the source code. Below is an opcode example that uses inherent addressing mode to clear the accumulator:

• CLRA    ; Clear Accumulator

There are many one byte opcodes that use inherent addressing modes. Some of these instructions include: LSRA, RORA, CLRX, DAA, SEI, CLI, COMA, COMX, NOP, DECA, DECX, INCA, or INCX, etc.

There are also few instructions that use inherent addressing mode to access the memory stack area. The example below saves register H, the high byte of H:X register onto the stack:

- PSHH  ; Push high order byte of X Register onto the stack

Additional instructions that use inherent addressing modes to push or pull registers from the stack: PSHA, PULA, PSHX, RTS, or RTI, etc.

### Immediate Addressing Mode (IMM)

To provide a method of loading data into registers, the immediate addressing mode is used. The size of this data is determined by the destination register specified in the operation. Here are some examples:

- LDA  #$56   ; Load Accumulator with 8-bit value of $56
- LDX  #$65   ; Load X with 8-bit value of $65
- LDHX #$1234 ; Load H:X with 16-bit value of $1234

### Direct Addressing Mode (DIR)

This addressing mode is used to access operands located in direct address space page '0' located at address ($0000 through $00FF). In this mode, the instruction consists of 2 bytes, the first byte is the opcode and the second is the direct address. This is a more efficient addressing mode than extended addressing because the upper 8 bits of the address are implied rather than being explicitly provided in the instruction. This saves a byte of program space and the bus cycle that would have been needed to fetch this byte. Here is an example of directly loading Accumulator:

- LDA  $50   ; Load Accumulator from memory location $50

There are many instructions that use this addressing mode, here is a list of some of these opcodes: **STA, LDX, STX, LDHX, STHX, ADD, ADC, INC, DEC, BSET, BCLR or BCLR,** etc.

### Extended Addressing Mode (EXT)

The extended is similar to direct addressing mode but requires a 16 bit (2 bytes) address for the memory operand. In this mode, the first byte is the opcode and the second two bytes are the 16-bit address. Extended addressing allows instructions to access any memory location in the 64K memory map. Below are two examples using extended addressing modes:

- STX   $0800   ; Store X register (low byte of H:X) into memory location $800
- LDHX  $1000  ; Load 16-bits into H:X register from address $1000

As with the direct addressing mode examples above, extended addressing use the same instructions but with 2 bytes address instead of one. Some instruction include: **STA, LDX, STX, STHX, ADD, ADC, SUB, or SBC,** etc.

## Indexed Addressing Modes

These modes are very powerful and provide and easy mechanism to access system memory in variety of ways. There are various indexed addressing modes available in the HCS08 core implementation. Below is a brief description on each modes:

* Indexed, No Offset (IX)

In this addressing mode, the contents of H:X register contains the effective address of the memory operand needed by the instruction. This mode is very efficient since the operation requires only one byte opcode. Two examples of this mode are shown below:

* LDA , X  ; Load Accumulator from memory location pointed by H:X register
* CMP , X   ; Compare memory location pointed by H:X to Accumulator

Here are some of instructions that use Indexed with no offset addressing: **STA, STX, ADC, ADD, SUB, SBC, DEC, INC, LDX, LDHX,** etc.

## Indexed, NO Offset with Post Increment (IX+)

This mode is a similar to the previous one but allows the index register to automatically increment to the next entry in memory after the data is accessed. Incrementing the index register does not increase instruction execution time, while providing a convenient and efficient method to update the index register to point to the next entry. Using this mode, the application software does not need to increment the memory address contained in H:X register manually. An example of indexed register with post increment mode is shown below:

* MOV   SCID, X+ ; Read SCI Data register & store in memory location pointed by X register. The Index register is incremented to point to the next entry in memory

Note that above instruction combines two addressing modes, the first is direct and the second is indexed with post increment. Here is another example using conditional branch and index with post increment addressing mode.

* CBNE   X+, LOOP ; Compare memory location pointed by H:X register to accumulator A and branch if not equal to loop. Increment X by 1 after. The index register points to the next data entry in memory.

### Index Addressing with 8-bit offset (IX1)

This addressing mode is similar to indexed addressing mode but an 8 bit offset is added to the value of the H:X register to form the effective address. With an 8 bit, the H:X register can be offset by up to 255 memory locations from the current address held in H:X register. An example of this mode is shown below:

- LDA  $50, X  ; Load ACC A from memory location pointed by H:X plus $50

Here are some instructions that use indexed with 8-bit offset addressing mode: **INC, DEC, NEG, TST, AND, ORA, or XORA,** etc.

### Indexed Addressing with 8-Bit Offset and Post Increment (IX1+)

Similar to the previous mode is 8-bit offset with X post incremented. The first byte is the opcode and the second byte is the 8-bit offset that is added to the index register value to form the effective address. In this mode, after the memory operand is accessed the index register is automatically incremented to the next memory address. Below is an example that shows indexed addressing with 8-bit offset post incremented mode:

- Loop: CBEQ $50, X+, Exit  ; This instruction compares the memory location pointed by X register with an offset value of $50. If the two operands are equal, the program will branch to Exit, else the next instruction will execute sequentially. In either case, the X register will automatically increment to the next memory location.

### Indexed Addressing with 16-Bit Offset (IX2)

This addressing mode is similar to IX1 mode but requires 16-bit offset instead of an 8-bit. With 16 bit, the H:X register can be offset by up to 64K memory locations from the current address held in H:X register.

Below is a block move program segment that shows examples of indexed addressing modes with 8 bit offset and 16 bit offset:

```
        ORG $100            ; Originate data at address $100
SRC     FCC 'Hello! World'  ; Source data
        FCB 0               ; Last byte to move is'0' (loop terminator)
DEST    EQU $80             ; Destination start address is $80

        ORG $C000           ; Originate program at address $C000
        CLRX                ; Clear the X register.
LOOP    LDA SRC, X          ; Read data byte from SRC addr. (16 bit offset)
        STA DEST, X         ; Write data byte to DEST addr. (8 bit offset)
        BEQ DONE            ; If data byte moved = zero, go to DONE.
        INCX                ; Increment X pointer to next byte location.
BRA     LOOP                ; Branch to LOOP to move next byte.
DONE    BRA DONE            ; Done, stay here.
```

In the program example above, the source is originated at address $100 and the destination data is originated at address $80. In order to use indexed addressing with offset to access the source data, 2 bytes will be required. Likewise, to access the destination data, only one byte is required since address $80 falls in page 0.

### Stack Pointer Relative, 8-Bit Offset (SP) and 16-Bit Offset (SP2)

These two modes are similar to IX1 and IX2 addressing modes but use the stack pointer instead of the index register. Stack pointer relative addressing is most commonly used to access parameters and local variables on the stack. The two examples below show 8-bit and 16-bit offset.

- LDA $10, SP ;   Load Accumulator from stack area at address SP + 16
- STX $200, SP ;   Store Index register to stack area at address SP + 512

### Relative Addressing Mode (REL)

This mode is used to specify the destination address for branch instructions. Typically, the programmer specifies the destination with a program label or an expression in the operand field of the branch instruction. The assembler calculates the difference between the program counter (which points at the next address after the branch instruction at the time) and the address represented by the label or expression in the operand field. This difference is called the offset and is an 8-bit two's complement number. The assembler stores this offset in the object code for the branch instruction.

During execution, the CPU evaluates the condition that controls the branch. If the branch condition is true, the CPU sign-extends the offset to a 16-bit value, adds the offset to the current PC, and uses this as the address where it will fetch the next instruction and continue execution rather than continuing execution with the next instruction after the branch. Since the offset is an 8-bit two's complement value, the destination must be within the range −128 to +127 locations from the address that follows the last byte of object code for the branch instruction. Here is an example using branch relative addressing mode:

- BNE   Loop   ; Branch to Loop if evaluated condition is not equal to '0'.

### Memory-to-Memory Addressing Modes

There are four other addressing modes that allow memory-to-memory accesses without loading or storing the operand into the core registers. These modes are called memory-to-memory addressing.

## Memory-to-Memory Immediate to Direct addressing

This mode combines immediate with direct addressing. An example is shown below:

- MOV    #$12, $56    ; Move the immediate value of $12 to memory location $56

## Memory-to-Memory Direct-to-Direct addressing

This mode allows moving data from a memory direct address to another memory direct address. This instruction consists of 3 bytes. The first is opcode, and the second is the source direct memory address and the third is the destination direct address.

- MOV    $20, $50   ; Move source data from address $20 into destination memory at
                           address $50

## Memory-to-Memory Indexed to Direct with Post increment

Instruction using this mode allow accessing an operand in memory using indexing with post increment and move the operand to a direct memory address. In this mode, the indexed is the source address and the destination is the direct address.

- MOV    X+, SPDR ;Move a data byte from memory pointed by H:X to SPI data regis-
                        ter. After the operand is accessed, H:X register is incremented by 1

## Memory-to-Memory Direct to Indexed with Post increment

This addressing mode is very similar to the previous one, but the source is the direct address and the destination is the indexed address.

- MOV    SPDR, X+ ;  Move a data byte from the SPI data register into the memory
                         location pointed by H:X. After the operand is transferred to desti-
                         nation memory, increment H:X by 1

This concludes our coverage of the HCS08 addressing modes with some instruction examples. For additional information and complete listings of the instruction set, refer to the HCS08 Family Reference Manual which can be found at the Freescale website:

*www.freescale.com*

## *HCS08 Memory Expansion*

Memory size on the previous MC68HC08 family was a limiting factor because it could not be extended to more than 64K of address space. To maintain backward compatibility with the HC08, HCS08 core also limits the CPU address space to 64K. To overcome the address space limitations, some HCS08 based MCU integrate additional hardware to allow for increase in the memory size. For example, devices with the suffix QE128 have 128K of memory space and allow addressing of 64K additional Flash memory provided on-chip. The expansion gives the system designer and software developer additional memory space for code development. Devices that offer memory expansion integrate a memory management unit to support the additional program or data memory space needed. The following section will focus on the additional hardware implemented to support memory expansion.

## *Memory Management Unit*

The implementation of the MMU on HCS08 devices gives system designers the additional memory that may be needed by some applications. In fact, the MMU allows memory expansion of up to 4 mega-bytes of address space but currently, only 64k byte expansion is offered totaling 128K of memory address space.

HCS08 CPU has a linear address of 64K byte space. All resources, including on-chip I/O peripherals control and status registers as well as memory arrays are mapped into this address space. As previously stated, certain number of HCS08 based MCUs having QE128 suffix, incorporate an MMU to support addressing of larger memory space than the standard 64K. The expanded memory hardware or the MMU uses fast on-chip logic to implement a transparent paged memory or bank switching mechanism.

The MMU allows the HCS08 CPU to access the additional memory space either through the expansion window or through linear addressing. The expansion window is located between addresses $8000 - $BFFF provides access to multiple 16K byte pages. Whereas, linear addressing allows accessing any address in the extended flash memory space. Accessing the additional memory is described in the following section.

### Expansion Via the Program Page Window

Through the program page window, one of eight 16K byte pages can be selected. This expansion window falls between addresses $8000 through $BFFF. On QE128 devices, the limit is eight 16K pages. These pages can be uniquely addressed through the program page (PPAGE) register. This register is a memory mapped I/O located at address $0078. The PPAGE register is not used to access the program page window unless the CALL or RTC instructions are executed. Without the execution of these instructions, the CPU continues to access the 64K byte linear space starting at address $0000 through $FFFF. Note that the

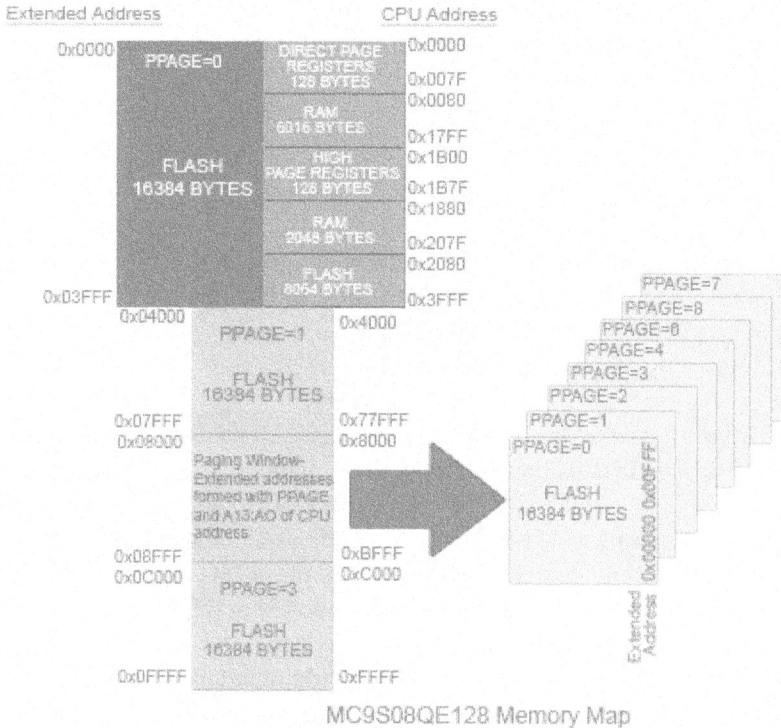

MC9S08QE128 Memory Map

**FIGURE 5.4. HCS08 Memory Map and Window Expansion**

64K linear memory map consists of four pages, 0, 1, 2 and 3. In other words, address $0000 - $3FFF is considered to be page '0', address $4000 - $7FFF is page '1', address $8000 - $BFFF is page '2' and finally address $C000 - $FFFF is considered to be page 3. However, these four 16K byte pages may still be accessed in the expansion window as page 0, 1, 2 and 3 when the CALL instruction is executed and the PPAGE register contains a value of 0, 1, 2 or 3, respectively. Since the PPAGE register is currently implemented with only 3 bits, the MMU allows selection of one of 8 different pages in the memory expansion window. For example, when the CPU accesses the expansion window at addresses $8000 through $BFFF, it simply accesses page '2' in the linear address space regardless the value written in the PPAGE register. If the PPAGE register is written with a different value, for example page 5, then when the CPU accesses the expansion window at the same address, page 2 will still be selected, not page 5. However, if the CALL Instruction is executed, then page 5 will be selected, assuming the PPAGE register still contains the value 5. Figure 5.4 depicts the graphical details of the memory map and the expansion window. As can be seen from the discussion above, the CALL instruction provides automatic page switching when executed. Also, to be able to switch back to the previous page, the return-from-call (RTC) instruction must be used. A closer look at these two instructions follows.

- CALL and RTC Instructions

The CALL is similar to a jump-to-subroutine (JSR) instruction. When executed, the CALL instruction will stack the return address, then selects any of the eight 16K pages in the expansion window at address $8000 - $BFFF. The specific page number is selected based on the value programmed into the PPAGE register. A value of '000'$_2$ selects page 0, a value of '001'$_2$ selects page 1, a value of '010'$_2$ selects page 2 and so on. Once executed, the CALL instruction pushes the current value of the PPAGE register and the 16-bit return address onto the stack. After the PPAGE is pushed, the new page number supplied by the CALL instruction is written into the PPAGE register. Finally, control is transferred to subroutine address specified in the instruction.

To return from the CALL, an RTC instruction is executed at the end of the subroutine. This instruction pulls the previously pushed PPAGE register as well as the 16-bit return address (PC value) from the stack. Lastly, normal operation is resumed to the return address which is the next instruction that following the CALL instruction.

### Expansion Via Linear Addressing

Linear addressing provides the user a mechanism to read or write any memory location in the extended address space. Besides the PPAGE register, the MMU includes 5 additional registers to allow for easy management and fast access to the extended address. A brief description of these registers is provided below.

- Linear Address Pointer Registers 2:0 (LAP2:LAP0)

The linear address pointers are written by the application software with a 17-bit address into 3 byte registers to select the extended address space. When written along with the linear data register, this allows read or write access to any location in the 128K linear memory space. A data byte or a 16-bit word may be accessed from the linear address through one of three data registers. These register are: linear byte (LB), linear byte post increment (LBP) and linear word post increment (LWP). The linear address pointer may be post incremented by either 1 or 2 after the data is accessed depending on the register accessed.

- Linear Byte Register (LB)

Reading from this register returns a data byte from the linear address space specified by LAP2:0 registers. Likewise, writing to the LB register will write a data byte at the linear memory address specified by the LAP2:0 registers.

• Linear Byte Post Increment Register (LBP)

Reading from this register returns a data byte from the linear address space specified by LAP2:0 registers. Likewise, writing to the LBP register will write a data byte at the linear memory address specified by the LAP2:0 registers. After the date byte is accessed, LAP is incremented by one to point to the next data entry.

• Linear Word Post Increment Register (LWP)

To allow for 16-bit word accesses, LWP and LBP are located sequentially in the MMU map. Reading a 16-bit word from the LWP register returns a data word from the linear address space specified by LAP2:0 registers. Likewise, writing to LWP register will write a data word at the linear memory address specified by the LAP2:0 registers. After the word is accessed, LAP is incremented by two to point to the next word entry.

Note that reading or writing to the linear byte (LB) register does not increment the address contained in LAP but reading or writing LBP or LWP will.

• Linear Address Pointer Add Byte Register (LAPAB)

Often, the application software may need to access a memory location with an offset from the base address contained in the LAP2:0 registers. It would involve some overhead to rewrite the LAP 2:0 every time we need to index memory from the base address. For this reason, the MMU is equipped with an 8-bit register called linear address pointer add byte (LAPAB) register, to do just that. The 2's complement value is added to the LAP2:0 register when LAPAB register is written, allowing an increase of up to 127 bytes or a decrease of up to 128 bytes from the base address contained in the LAP 2:0. For example, writing a value of 8 to LAPAB register causes the current address to increase by 8 from the current address pointed to by LAP, writing a value of $AA causes the current address to decrease by a value of $56, etc.

As can be seen from the discussion above, the benefits of the LAPAB register the application software needs only to write a one byte register to index or offset linear address by as much as +127 and -128 bytes without having to rewrite all 3 LAP registers.

Aside from the general use of accessing linear address space at run-time, these 3 registers, LB, LBP, and LWP are typically used during flash programming.

## Introduction to System Resets and Interrupts

Resets and interrupt operations are discussed together here because they share the same concept of vector fetching to force the CPU to a new starting point for further processing. HCS08 processors have several events that cause them to abort or suspend program execution. Each of these events will be covered in details in the following section. Before beginning coverage of interrupts and interrupt handling, a detailed description of all the HCS08 resets is provided first.

System resets have the highest priority and can not be masked. Once a reset is detected, the CPU immediately suspends main program execution and transitions into the reset handler via vector at address $FFFE and FFFF.

There can be up to 8 reset sources on HCS08 devices with power-on and external resets being the most common. Once detected, the CPU is forced into the reset state until the signal is negated. Resets can be externally or internally generated depending on the cause or the nature of the reset condition. Since there are a number of reset sources implemented on this architecture, often the application software must determine the source so it can handle the reset condition appropriately.

The HCS08 is equipped with a reset module which determines the cause of reset, as well as records the source of the reset. This information is stored in the system reset status register (SRS). The system reset status register format is shown in Figure 5.5. A brief cover-

| | Bit 7 | 6 | 5 | 4 | 3 | 2 | 1 | 0 |
|---|---|---|---|---|---|---|---|---|
| | POR | PIN | COP | ILOP | 0 | 0 | LVD | 0 |
| POR: | 1 | 0 | 0 | 0 | 0 | 0 | 1 | 0 |

FIGURE 5.5. System Reset Status Register (SRS)

age of each reset source follows:

### Power-on Reset

The power-on reset is considered a normal start up sequence when power is applied to the MCU for the first time. The reset condition forces the MCU to initial values and disables all on-chip peripherals. Once the reset signal is negated, HCS08 transitions into the reset handler by fetching the reset vector into the program counter located at address $FFFE and $FFFF. The contents of this vector is a pointer to the first instruction in the reset handler. The reset handler is a routine that configures the system to known state to satisfy a particular application's requirements. Coming out of reset, the MCU defaults to using the internally generated clock (self clocked mode) at approximately 8MHz bus clock speed. This allows the CPU to start program execution without having to wait for the external oscillator stabilization delay. Typically, the application software during initialization configures the clock module to operate at the desired system clock frequency.

### External Pin Reset

This reset is very similar to power-on and sometimes referred to as warm start. The MCU enter the reset state when the external reset pin is asserted. As soon as the reset signal is negated, the MCU fetches the reset vector at address $FFFE and $FFFF. Often, external reset may be used to abort program execution and restart the MCU due to some unusual event.

### COP Watchdog Reset

To ensure the MCU is fault tolerant, the computer operating properly (COP) timer prevents system lockup if the software becomes trapped in a loop with no controlled exit. To prevent the COP from timing out and generating a system reset, the application software must service the COP on a periodic basis by writing any value to the reset status register (SRS). Some HCS08s require writing a $55 then $AA sequence to clear the COP counter. For other HCS08s, any value is acceptable. If the application software fails to service the COP timer, assuming it is enabled, it will time out and will generate a reset condition to the MCU. The MCU immediately transitions into the reset handler by fetching the reset vector at address $FFFE and FFFF.

By default, after any reset, the COP timer is enabled. If the application does not need the COP function, it can be disabled by clearing the COP Enable (COPE) bit in the write-once system option register (SOPT).

### Illegal Opcode Reset

This condition is detected when an attempt is made to execute an instruction that is not defined by HCS08 processor instruction set architecture. An illegal[2] instruction causes the CPU to transition to the reset handler via the reset vector at address $FFFE and $FFFF.

---

2. The Stop instruction can also be considered an illegal opcode if it has been disabled in the SOPT register.

### Illegal Address Reset

An illegal address detect feature on some HCS08 derivatives forces the MCU to reset if the CPU attempts to access data or execute an instruction from any address that is identified as an illegal address in the memory map. For example, access to non-resident memory will cause a reset signaling to the processor. Once detected, the MCU transitions into reset handler via the reset vector at address $FFFE and FFFF.

### Loss-Of-Clock and Loss-Of-Lock Resets

Depending on the HCS08 derivative, these two signals can generate reset conditions or interrupts to the MCU depending on the application requirements. When programmed to generate resets, the MCU vectors to the reset vector at address $FFFE and $FFFF, and when programmed to generate interrupt, the loss-of-clock and loss-of-lock use vector $FFF6 and $FFF7.

### Low-Voltage Detect (LVD) Reset

HCS08 devices are equipped with a circuitry that detects low voltage conditions to protect memory contents from being corrupted to control the MCU states during supply voltage variations. The system comprises a power-on reset (POR) circuit and an LVD circuit with a user selectable trip voltage, either high (VLVDH) or low (VLVDL). The LVD circuit is enabled when the low voltage detect enable (LVDE) in the system power management status and control 1 register (SPMSC1) is high and the trip voltage is selected by low voltage detect voltage select (LVDV) control bit in the system power management status and control 2 register (SPMSC2). The LVD is disabled upon entering any of the stop modes unless the LVDSE bit is set.

The LVD can be configured to generate a reset upon detection of a low voltage condition by setting low voltage detect reset enable (LVDRE) to 1. After an LVD reset has occurred, the LVD system will hold the MCU in reset until the supply voltage has risen above the level determined by LVDV. If low-voltage detect circuitry is programmed to generate reset, the CPU uses the reset vector at address $FFFE and $FFFF, else will use the interrupt vector at address $FFF8 and $FFF9.

Resets are the only exceptions that do not save any state information or pushes anything on the stack. In other words, there is no recovery back to the main program when a reset is initiated since the machine state information is not saved. A few things should be kept in mind on the initial state of the core when a reset occurs:

1. Most MCU control and status register are forced to their initial default values.
2. The program counter is loaded from the reset vector at address $FFFE and $FFFF.

3. The Condition Code Register (CCR) interrupt mask (I) bit is forced to '1', masking all interrupts.

4. The System Reset Status Register (SRS) records the source of reset condition.

The information in the SRS register provides the application software an easy and quick method of determining the cause of reset and allows it to immediately take the appropriate action to handle the reset condition.

## HCS08 Interrupts

Often, it is necessary to execute sets of instructions in response to requests from various I/O peripheral devices. These request are asynchronous events with respect to main program execution. The instructions executed in response to the asynchronous events from I/O devices are called interrupt service routine or (ISR).

There are many interrupt sources in a particular HCS08 system. These interrupt sources originate from the on-chip and/or possibly off-chip I/O peripherals or pins. An interrupt is considered a normal condition which signals the CPU to handle an input or output event. When acknowledged, the CPU suspends normal program execution, creates a stack frame, saves its registers on the system stack, updates the condition code register and vectors to the appropriate interrupt service routine. Once the interrupt service routine handles the I/O event, the MCU restores its register from the stack frame, returns to the suspended program and then continues normal program execution.

### TABLE 5.1  HCS08 Vector Table

| Vector Priority | Address (High/Low) | Vector Name | Module | Source | Enable | Description |
|---|---|---|---|---|---|---|
| Lower | $FFC0/FFC1 through $FFCA/FFCB | | Unused Vector Space (available for user program) | | | |
| | $FFCC/FFCD | Vrti | System control | RTIF | RTIE | Real-time interrupt |
| | $FFCE/FFCF | Viic | IIC | IICIS | IICIE | IIC control |
| | $FFD0/FFD1 | Vatd | ATD | COCO | AIEN | AD conversion complete |
| | $FFD2/FFD3 | Vkeyboard | KBI | KBF | KBIE | Keyboard pins |
| | $FFD4/FFD5 | Vsci2tx | SCI2 | TDRE TC | TIE TCIE | SCI2 transmit |
| | $FFD6/FFD7 | Vsci2rx | SCI2 | IDLE RDRF | ILIE RIE | SCI2 receive |
| | $FFD8/FFD9 | Vsci2err | SCI2 | OR NF FE PF | ORIE NFIE FEIE PFIE | SCI2 error |
| | $FFDA/FFDB | Vsci1tx | SCI1 | TDRE TC | TIE TCIE | SCI1 transmit |
| | $FFDC/FFDD | Vsci1rx | SCI1 | IDLE RDRF | ILIE RIE | SCI1 receive |
| | $FFDE/FFDF | Vsci1err | SCI1 | OR NF FE PF | ORIE NFIE FEIE PFIE | SCI1 error |
| | $FFE0/FFE1 | Vspi | SPI | SPIF MODF SPTEF | SPIE SPIE SPTIE | SPI |
| | $FFE2/FFE3 | Vtpm2ovf | TPM2 | TOF | TOIE | TPM2 overflow |
| | $FFE4/FFE5 | Vtpm2ch4 | TPM2 | CH4F | CH4IE | TPM2 channel 4 |
| | $FFE6/FFE7 | Vtpm2ch3 | TPM2 | CH3F | CH3IE | TPM2 channel 3 |
| | $FFE8/FFE9 | Vtpm2ch2 | TPM2 | CH2F | CH2IE | TPM2 channel 2 |
| | $FFEA/FFEB | Vtpm2ch1 | TPM2 | CH1F | CH1IE | TPM2 channel 1 |
| | $FFEC/FFED | Vtpm2ch0 | TPM2 | CH0F | CH0IE | TPM2 channel 0 |
| | $FFEE/FFEF | Vtpm1ovf | TPM1 | TOF | TOIE | TPM1 overflow |
| | $FFF0/FFF1 | Vtpm1ch2 | TPM1 | CH2F | CH2IE | TPM1 channel 2 |
| | $FFF2/FFF3 | Vtpm1ch1 | TPM1 | CH1F | CH1IE | TPM1 channel 1 |
| | $FFF4/FFF5 | Vtpm1ch0 | TPM1 | CH0F | CH0IE | TPM1 channel 0 |
| | $FFF6/FFF7 | Vicg | ICG | ICGIF (LOLS/LOCS) | LOLRE/LOCRE | ICG |
| | $FFF8/FFF9 | Vlvd | System control | LVDF | LVDIE | Low-voltage detect |
| | $FFFA/FFFB | Virq | IRQ | IRQF | IRQIE | IRQ pin |
| | $FFFC/FFFD | Vswi | Core | SWI Instruction | — | Software interrupt |
| Higher | $FFFE/FFFF | Vreset | Systemcontrol | COP LVD RESET pin Illegal opcode | COPE LVDRE — — | Watchdog timer Low-voltage detect External pin Illegal opcode |

Table 5.1 on page 119 above provides a summary of all HCS08 vectors. It shows the reset and interrupt vectors, interrupt sources and local masks. Each interrupt source is enabled by its local mask bit and eventually by the global interrupt mask (I) bit in the condition code register (CCR). If multiple interrupts assert simultaneously, the higher priority request will be serviced first. The vector table shows the order of interrupt request priorities. The vector at the highest address has the highest priority, and the vector at the lowest address has the lowest priority. For example, in the vector table shown above, Software

Interrupt (SWI) has the highest interrupt priority, interrupt request (IRQ) pin has the second highest, and finally real-time interrupt (RTI) has the lowest priority.

### Interrupt Processing

When an enabled interrupt request is signaled, the CPU first completes the current instruction being executed before it responds to the request. Once acknowledged, the CPU hardware executes the following steps:

1. Creates an interrupt stack frame and saves all its registers on the stack, as shown in Figure 5.2. For backward compatibility with the M68HC05 family, the H register is not automatically saved or restored. It is good programming practice to push H onto the stack at the start of the interrupt service routine (ISR) and restore it just before the RTI instruction is executed at the end of the ISR. When programming in C, most compilers may automatically include push and pull H register.

2. While the CPU is responding to the interrupt, the 'I' bit in the CCR is automatically set, and thus masking all interrupts to avoid the possibility of another interrupt from being acknowledged inside an interrupt service routine.

3. Fetches the interrupt vector for the highest priority interrupt that is currently pending.

4. Fills the instruction queue with the first three bytes of program information starting with the address fetched from the interrupt vector location.

5. Begins program execution in the interrupt handler or ISR.

Once, the CPU handles the interrupt event, it requires an RTI instruction to restored all of its registers from the stack. The restored value of the 'I' bit in the CCR will reenable interrupts. RTI instruction also restores the CPU registers to their pre-interrupt values by pulling all previously saved information from the stack.

### Interrupt Nesting

Interrupt nesting is not recommended with the HCS08 architecture, but some applications may require nesting in order to process incoming events quickly. As mentioned earlier, the S08 can not process more than one interrupt request at a time. In other words, the CPU hardware is unable to nest multiple interrupts but it is possible to nest in software. Interrupt nesting or enabling is achieved by clearing the 'I' bit in the CCR with a clear-interrupt-mask (CLI) instruction during the execution of the ISR. The normal procedure for re-enabling interrupts inside the ISR is to first ensure that the original request is cleared. If the clear interrupt (CLI) instruction is executed prior to clearing the interrupt flag that caused the original request, the CPU will most likely see the same event requesting again. This condition is undesirable and should be avoided. Enabling interrupt nesting allow other incoming events to be serviced without waiting for the first service routine to complete.

The interrupt service routine normally ends with a return-from-interrupt (RTI) instruction. This instruction restores the CCR, A, X, and PC registers to their pre-interrupt values by popping the previously saved information from the stack.

**Device Modes of Operation**

HCS08 devices have a number of low power saving modes, as well as the normal run and background debug mode. Run mode is the normal operating mode and it is selected when the application needs to process events at relatively high speed. Background mode is used for application debug and development.

To reduce power consumption, numerous low power modes are available and may be invoked by the user code when the application is partially or fully idle. The following section discusses normal run mode followed by debug and low power saving modes.

• Run Mode

From power-on or external pin reset, this mode is selected if the BKGD/MS pin is high at the rising edge of the reset signal. In this mode, the CPU executes instruction from program memory starting at the address contents of the reset vector at $FFFE and $FFFF. This vector usually points the program counter (PC) to the start-up routine in order to initialize and configure the MCU to a particular application requirements.

• Background Debug Mode

This mode is an auxiliary operating mode used for software debug and development which is managed by the Background Debug Controller (BDC). The HCS08 based MCU is equipped with debug module (DBG) implemented in on-chip hardware that offers means of analyzing MCU operation during software development.

The BDC provides a single-wire debug interface via a single open-drain bidirectional pin to the target MCU. This interface offers a convenient means for programming the on-chip FLASH or EEPROM memories. Also, the BDC is the primary debug interface for development which provides non-intrusive access to memory data and traditional debug features. For example, the debug mode allows access to view and modify CPU registers, view and modify memory, set breakpoints, and invoke single instructions trace commands.

This mode is entered in any of five ways:

1. If BKGD/MS pin is asserted low at the rising edge of reset
2. When a BACKGROUND command is received through the BKGD pin

3. When a BGND instruction is executed

4. When encountering a BDC breakpoint

5. When encountering a DBG breakpoint

Once entered, the CPU is held suspended waiting for serial commands. For additional information, refer to Chapter 8.

- Wait Mode

This mode is entered when the CPU executes a WAIT instruction. This instruction causes the CPU to clear 'I' bit in CCR to enable interrupts before it enters the wait mode. Typically, the application software executes this instruction when there is a need to save power while waiting for some internal or external interrupt event before the program resumes. In this mode, the system clocks are on and applied to all enabled on-chip peripherals but the but CPU is not clocked and awaiting interrupt occurrence. Once signaled with an interrupt, the CPU exits the wait mode, saves all of its registers onto the stack, and begins program execution at the memory location pointed by the appropriate interrupt vector.

- Stop Modes

HCS08 have 3 additional lower power modes entered by executing a STOP instruction. In these modes, the CPU clock and most or all peripheral clocks are shut down. When all internal peripherals are shutdown, they are not able to wake up the CPU with an interrupt event rather, the interrupt must originate from either an external reset source or the real-time-interrupt timer (RTI), if enabled. The interrupt request (IRQ) pin is one of three possible sources that can wake up the CPU from Stop. A second possibility is a keyboard interrupt which is considered an external source and the third is the RTI timer.

To properly implement the stop modes, the Stop Enable (STOPE) bit in the System Option (SOPT) must be set to a logic '1'. If cleared, the MCU will not enter stop but instead will be treated as Illegal Opcode and forces a system reset. Since there are 3 low power stop mode options, each have advantages and disadvantages. The modes are described below:

1. Stop1

This is the lowest power consumption mode which is entered upon the execution of STOP instruction while the system management status and control register 2 (SPMC2) configuration bits, PDC and PPDC are set to '10'$_2$. This mode shuts down the entire system clocks including the RTI timer, and the only way to exit this mode is through the external interrupt request (IRQ) pin or Reset. This mode is not available on all HCS08 devices.

2. Stop2

This mode consumes a little more power than Stop1. It is entered upon the execution of STOP instruction while the SPMC2 configuration bits, PDC and PPDC are set to '11'$_2$ . This mode shuts down the entire system clocks except the RTI clock, if enabled. This mode may be exited by an IRQ, RTI or Reset.

3. Stop3

This mode consumes a little more power than Stop2. Again, this mode is entered upon the execution of STOP instruction while the SPMC2 register configuration bits, PDC and PPDC are set to '1X'$_2$ . This mode has the most options to exit. It is possible to wake from the analog comparator, the analog-to-digital converter, the RTI, a Keyboard interrupt or Reset.

When the application software executes a WAIT or STOP instruction, the CPU automatically clears the 'I' bit in the Condition Code Register to ensure interrupts are enabled prior to entering the low power mode, since interrupts are the normal way to wake the CPU from these low power modes.

- Conclusion

The HCS08 is increasingly becoming more popular among industrial and automotive designers. Its low power, reduced cost and flexibility make the HCS08 family of devices ideal for many applications. Since the HCS08 takes part in the Flexis family, system designers that are currently using this 8-bit architecture can very easily upgrade to 32-bit architecture with little or no change.

# CHAPTER 6

*Memory*

## Introduction to Memory

Products based on the S08 and ColdFire V1 cores, including Flexis, are single chip microcontrollers with all memory systems integrated, and generally without the option of an external bus. This makes the microcontroller easy to use, and allows the user to focus on the target application without having to be concerned about complex bus loading and timing specifications. However, since resource extension with external memories is very restricted, the application microcontroller must be chosen carefully to ensure that the application will have enough internal memory resources, even allowing for possible software or data size overruns and future product extensions. A major advantage of Freescale's Controller Continuum, and the Flexis product families in particular, is the availability of a wide range of memory size options within a pin compatible product family.

Both the HCS08 and ColdFire V1 CPUs are built on the Von Neuman architecture, which has a single address space for both data and program, with a unified data bus for transferring both instructions and data between the memory system and the CPU core. This makes it possible to store data in the system ROM, such as calibration tables or system configuration, and to execute programs from RAM, which is useful for some debug and monitoring functions, and for Flash programming.

Besides the CPUs, the most significant difference between the HCS08 and ColdFire V1 microcontrollers is the memory map organization. The HCS08 locates RAM and system registers at low addresses starting at address 0x00000 and stretching up in the memory map. This is to allow the use of efficient 8-bit addressing and the bit manipulation instructions, which is only available for page 0 of the memory map (addresses 0x0000 to 0x000FF). Flash ROM memory on the HCS08 is located at the top of the memory map 0xFFFFF and stretches down. This is to accommodate the reset and interrupt vectors which are located at the very top of the memory map.

In ColdFire V1 systems, however, the flash ROM memory is located at the bottom of the memory map, starting at address 0x(00)00_0000 and stretching up. This is to support the ColdFire processor exit from reset, where it fetches the initial 32-bit values for the supervisor stack pointer and program counter from locations 0x(00)00_0000 and 0x(00)00_0004 respectively. Compared to the HCS08, RAM is shifted up in the memory map to start at 0x(00)80_0000. Peripheral registers are located at the top of the ColdFire V1 memory map, 0x(00)FF_FFFF and stretch down.

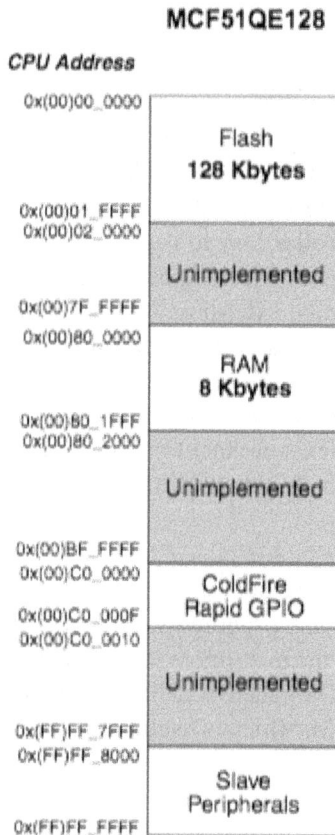

FIGURE 6.1. **Example ColdFire Memory Map**

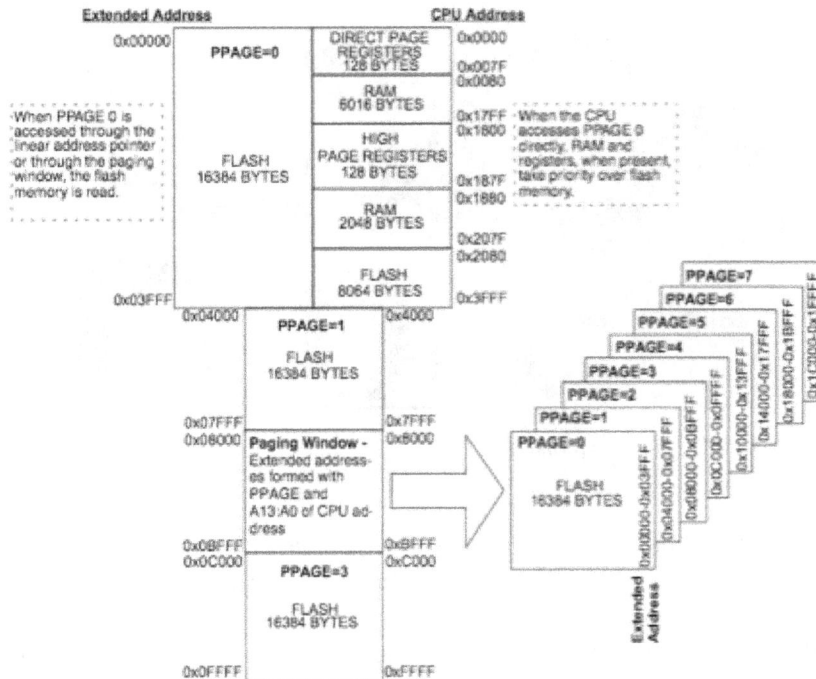

FIGURE 6.2. **Example HCS08 Memory Map**

## External Memories

A few ColdFire V1 products offer the option of an external bus, the Mini-FlexBus (see Chapter 9). As the name implies, this is a reduced functionality version of the standard ColdFire FlexBus, intended for interfacing to slave-only external devices such as ROMs, Flash, SRAM, programmable logic devices, and external peripherals. The Mini-FlexBus supports up to 20 bit addresses, and both 8 and 16 bit data ports in the following three modes:

- Up to a 20-bit address (non-multiplexed) with 8-bit data
- Up to a 20-bit address (multiplexed) with 16-bit data (write masking of upper/lower bytes not supported)
- Up to a 20-bit address (multiplexed) with 8-bit data

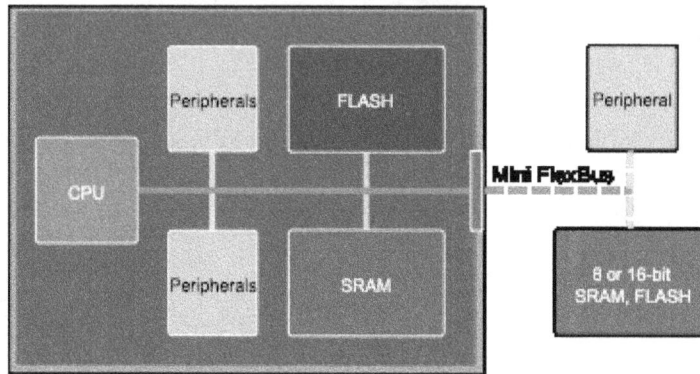

FIGURE 6.3. Mini Flexbus Expansion

With all S08 and ColdFire V1 microcontrollers without the Mini-FlexBus option, it is not practical to implement external program or working data memories. However, a number of specialized memories with synchronous serial interfaces such as SPI and I²C are available from a number of suppliers. These tend to offer non-volatile storage options for data that needs to be retained when the microcontroller is powered down. This may be system configuration data, or local application data. An example are the stored frequencies of selected radio stations in a tuner, portable or car radio. The following two types of memories are commonly used:

### Electrically Erasable Programmable Read Only Memory (EEPROM)

EEPROMs requires no power to retain data, and is popular due to the ability to reprogram a single byte, or even a bit, of data. The possible disadvantages of EEPROMs are the time and sequence required to reprogram a data location, and the limited number of write erase cycles, although modern devices support in excess of 1 million write erase cycles. Some EEPROM devices may also require multiple supply voltages to operate.

### Battery Backed-up Serial RAM

The advantages of RAM are the fast write times, without the need for an erase write control sequence, and unlimited number of read and write cycles. Some devices are also available with an integrated Real Time Clock (RTC), which will keep time at very low power consumption when the microcontroller is powered down. The main disadvantage or battery backed-up RAM is the requirement for a battery, or a super capacitor. For applications that may remain powered down for long periods of time, and do not require an RTC, a

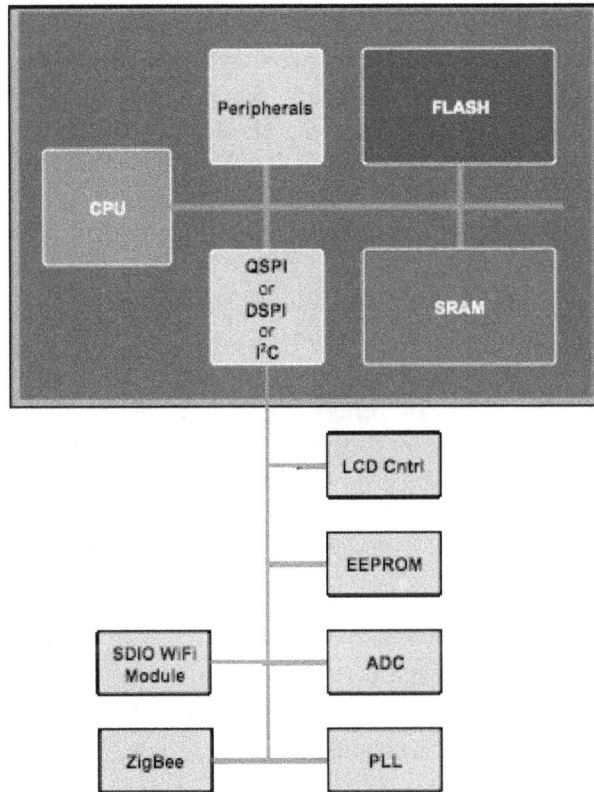

FIGURE 6.4. **Typical SPI or I2C Serial Bus Expansion**

long term non rechargable batteries such a Lithium based batteries are a good choice. The right combination of RAM and battery can provide reliable data storage for many years.

## *RAM*

The S08 and ColdFire V1 based microcontrollers include various amounts of integrated Static RAM (SRAM), which is primarily used for run time data storage, including the system stack. It is also possible to execute instructions from RAM, but generally this is only used during some debug operations or while re-programming the Flash memory.

SRAM data can be accessed as bytes with most instructions on the S08 and ColdFire V1 cores, in addition the ColdFire V1 core may access data as a 16-bit word or a 32-bit long word. Data may be addressed using the available direct and indirect addressing modes of each core. The HCS08 includes a very efficient 8-bit direct addressing mode, which is limited to the lowest 256 address locations, 0x0000 to 0x00FF. This HCS08 direct page includes the system and peripheral control registers, and up to 128 bytes of direct page RAM, that should be used for the most frequently accessed data and any bit based software flags or status bits. Both the HCS08 and ColdFire V1 cores can access data as bits using the corresponding bit manipulation instructions, but the HCS08 bit manipulation addressing is restricted to an 8-bit address and the direct address page, so any data that needs to be bit addressable must be placed in the direct page RAM.

At power-on, the contents of RAM are un-initialized, and are unaffected by reset. It is possible retain RAM data during power down, by maintaining the required minimum voltage on the $V_{RAM}$ pin. This may be done by using a back-up battery or a super capacitor. For compatibility with M68HC05 microcontrollers, the HCS08 core resets the stack pointer to 0x00FF, however, it is usually best to re-initialize the stack pointer to the top of the RAM so that the direct page of RAM may be used for frequently accessed and bit addressable data. The following instruction sequence inserted into the initialization code will reset the stack pointer to the top of memory. The RamLast parameter is part of the Freescale provided equate file.

```
LDHX        RamLast+1        ;point one past top of RAM

TXS                          ;SP<- (H:X-1)
```

When security is enabled, the RAM is considered a secure memory and is not accessible through the background debug mode.

## Flash

Flash memory is primarily intended for program storage, but may also be used for storing data that is not frequently changed, such as calibration and system configuration. It may be electrically erased and reprogrammed, allowing for product firmware updates, parameter re-calibration, and system re-configuration. Although the number of write erase cycles is limited, this is unlikely to be an issue with system software and stable data, such as calibration. The typical Freescale Flash based microcontrollers are specified for 100,000 write erase cycles, so if this not sufficient for your firmware updates, you need to have a word with your software development team. However, in many applications there is a need to store temporary values in a non-volatile memory, and with the right software the Flash memory may be used as an effective EEPROM.

**Flash Memory Organization**

Flash memory is divided into a number of pages or sectors, for HCS08 devices these are referred to as pages and typically each page contains 512 bytes, for the ColdFire V1 these are referred to as sectors and each sector typically contains 1024 bytes. In addition to pages and sectors, the Flash may also be divided into blocks, where each block is made up of a binary multiple of pages or sectors. For example, the HCS08 60K byte Flash consists of 120 pages of 512 bytes each, and a block may contain 512, 1K, 2K, 4K, 8K, 16K, or 32K bytes. An erased bit reads as a 1, and a programmed bit reads as a 0.

Like most 8-bit architectures, the HCS08 actually supports up to 16-bit address space, but even this has the inherent limit of 64K bytes. HCS08 products with more than 60K bytes of Flash use a Memory Management Unit (MMU) to extend the address range beyond 64K bytes. The MMU is used to access one of a number of 16K byte Flash pages that appear in a reserved 16K byte memory space reserved in the base memory map as shown in Figure 6.2, "Example HCS08 Memory Map.

ColdFire V1 based products in 0.25 micron technology (all families introduced at the time of writing this book) implement two interleaved 16-bit wide Flash blocks to improve performance. Although each block has a two cycle access time, by interleaving two blocks the memory system supports single cycle sequential accesses. Beyond the awareness of the impact on execution times during software changes of flow and non sequential data access, the user does not need to be concerned with the Flash memory organization, as it is taken care of by the system hardware.

Some ColdFire V1 products, such as the MCF51EM and MCF51MM families, separate the Flash memory into two distinct blocks with independent controllers. This allows code execution from one block, while the other is being erased or programmed, without the need to load and execute a routine from RAM. This may be used for storing data in the application, and is also used to support the robust firmware update function. To use the robust firmware update, the application software should fit into just one of the Flash blocks, or half the total Flash available. This leaves the other block available for programming with a completely new version of the software, while the application executes out of the first block. This is particularly useful for remote firmware/software updates, as the download may take a long time, or be interrupted, and several attempts may be needed for a successful transfer and validation. Where available, the hardware Cyclic Redundancy Check (CRC) may be used for validation. Once the new firmware is programmed and validated, the program can switch to the new firmware by programming a single control bit, which swaps the location of the two Flash blocks in the memory map. This also makes it possible to revert to the older firmware, if required, as long as it has not been erased or overwritten. The robust firmware update function is described in detail in "Robust Firmware Updates" on page 139.

Please note that products such as the MCF51EM and MCF51MM with the separate Flash memory blocks and independent controllers, also have separate control and status registers for each block.

**Flash Erasing and Programming**

While read accesses to the Flash memory require no special setup actions, erasing, writing and verifying of the Flash memory array is controlled with a set of Command Operations. The following Flash memory commands are available:

- Mass Erase
- Erase Verify (mass)
- Sector Erase
- Program
- Burst Program

Flash memory operations use a separate clock signal for timing operations, the Flash CLocK (FLCK). To function correctly, and avoid damage to the Flash memory array, the FCLK must be set to a frequency between 150 to 200 kHz. This is accomplished by programming the appropriate value into the Flash Clock DIVide (FCDIV) register. Until this register has been written, no Flash commands will be executed. Also, further changes of the FCDIV register may be prevented by writing the FDIVLD bit (7) of the FCDIV register to 0. The source system clock frequency, typically 25 MHz, is divided by a six bit divider and an optional by 8 bit prescaler, as programmed in the bit fields FDIV and PRDIV8 of the FCDIV register.

For example, if the typical system clock frequency of 25 MHz is used, the FCDIV field should be set to 0x0F (001111) and the PRDIV8 bit should be set to 1. The resulting FCLK frequency will then be 195 kHz, within the allowable range of 150 to 200kHz.

The Flash memory requires a higher voltage for erasing and programming than the microcontroller supply voltage. To provide the convenience of a single system voltage and reduce the target system cost, this higher voltage is generated by an integrated high voltage generator. If the system voltage falls beyond a given threshold, this high voltage generator is no longer able to supply sufficient current to program an entire 32-bit long word, and programming must be restricted to 16-bits per programming command. The status of the available system voltage is indicated by the Low Voltage Detect Flag (LVDF) in the System Power Management Status and Control 1 register (SPMSC1). If the LVDF is clear the Flash memory must be programmed 32-bits at a time, when the LVDF is set, the programming sequence must be modified such that odd and even bytes are written separately. This is achieved by programming the same location twice, with alternate bytes being set to all ones (the erased state), as follows:

1. First write, Address 0xaaaa_aaaa, Data 0xddFF_ddFF
2. Second write, Address 0xaaaa_aaaa, Data 0xFFdd_FFdd

## Mass Erase Command

The mass erase operation erases the entire Flash memory array, and is initiated with the following sequence:

1. Write to an aligned Flash block address (e.g. 0x(00)00_0000) to start the command sequence for the mass erase command. The actual address and data written will be ignored.
2. Write the mass erase command, 0x41, to the Flash Command (FCMD) register.
3. Clear the Flash Command Buffer Empty Flag (FCBEF) flag in the Flash Status (FSTAT) register by writing a 1 to FCBEF to launch the mass erase command.

When the mass erase command is successfully executed the Flash Command Complete Flag bit (FCCF) in the Flash STATus register will be set.If any portion of the Flash memory is protected, the Protection Violation Flag (FPVIOL) bit in the FSTAT register will be set, and the mass erase operation will not be carried out.

## Erase Verify Command

The erase verify operation verifies that the entire Flash memory array is erased., and is initiated with the following sequence:

1. Write to an aligned Flash block address (e.g. 0x(00)00_0000) to start the command sequence for the erase verify command. The actual address and data written will be ignored.
2. Write the erase verify command, 0x05, to the Flash Command (FCMD) register.
3. Clear the Flash Command Buffer Empty Flag (FCBEF) flag in the Flash Status (FSTAT) register by writing a 1 to FCBEF to launch the erase verification command.

When the erase verify operation is complete the Flash Command Complete Flag bit (FCCF) in the Flash STATus register will be set.If all Flash memory locations were successfully erased the Flash BLANK (FBLANK) bit in the FSTAT register will be set, else it will remain cleared. The erase verify operation will require number of bus cycles to execute that is equal to the number of addresses in the Flash array memory plus several overhead bus cycles.

### Sector Erase Command

The sector erase operation erases a 1 K byte sector of the Flash memory array, and is initiated with the following sequence:

1. Write to an aligned Flash block address in the sector to be erased (e.g. 0x(00)00_0400 for sector 1) to start the command sequence for the sector erase command. The actual address within the sector and data written will be ignored.

2. Write the sector erase command, 0x40, to the Flash Command (FCMD) register.

3. Clear the Flash Command Buffer Empty Flag (FCBEF) flag in the Flash Status (FSTAT) register by writing a 1 to FCBEF to launch the sector erase command.

When the sector erase command is successfully executed the Flash Command Complete Flag bit (FCCF) in the Flash STATus register will be set.If the Flash memory sector is protected, the Protection Violation Flag (FPVIOL) bit in the FSTAT register will be set, and the sector erase operation will not be carried out.

### Program Command

The program operation programs a previously erased address in the Flash memory using an embedded algorithm, and is initiated with the following sequence:

1. Write to a Flash block address to start the command write sequence for the program command. The data written will be programmed to the address written.

2. Write the program command, 0x20, to the Flash Command (FCMD) register.

3. Clear the Flash Command Buffer Empty Flag (FCBEF) flag in the Flash Status (FSTAT) register by writing a 1 to FCBEF to launch the program command.

When the program command is successfully executed the Flash Command Complete Flag bit (FCCF) in the Flash STATus register will be set.If the address written is protected, the Protection Violation Flag (FPVIOL) bit in the FSTAT register will be set, and the program operation will not be carried out.

### Burst Program Command

The burst program operation programs a previously erased sequential data in the Flash memory using an embedded algorithm. The pipelining of programing operations allows the necessary high voltage generation to remain operational between programming commands, thus reducing the time required to program multiple sequential locations, by greater than 50% for the entire array. The following sequence is used:

1. Write to a Flash block address to start the command write sequence for the program command. The data written will be programmed to the address written.

2. Write the burst program command, 0x25, to the Flash Command (FCMD) register.

3. Clear the Flash Command Buffer Empty Flag (FCBEF) flag in the Flash Status (FSTAT) register by writing a 1 to FCBEF to launch the burst program command.

4. After the FCBEF flag in the FSTAT register returns to a 1, repeat steps 1 through 3. The address written is ignored but is incremented internally.

When the burst program command is successfully executed the Flash Command Complete Flag bit (FCCF) in the Flash STATus register will be set.If an address to be programmed falls within a protected area, the Protection Violation Flag (FPVIOL) bit in the FSTAT register will be set, and the burst program operation will not be carried out.

### Flash Configuration and Security

The Flash memory is configured, controlled and monitored via the memory mapped registers shown in Figure 6.1.

TABLE 6.1

| Register Acronym | Register Name | Register Function |
|---|---|---|
| FCDIV | Flash Clock Divider | Controls the divider that generates the Flash memory clock for erase and program functions, from the system clock |
| NVOPT | Flash Options | A reserved Flash location (0x00_040C) stores the selected security options. On reset, the data from NVOPT is transferred to FOPT |
| FOPT | Flash Options | Read only register shows the status of security options |
| FCNFG | Flash Configuration | Controls writes to the security backdoor comparator |
| NVPROT | Flash Protection | A reserved Flash location (0x00_040C) stores the selected security options. On reset, the data from NVPROT is transferred to FPROT |
| FPROT | Flash Protection | Flash protection configuration and status |
| FSTAT | Flash Status | Shows the operational status of the Flash memory module |
| FCMD | Flash Command | Command register for controlling erase, program and verify operations |

The Flash Clock Divider register must be initialized before any erase or program operations can be carried out, as described earlier in "Flash Erasing and Programming" on page 132. The Flash Status register includes the following status flags:

- Command Buffer Empty (FCBEF), indicates that a new command may be written.
- Command Complete (FCCF), indicates that a command operation has completed
- Protection Violation (FPVIOL), indicates an attempt to program or write to a protected Flash area
- Access Error (FACCERR), indicates a violation of the command sequence, or the execution of a CPU STOP instruction while a Flash command operation was in progress

## Flash Memory Protection

Flash protection allows selected Flash areas to be protected against erasing and programming. The area of Flash to be protected is controlled through the Flash Protection (FPROT) register. Since this is a volatile register, it's configuration must be stored in a non-volatile location, a reserved Non Volatile Protection (NVPROT) location in the flash memory, and it's value is automatically loaded into the Flash Protection (FPROT) register at reset. The exact address of the Non Volatile Protection (NVPROT) location will vary according to the microcontroller used, and will be defined in the relevant CodeWarrior define file and the product Reference Manual.

The Flash Protection Open bit (FPOPEN) selects if the entire Flash memory should be protected, if only partial protection is required, then the area to be protected is selected in Flash Protection Size field (FPS). For partial protection, various sizes from 2 K bytes to 128K bytes are selectable.

On reset, the data stored in FPS and FPOPEN fields of Non Volatile Protection location (NVPROT) is transferred into the corresponding fields of Flash Protection (FPROT) register. FPROT may be read or written, as long as the protected area is increased. Any attempt to decrease the size of protected memory will be ignored. The selected protection stored in the NVPROT location may only be changed if the respective Flash area is not protected, and will take effect following the next reset. The respective product user manual shows the available options for Flash memory protection.

## Flash and Microcontroller Security

It may be desired to secure a microcontroller application for a number of different reasons. A popular motivation is to prevent equipment and or software cloning. As the complexity of software increases the portion of development investment going to software keeps increasing, and therefore becomes more valuable and more desirable to protect. Other reasons for security could be data protection, such as patient or customer data, product authentication codes or cryptography keys. The Flash memory on Freescale microcontrollers includes a Security Bit that may be used to lock out all external accesses, thus providing a high level of protection for internally stored programs and data. However, no security system is absolute, and care must also be taken that the application program itself does not introduce vulnerabilities.

Setting the security bit generally disables all external access, by disabling the background debug mode, the JTAG mode, and any external parallel buses. On products with the Mini-FlexBus, it is possible to enable the on chip security while allowing use of the external bus for data accesses only. On **ColdFire V1** the security "bit" is in fact implemented as two bits, to help prevent accidental securing of the device, on HCS08 it is a single bit. On **ColdFire V1** devices security is engaged by setting the two nonvolatile register bits (SEC01, SEC00) in the Flash Options (FOPT) register to 10. Any other combination will leave security disengaged. **On HCS08 devices it's the inverse, combination 10 in SEC01 and SEC00 disengages security, and any other combination engages it**. During reset, the contents of the Non-volatile Options (NVOPT) location in the flash memory, are copied into the working Flash Options (FOPT) register in the high-page register space. Security is engaged by programming the NVOPT location, and this is generally be done at the same time the Flash memory is programmed. Security implementation varies slightly between the pin-compatible HCS08 and ColdFire V1 family of devices. This is a result of differences inherent in the HCS08 and ColdFire microcontroller architectures. This chapter describes the ColdFire V1 implementation and notes the significant differences for the HCS08.

Care must be taken that security is not enabled during development, as this will seriously hinder most debugging. Another consideration is the evaluation of any field returns once the final product is shipping. For these reasons Freescale offers two options with the secure function:

- Block Erase
- Backdoor Access

Even when security is engaged, it is possible to initiate the flash Block Erase operation through the background debug mode (BDM). After the entire the Flash memory is erased, the security bits in the NVOPT location will also be erased, but by then it is not possible to recover data or program information. It does, however, allow the microcontroller to be reprogrammed with new software, or evaluated for any possible system failures.

The backdoor access is user selectable, and allows the user to program a private backdoor access key into the device. The user must also include a backdoor access request function in the application, that allows the requester to enter the backdoor key for comparison. The microcontroller security system will carry out the comparison of the entered and the stored keys, and if matched, will change the security bits into the unsecured state. If it is required that the microcontroller remains unsecured past the next reset, then the corresponding security bits in the Non Volatile Protection location (NVPROT) must also be re-programmed to the unsecured state.

The backdoor access is enabled by setting the Key Enable (KEYEN) bits in the Flash Options (FOPT) register, and the corresponding non-volatile bits in the Non Volatile Protection location (NVPROT), to the Enabled combination (10), and programming an eight byte Backdoor Comparison Key into the reserved Flash memory locations, 0x(00)00_0400 - 0x(00)00_040C on ColdFire V1 microcontrollers. Security can then be temporarily disabled by the user code, setting the Key Access (KEYACC) bit in the Flash Configuration (FCNFG) register enables the key comparison system, and writing the backdoor key to the designated Flash address starts the comparison. The KEYACC bit redirects the writes to the backdoor key comparison logic, that will temporarily disable security on a successful match. It is highly recommended that the backdoor key is not stored as part of the user program, but must be entered externally when so required by the user program for unlocking. If required the microcontroller security may be disabled for longer term by reprogramming the Non Volatile Protection location (NVPROT) field in the reserved Flash location.

The following is the exact sequence required for temporarily disabling security using the backdoor key:

1. Set the Key Access (KEYACC) bit in the Flash Configuration (FCNFG) register
2. Execute three NOP instructions to provide time for the backdoor state machine to load the starting address and number of keys required into the flash state machine
3. Sequentially write the correct longwords to the flash address(es) containing the backdoor keys
4. Clear the Key Access (KEYACC) bit. Depending on the user code used to write the backdoor keys, a wait cycle (NOP) may be required before clearing the KEYACC bit.
5. If all data written match the stored backdoor keys, the microcontroller is unsecured and the SEC[1:0] bits in the Flash Options (FOPT) register are forced to an unsecured state. If required, the Security Bits SEC may now also be reprogrammed the Non Volatile Protection location (NVPROT) field in the reserved Flash location.

The backdoor key access sequence is monitored by an internal security state machine. An illegal operation during the backdoor key access sequence causes the security state machine to lock, leaving the MCU in the secured state. A reset of the MCU causes the security state machine to exit the lock state and allows a new backdoor key access sequence to be attempted. The following operations during the backdoor key access sequence lock the security state machine:

1. If any of the keys written does not match the backdoor keys programmed in the Flash array

2. If the keys are written in the wrong sequence

3. If any of the keys written are all 0's or all 1's

4. If the Key Access (KEYACC) bit does not remain set while the keys are written

5. If any of the keys are written on successive CPU clock cycles.

6. Executing a STOP instruction before all keys have been written

After the backdoor keys have been correctly matched, security is temporarily turned disabled (until the next reset). After security is temporarily disabled, the Flash security byte in the Non Volatile Protection location (NVPROT) location may be programmed to the unsecured state, if it is desired for the microcontroller to remain unsecured following the next reset.

## Robust Firmware Updates

As firmware updating becomes more common, even for microcontroller systems, and as it is increasingly desirable to carry out these firmware updates remotely, it is necessary to provide a robust method to support such firmware updates. Although it is possible to design robust firmware updates with most existing microcontrollers through careful software design, this tends to be complex and time consuming. Some ColdFire V1 microcontrollers, such as the MCF51EM and MCF51MM, now include hardware support that make robust firmware updating much easier and highly reliable.

This is achieved by dividing the integrated Flash array into two blocks, with separate Flash controllers, and then providing a means to swap the two arrays in the memory map. Since ColdFire microcontrollers fetch the initial 32-bit values for the supervisor stack pointer and program counter from locations 0x(00)00_0000 and 0x(00)00_0004, this also swaps the array that the system effectively boots from.As with any robust firmware update scheme, the cost is that only half of the total Flash memory is available for the application firmware. A fringe advantage is, that it is possible to retain the previous version of firmware until the next update, so in case of unexpected problems, it is very easy to revert back one version.

This allows the application to run from the Flash array currently starting at address 0x(00)00_0000, while erasing and reprogramming the upper Flash array. It is recommended to carry out a cyclic redundancy check (CRC), using the CRC hardware accelerator, to verify that the transfer and programming is complete and error free, before carrying out the flash block and firmware swap. Should the firmware update be interrupted for any reason, or an error be discovered, the firmware update process can be restarted, while the application continues to function with the previous firmware.

In microcontrollers without the Independent Real Time Clock (IRTC) module, like the MCF51MM, firmware update is applied by toggling the Array Select (ARRAYSEL) bit in the System Options 3 (SOPT3) register. At reset this control bit is loaded with the Array Select (ARRAYSEL) bit 0 in the reserved non-volatile register Flash Array Select (FLASHAS) located at address 0x(00)00_0410, providing that bit 1 of ARRAYSEL in FLASHAS is 0. This is to prevent unintentional block swapping when the flash is fully erased, all 1s'.

For microcontrollers with the Independent Real Time Clock (IRTC) module, like the MCF51EM, firmware update is applied by toggling the IRTC Configuration Data Register Configuration bit 0 (IRTC_CFG_DATA[CFG0]). Please note that this is located in the Independent Real Time Clock (IRTC), described in Chapter 16, and this configuration register is maintained by the backup battery when the main system power is not available. The tamper detect circuit monitors the backup battery, and indicates if battery power was lost. If a backup battery is not used in the application, then Flash array 0 must always include an initialization routine that determines which array has the most up to date firmware, and executes a swap if necessary. This can be easily implemented by saving the firmware version number in a specified Flash location for comparison.

The recommended procedure for swapping Flash arrays, is to load a short software routine into RAM, that implements the following steps:

1. Disable Flash speculation by the V1 ColdFire CPU by setting the CPU Configuration Register Flash Speculation Disable bit (CPUCR[FSD]) to 1
2. Disable all interrupts
3. Depending on the microcontroller used, toggle either the Array Select (ARRAYSEL) bit in the System Options 3 (SOPT3) register, or the IRTC Configuration Data Register Configuration bit 0 (IRTC_CFG_DATA[CFG0])
4. Re-enable interrupts
5. If desired, re-enable Flash speculation by the V1 ColdFire CPU by re-setting the CPU Configuration Register Flash Speculation Disable bit (CPUCR[FSD]) to 0
6. Jump back (using an indirect jump) to main application code residing in the Flash memory

Note that swapping the Flash arrays does NOT change the location of the Flash controllers in the memory map. Registers in FTSR1 and FTSR2 are fixed in the memory map. Only the Flash arrays are impacted by changing ARRAYSEL or IRTC_CFG_DATA[CFG0].

It is possible to use an alternative flash array swapping procedure that does not involve having to load a software routine into RAM and executing this. This may be desirable due to the inherent risks in executing any code from RAM. Since the array swap is implemented as a change in the decoding of the high order addresses, and does not actually change the flash array directly, the swap is accomplished in a single system clock cycle. It is therefore possible to implement the flash array swap in flash, but care must be taken that the correct software is correctly located at the immediately sequential addresses in the updated flash array following the instruction that executes the swap. It is also advisable to insert several No Operation (NOP) instructions following the swap, in both arrays. It is also good practice to locate the array swap routine in all firmware versions at exactly the same addresses, so that it is always present in the same location in both arrays. The following procedure should then be followed:

Located in original flash array;

1. Disable Flash speculation by the V1 ColdFire CPU by setting the CPU Configuration Register Flash Speculation Disable bit (CPUCR[FSD]) to 1
2. Disable all interrupts
3. Depending on the microcontroller used, toggle either the Array Select (ARRAYSEL) bit in the System Options 3 (SOPT3) register, or the IRTC Configuration Data Register Configuration bit 0 (IRTC_CFG_DATA[CFG0])
4. Several No Operation (NOP) instructions

Located in updated flash array;

5. At the array addresses as 4. above, several No Operation (NOP) instructions
6. Re-enable interrupts
7. If desired, re-enable Flash speculation by the V1 ColdFire CPU by re-setting the CPU Configuration Register Flash Speculation Disable bit (CPUCR[FSD]) to 0
8. Main application code, or a Jump (using an indirect jump) to it.

In microcontrollers with the Independent Real Time Clock (IRTC) module, like the MCF51EM, the correct selection of the flash array to be used is dependent on the IRTC backup power being maintained while the main microcontroller power is off. In case all power is lost, the data in the IRTC Configuration Data Register Configuration (IRTC_CFG_DATA) will also be lost, fortunately this is detected and indicated by the IRTC tamper detection mechanism as a tamper interrupt with the Tamper detect flag (TMPR) bit 0 in the IRTC Interrupt Status Register (IRTC_ISR). In applications using a microcontroller with an IRTC, but where loss of battery backup is of no consequence, or where no backup battery is used, robust firmware updates may still be used, but the correct flash array configuration must be determined by software after reset. The recommended approach is to include a firmware version number at a predetermined address, and include a software routine in the reset initialization software that checks the firmware version numbers in both flash arrays, and then configures the IRTC Configuration Data Register Configuration bit 0 (IRTC_CFG_DATA[CFG0]) accordingly. Although this routine should only be needed in array 0, as this will be located at start address 0x(00)00_0000 by default, it is prudent to also include it in an initialization routine in array 1, for the possible case where the tamper detect may be triggered, but sufficient power was retained up to reset to retain data in the IRTC Configuration Data Register Configuration bit 0 (IRTC_CFG_DATA[CFG0]).

# CHAPTER 7

# *Interrupt Controller (INTC)*

## *Introduction to ColdFire V1 Interrupt Controller*

Interrupts provide a mechanism for the processor core to react to real-time events that may be generated internally from the on-chip I/O peripherals or from an external source such as an interrupt request pin or port. Once acknowledged, the CPU saves the current machine state onto the system stack, executes an interrupt service routine (ISR) to service the event, and then returns to the interrupted program by restoring the machine state information to allow resumption of the interrupted program.

Each interrupt source supplies a unique vector number when acknowledged by the core to reduce interrupt handling overhead by eliminating the need to poll I/O status inside the interrupt service routines.

Due to a large number of I/O peripherals integrated into each member of the Cold-Fire V1 family derivatives, an interrupt controller (INTC) is integrated on the all derivatives and designed to handle up to 192 interrupt sources.

Currently, only 32 interrupt sources are implemented on both ColdFire V1 and the HCS08 MCUs (Flexis) to provide complete hardware compatibility between the 8-bit and 32-bit MCU families.

It is important to mention here that the HCS08 architecture does not integrate an interrupt controller such as the one found in the ColdFire V1 because historically Freescale 8-bit MCUs were designed to handle incoming interrupt request automatically by the core hardware.

To provide applications that demand the least amount of latency to servicing events, the (INTC) supports interrupt nesting of up to seven priority levels, permitting a higher priority level to preempt lower priority ones.

The INTC implementation on CFV1 provides a unique vector and pre-assigned level to each interrupt source. The levels assigned are from '1' to '7', level '1' is the lowest priority, and '7' is the highest. Since there are 39 interrupt sources and only seven levels supported, it means that a group of interrupt sources must share the same level.

For example, interrupt level 4 is assigned by the INTC to interrupt source number 10, 11, 12, 13, 14, and INTC_FRC 35, within this level there are actually 9 INTC priorities, 7 - 0 plus a MID priority, where 0 is lowest and 7 is the highest. The MID priority falls between priority 3 and 4. Refer to Table 1 for the INTC priority assignments.

To fully understand the interrupt levels and priority assignments for all 39 sources, let us use level 5 shown in Table 1 below as an example of how interrupt sources are prioritized. Within this level, there are 5 interrupt sources: TPM2_CH0 is assigned to the highest, TPM2_CH1 is assigned to the second highest, TPM2_CH1 is assigned to the third highest, and so on, while interrupt source Force_LVL5 is assigned to the lowest priority. Notice that priority 2, 3 and 4 are not used and are not assigned.

Again, let us use level '1' as another example, Timer 3 has 8 requests and is assigned to level 1. Within this level, TPM_CH0 is assigned to the highest, priority 7, TPM_CH1 assigned to the second highest, priority 6, and so on. And finally, Force_LVL1 is assigned to the lowest priority. The question that comes to mind, what happens if multiple requests at the same level are pending simultaneously? the answer is the highest priority request will be acknowledged and serviced first. Let us assume that TPM3_CH2, TPM3_CH4, TPM3_CH5 and TPM_OVFL are requesting at the same time, since TPM3_CH2 has the highest priority, it will be acknowledged and serviced before all other requests at this level. Keep in mind that if a higher level is currently requesting at the same time as level 1, then the highest priority within the higher level will be acknowledged and service first.

**Interrupt Priority Elevation**

As previously stated, the INTC interrupt assignment is fixed. The fix assignment could pose a potential problem for some applications that often require that a low priority request be temporarily or permanently elevated to a higher priority. To overcome this problem, the INTC allows elevation or remapping of two interrupt sources to the highest level and highest maskable priority. This can be achieved through two registers implemented in the interrupt controller's programming model. The two registers are: the INTC_programmable level 6 priority 7 (INTC_PL6P7), and the INTC_programmable level 6 priority 6 (INTC_PL6P6).

For example, to elevate the priority of any interrupt source[1] to level 6 priority 7, simply write the request source number into INTC_PL6P7. Likewise, to elevate another interrupt source to level 6 priority 6, write the interrupt source number into INTC_PL6P6. Again, if you refer to the table below, it is possible to remap one request to level 6 priority 7, and another to level 6 priority 6. Below are two examples that depict elevation of 2 requests to the highest maskable priorities:

**Example 7.1.** **To elevate SCI2 Receive (SCI2_RX) request currently at level 2 priority 4, to level 6 priority 7, write $20_{10}$ to PLP6P7 register.**

**Example 7.2.** **To elevate Timer3 Channel 4 (TPM3_CH4) request currently at level 1 priority 3, to level 6 priority 6, write $27_{10}$ to PLP6P6 register.**

TABLE 1. **INTC Interrupt Level Assignment**

| IRQ Source | Level | Priority within Level | Interrupt Source Number | Vector Number |
|---|---|---|---|---|
| IRQ_Pin | 7 | Mid | 0 | 64 |
| Low_Voltage[a] | 7 | 3 | 1 | 65 |
| Force_LVL7 | 7 | 0 | INTC_FRC[32] | 96 |
| Remapped_I6P7 | 6 | 7 | INTC_PL6P7 | *[b] |
| Remapped_I6P6 | 6 | 6 | INTC_PL6P6 | * |
| TPM1_CH0 | 6 | 5 | 2 | 66 |
| TPM1_CH1 | 6 | 4 | 3 | 67 |
| TPM1_CH2 | 6 | 3 | 4 | 68 |
| TPM1_OVFL | 6 | 1 | 5 | 69 |
| Force_LVL6 | 6 | 0 | INTC_FRC[33] | 97 |
| TPM2_CH0 | 5 | 7 | 6 | 70 |
| TPM2_CH1 | 5 | 6 | 7 | 71 |
| TPM2_CH2 | 5 | 5 | 8 | 72 |
| TPM2_OVFL | 5 | 1 | 9 | 73 |
| Force_LVL5 | 5 | 0 | INTC_FRC[34] | 98 |
| SPI2 | 4 | 7 | 10 | 74 |
| SPI1 | 4 | 6 | 11 | 75 |
| SCI1_ERR | 4 | 5 | 12 | 76 |

---

1. Force levels can not be assigned to higher priorities.

**Interrupt Controller (INTC)**

TABLE 1. INTC Interrupt Level Assignment

| IRQ Source | Level | Priority within Level | Interrupt Source Number | Vector Number |
|---|---|---|---|---|
| SCI1_RX | 4 | 4 | 13 | 77 |
| SCI1_TX | 4 | 3 | 14 | 78 |
| Force_LVL4 | 4 | 0 | INTC_FRC[35] | 99 |
| I²Cx | 3 | 7 | 15 | 79 |
| KBIx | 3 | 6 | 16 | 80 |
| ADC | 3 | 5 | 17 | 81 |
| ACMPx | 3 | 4 | 18 | 82 |
| Force_LVL3 | 3 | 0 | INTC_FRC[36] | 100 |
| SCI2_ERR | 2 | 5 | 19 | 83 |
| SCI2_RX | 2 | 4 | 20 | 84 |
| SCI2_TX | 2 | 3 | 21 | 85 |
| RTC | 2 | 2 | 22 | 86 |
| Force_LVL2 | 2 | 0 | INTC_FRC[37] | 101 |
| TPM3_CH0 | 1 | 7 | 23 | 87 |
| TPM3_CH1 | 1 | 6 | 24 | 88 |
| TPM3_CH2 | 1 | 5 | 25 | 89 |
| TPM3_CH3 | 1 | 4 | 26 | 90 |
| TPM3_CH4 | 1 | 3 | 27 | 91 |
| TPM3_CH5 | 1 | 2 | 28 | 92 |
| TPM3_OVFL | 1 | 1 | 29 | 93 |
| Force_LVL1 | 1 | 0 | INTC_FRC[38] | 102 |

a.  Low Voltage Warning and Low Voltage Detect share the same vector when interrupts are enabled. Alternatively, the Low Voltage Detect circuitry can be enabled to generate reset.

b.  Programmable (Any I/O vector)

## Level 7 assignments

Most ColdFire V1 MCUs integrate an external interrupt request (IRQ) pin. Typically, the IRQ pin may be used to signal the CPU of some unusual or critical event condition. To provide a quick recognition of this input signal with the least amount of latency, the IRQ pin is assigned to the highest level (level 7) middle (Mid) priority. Mid assignment is a fixed priority that lies between 3 and 4.

Also, the low-voltage detect (LVD) circuitry and the software interrupt Force-LVL7 share the same level with the IRQ pin. However, LVD is assigned priority 3 and Force_LVL7 is assigned priority 0.

### Software Interrupts

The CFV1 interrupt controller (INTC) allows the application software to schedule a unique interrupt for each possible level at the lowest priority within the level. Scheduling or forcing a software interrupt may be used for functional or debug purposes.

In some case, an interrupt service routine may have a critical portion that requires a high priority and a lesser-critical portion that can be scheduled to execute later at lower priority. The critical portion of the interrupt service routine can schedule a lower priority request for the lesser-critical portion with a software interrupt. The software interrupt request can be scheduled by writing the INTC_FRC register.

These interrupts may be self-scheduled by setting one or more bits in the INTC_FRC register. The INTC_FRC register contain one bit for each interrupt level and may be modified directly using a read-modify-write sequence. If a read-modify-write sequence is not desirable, the INTC has two register pairs that allow a simple write operation to force or clear a software interrupt. The two registers are: INTC set interrupt force (INTC_SFRC) and INTC clear interrupt force (INTC_CFRC). To schedule a software interrupt, the application software writes INTC set interrupt force (INTC_SFRC) register with a value between 0x20 and 0x26 to schedule one of the seven software interrupts. Below are examples that may be used to force any of the seven software interrupt requests:

**Example 7.3. Scheduling Software Interrupts**

To force level 7 request, write 0x20 ($32_{10}$) into INTC_SFRC register
To force level 6 request, write 0x21 ($33_{10}$) into INTC_SFRC register

--
--
--

To force level 1 request, write 0x26 ($38_{10}$) into INTC_SFRC register

Only values 0x20, 0x21, 0x22, 0x23, 0x24, 0x25 and 0x26 should be used. Other values outside 0x20-0x26 range will not have any effect nor will cause error conditions.

To clear a software interrupt request, the interrupt handler writes the INTC clear interrupt force (INTC_CFRC) with a value between 0x20 and 0x26 as shown in the example below:

**Example 7.4. Scheduling Software Interrupts**

To clear level 7 request, write 0x20 ($32_{10}$) into INTC_CFRC register
To clear level 6 request, write 0x21 ($33_{10}$) into INTC_CFRC register

--
--
--

To clear level 1 request, write 0x26 ($38_{10}$) into INTC_CFRC register

-------------------------------------------------------------------------------

## Software Interrupt Acknowledge

To further improve performance with a system having high rates of interrupt activity, the overhead of a context restore and return-from-exception (RTE) instruction execution can be avoided. This may be accomplished by enabling the core to execute a software interrupt acknowledge (IACK) cycle at the end of the interrupt service routine.

Basically, there are two methods to enable INTC query at the end of the service routine to find whether or not there are other requests currently pending. These two methods are described below:

• Automatic IACK Cycle Enabling

The interrupt controller design supports the concept of automatic vectoring to another request service routine during the execution of an interrupt handler. Automatic vectoring is a useful concept that allows the interrupt controller during the execution of an interrupt service routine examine if there are other pending requests. If a request is pending, the overhead associated with interrupt exception processing (including machine state save/restore functions) are avoided. In general, IACK cycle is performed near the end of an interrupt service routine. If the INTC detects active interrupt sources pending, the current ISR passes control to the appropriate service routine, but without taking another exception processing.

The automatic vectoring concept is supported by implementing a software interrupt acknowledge (SWIACK) register. This register is read by the INTC and returns an 8-bit vector number associated with the highest unmasked interrupt source, if a request pending.

The CPU uses the returned value as the vector number and transitions to that request service routine without restoring context. But if there are no pending requests, the interrupt controller returns an all-zero value as the operand from the SWIACK register and the CPU simply restores the context from the stack then returns to the interrupted program.

Automatic query of the interrupt controller can be enabled by setting the CPU configuration interrupt acknowledge enable bit (CPUCR$_{[IAE]}$). With this bit set, the interrupt controller is forced to execute an IACK cycle to read the INTC software interrupt acknowledge (SWIACK) register.

• Manual Software IACK Cycle

The interrupt controller also supports the concept of a software IACK. This method provides a mechanism to query the interrupt controller near the end of an interrupt service routine (after the current interrupt request has been negated) to determine if there are any pending requests within the current level being serviced. If the response to the software IACK's cycle is non-zero, the service routine uses the returned value as the vector number of the highest pending interrupt request and passes control to the appropriate handler. If the returned value is zero, this means there are no requests currently pending.

Each interrupt level has a level interrupt acknowledge (INTC_LVL$n$IACK) register associated with it, where $n$ is a value between 1 and 7. Each of these 8 bit registers hold the highest priority interrupt level vector for the group that share the same level. The eight (INTC_LVL$n$IACK) registers are implemented in the INTC programming model to provide a method to query the interrupt controller only for requests within the level currently being serviced, if so desired.

**Interrupt Vector Generation**

It was stated earlier that the ColdFire V1 vector table consists of 256 entries. For convenience we are showing the basic structure of the exception vector table here to help the reader understand how the interrupt vectors are generated by the INTC. Recall that the exception vector base address is determined by the value written into the VBR. As can be seen from Table 2, the first 64 exception vectors, vector number 0- 0x3F are reserved for the core to handle system exception such as, reset, access error, address error, etc. The last 192 entries are reserved for user's interrupt vectors. The interrupt controller is currently designed to support 30 interrupt sources, numbered 0 - 29, two remappable requests, plus 7 software requests. For each interrupt source, the INTC uses the following equation to obtain the vector number being acknowledged:

$$\text{Vector\_Number} = 64 + \text{Interrupt source number} \qquad \text{(EQ 1)}$$

Future ColdFire V1 derivative could support a lot more than 39 requests. This will be determined by the number on-chip I/O peripherals integrated.

**TABLE 2. Exception Vector Assignments**

| Vec # | Offset | Stacked PC | Assignment |
|-------|--------|------------|------------|
| 0x0 | 0x000 | - | Initial SP |
| 0x1 | 0x004 | - | Initial PC |
| 0x2 | 0x008 | Fault | Access Error |
| 0x3 | 0x00C | Fault | Address Error |
| --- | --- | --- | --- |
| --- | --- | --- | --- |
| 0x20-0x2F | 0x080- 0x0BC | Next | Trap # 1 - Trap # 15 |
| 0x30-0x3C | 0x080- 0x0BC | --- | Reserved |
| 0x3D | 0x0C0 | Fault | Unsupported Instructions |
| 0x3E-0x3F | 0x1F8-0x1FC | --- | Reserved |
| 0x40-0xFF | 0x100 - 0x3FC | Next | I/O Interrupts |

## Interrupt Nesting Prevention

ColdFire V1 processor guarantees that the first instruction in the interrupt handler is executed before sampling for interrupts is resumed because it is treated as part of exception processing sequence. This guarantee provides a way to raise the interrupt mask to the highest level, if so desired, to prevent a higher level from preempting the current level being serviced. The store/load status register (STLDSR) instruction is provided in the ISA to safely mask interrupts by raising the mask field to the desired level.

Nesting prevention can be used to provide compatibility with the HCS08 family. If nesting is disabled by raising the interrupt mask to the highest level in every interrupt service routine, the ColdFire V1 core can provide compatibility since there is no nesting support by the HCS08 architecture.

## Low Power Modes

Often when the application is idle, the CPU may enter one of the low power modes by executing either a wait or stop instruction for the purpose of reducing power consumption. The low power mode may be exited by the occurrence of an enabled unmasked interrupt request. The INTC_WCR register in the interrupt controller defines wake-up conditions for interrupt recognition during wait and stop modes. The wake-up mode set up can be configured by executing the following steps:

1. Set the ENB bit in the INTC_WCR to enable wakeup and define the interrupt mask level needed to force the core to exit the wait or stop mode. The mask level written can by a value between 0 and 6 in the (INTC_WCR[MASK]) field.

2. Execute a stop instruction to place the processor into wait or stop mode.

3. After the processor is stopped, the interrupt controller enables special logic that evaluates the incoming interrupt sources in a purely combinational path; no clocked storage elements are involved.

4. If an active interrupt request is asserted and the resulting interrupt level is greater than the mask value contained in INTC_WCR[MASK], the interrupt controller asserts the wake-up signal.

This signal is routed to the clock generation logic to exit the low-power mode to resume processing. Typically, the interrupt mask level loaded by the application software into the core status register field $SR_{[12:10]}$ before the execution of the STOP instruction should match the INTC_WCR[MASK] value.

The interrupt controller's wake-up signal is defined as:

$$\text{Wake-up}=\text{INTC\_WCR[ENB]}\&(\text{level of any asserted\_int\_request} > \text{INTC\_WCR}_{[MASK]}). \quad \textbf{(EQ 2)}$$

### ColdFire V1 and HCS08 compatibility

The hardware compatibility between the HCS08 and ColdFire V1 core is based on exact implementation of I/O peripherals, memory size and identical pin assignment. Whereas, the software compatibility is based on the similarities of the vector table and I/O requests assignments.

As with the ColdFire V1, the HCS08 core also supports a 32-entry exception vector table. The first two vectors are reserved for internal CPU/system exceptions and the remaining 30 are available for I/O interrupt requests. While the ColdFire V1 exception vector table supports the same I/O interrupt requests, but has the first 64 vectors reserved for system exceptions. The requirement for an exact match between the interrupt requests and priorities across the two architectures means the 30 sources are mapped to a sparsely-populated two-dimensional ColdFire V1 array of seven interrupt levels and nine priorities within each level. The following association between the HCS08 and ColdFire V1 vector numbers applies:

$$\text{ColdFire V1 Vector Number} = 64 + \text{HCS08 Vector Number} \quad \textbf{(EQ 3)}$$

The CFV1_INTC and HCS08 cores perform a cycle-by-cycle evaluation of the active requests and signal the highest-level, highest-priority request to their respective cores in the form of an encoded interrupt level and the exception vector associated with the request. Table 3 provides a high-level architectural comparison between HCS08 and

**TABLE 3. CFV1 and HCS08 Compatibility**

| Attributes | HCS08 | CFV1 |
|---|---|---|
| Vector Table | 32 Vectors, 16-bit entries located at upper end of memory (relocation of VCT is not possible) | 103 Vectors, 32-bit entries located at the beginning of memory. VCT can be relocated with the VBR |
| CPU Exception Vectors | only 2 vectors (0-1) 2-31 reserved for I/O | 64 vectors (0-63) 64-102 reserved for I/O |
| Stack Frame | 5 bytes (PC, A, X[a], and CCR) | 2 longwords (PC, stack format and core SR) |
| Interrupt Levels | Only one Level available (masked by CCR[I] bit) | Six levels masked by $SR_{[12:10]}$ |
| Interrupt Nesting | One level only | Hardware support for 7 level nesting |
| Non-Maskable IRQ Support | None | Supported (Level 7 only) |
| IRQ Sensitivity | All interrupt requests are level sensitive | Levels 1-6 are level sensitive, level 7 is edge sensitive |
| INTC Vectoring | Fixed priorities and vector assignments | Fixed priorities and vector assignments, but 2 interrupt request sources can be remapped to highest maskable priority |
| Software Interrupts | Software interrupt instruction (SWI) | Seven levels available |
| Software IACK | None | Available |
| Exit Instruction from ISR | Return-from-interrupt (RTI) | Return-from-exception (RTE) |

a. The high byte of H:X register is not saved onto stack to maintain compatibility with HC05 processors.

ColdFire V1 interrupt processing, as these differences are important in the definition of the CFV1_INTC module.

### Emulation of HCS08 Interrupts on CFV1

As shown in Table 3 above, the HCS08 architecture specifies only one level of nesting capability controlled by the interrupt mask bit in the HCS08 CPU condition code register ($CCR_{[I]}$). Clearing and setting the CCR[I] will respectively enable and disable interrupts. Whereas, the ColdFire V1 architecture defines seven interrupt levels, controlled by the 3-bit interrupt priority mask field in the core status register, SR[I], with an automatic supports of seven interrupt level nesting. ColdFire V1 core has two non-maskable interrupts,

(IRQ pin and Low voltage detect), both requests are treated as edge-sensitive interrupts, while HCS08 has none.

To emulate the HCS08's 1-level IRQ capabilities on CFV1 core, only two SR[I] settings are used: Writing 0 to $SR[I_2:I_0]$ enables all interrupt levels, while writing 7 disables all interrupts, except level 7.

To mimic the HCS08 interrupt mechanism, the application software needs to write the SR[I] bit field with 0 to enable interrupts, and the interrupt handler needs to write 7 as the first instruction to disable nesting of all maskable interrupts.

A better and more convenient way to accomplish this with no software overhead, is to set the CPU configuration interrupt masking enable (CPUCR[IME]). Setting this bit will automatically force the processor to raise the SR[I] field to level 7 as soon as the interrupt is acknowledged. If the latter method is used, the handler does not need to execute a STLDSR or MOVE to SR instruction.

### INTC Initialization

The CPU core and the interrupt controller's reset state has all requests masked via the core status (SR). Before any interrupt requests are enabled, the following steps must be taken:

1. The CPU configuration (CPUCR) register needs to be set to the desired system configuration as per the application requirements.
2. The appropriate interrupt vector table and interrupt service routines must be loaded into memory.
3. If desired, enable wakeup by writing the INTC wake-up control (INTC_WCR) register and select the required interrupt mask level to force the CPU exit the wait or stop instruction upon the appropriate level assertion. The following mask values depict which level(s) will be able to wake up the CPU:

   0 - enables all interrupt levels

   1 - enables level 2 - level 7

   2 - enables level 3 - level 7

   3 - enables level 4 - level 7

   4 - enables level 5 - level 7

   5 - enables level 6 - level 7

   6 - enables level 7 only.

The maximum value that can be written into the INTC_WCR mask field is 6.

4. Write the interrupt mask level in the core's status register $SR[I_2:I_0]$ to '$000_2$' to enable all interrupt levels.

## Conclusion

From our discussion in this chapter, it is clear that an HCS08 code can be ported into a ColdFire V1 system with no major efforts. Being part of the Flexis family, system designers currently using the HCS08 are provided with easy migration path from 8-bit to 32-bit performance with little or no change.

ColdFire V1 processors provide the engine for the industry's first 8- and 32-bit compatible devices, allowing easy migration between the architectures while delivering 32-bit performance at a lower cost. The Flexis family is designed to share the same peripheral set, leveraging the extensive I/O devices portfolio to enhance end-use applications with the latest features and capabilities.

Pin compatibility and a common CodeWarrior® integrated development environment (IDE) helps bridge the gap and complete the connection between the 8-bit and ColdFire V1 architectures.

# CHAPTER 8

# *Debug & Development*

## *Introduction to Debug*

### Why is debug needed?

Developing any kind of microcontroller system always involves a combination of software and hardware debug. As MCUs become more highly integrated, the debug process gets tougher. Software is never perfect, especially during the development phases. Its interaction with multiple hardware modules adds to the burden facing the developer. Hardware interfaces, especially to devices that are external to the MCU may also require debugging. Tools that can monitor and control the execution of software and also provide an easy method to exercise hardware interfaces reduce the time it takes to determine the cause of software and hardware problems. With these tools, the programmer can rapidly apply software patches as well as dynamically modify the run time environment to verify fixes before proceeding with a more formal method of eliminating errors. Many of these tools also provide methods to gather statistics on the real time performance of the system, which may in turn help designers understand the limitations and sensitivities of the microcontroller system and adapt the design to improve performance and/or cost.

### What kind of debug features are needed?

Two key aspects of debugging that the developer has to consider are:

- the level of intrusion that debugging causes on the real time behavior of the application.
- how much visibility of internal operations is required.

The ColdFire V1 and HCS08 processors support a variety of debug mechanisms which allow the developer the option to either implement software routines and interrupt handlers to trap code execution and data accesses, or alternatively use the external debug port and software tools to provide a more sophisticated debug capability. In both cases, no external bus is required to perform any of the debug operations. This significantly reduces the complexity of debugging, and eliminates the need to enable the external bus to control and monitor program execution and data accesses.

At a minimum a debugger has to be able to perform the following operations:

- Halt the core
- Run the core in real time
- Single step one line of code at a time
- Set a break point at which a running program will halt
- Read and write memory locations
- Read and write core registers

Additional debug features are often needed to ease software and hardware development. These features may include, but are not limited to:

- Setting a watch point which will record the data accessed by the running program
- Recording the execution of a program for later analysis
- Recording all data accesses within a memory range, or outside a memory range
- Defining a set of conditions that must be met before a break point occurs, or a program or data trace starts.

To maintain compatibility between the 8 and 32-bit architectures, the ColdFire Version 1 core supports BDM functionality using the HCS08's single-pin interface. The traditional 3-pin full-duplex BDM serial communication protocol based on 17-bit data packets is replaced with the HCS08 protocol where all communications are based on an 8-bit data packet using a single package pin (BKGD).

An on-chip trace buffer allows a stream of compressed processor execution status packets to be recorded for subsequent retrieval to provide program (and partial data) trace capabilities.

## Overview

Debug support is divided into three areas:

- Background debug mode (BDM) - Provides low-level debugging in the ColdFire V1 and HCS08 cores. In BDM, the core is halted and a variety of commands can be sent to the processor to access memory, registers, and peripherals. The external emulator uses a one-pin serial communication protocol.
- Real-time debug support - Use of the full BDM command set requires the processor to be halted, which many real-time embedded applications cannot support. The core

includes a variety of internal breakpoint registers which can be configured to trigger and generate a special interrupt.

The resulting debug interrupts let real-time systems execute a unique service routine that can quickly save the contents of key registers and variables and return the system to normal operation. The external development system can then access the saved data, because the hardware supports concurrent operation of the processor and BDM-initiated memory commands. In addition, the option is provided to allow interrupts to occur.

- Program trace support - The ability to determine the dynamic execution path through an application is fundamental for debugging. The ColdFire V1 solution implements a trace buffer that records processor execution status and data, which can be subsequently accessed by the external emulator system to provide program (and optional partial data) trace information.

**Main Features**

The ColdFire Version 1 debug definition supports the following features:

- Classic ColdFire V1 DEBUG_B+ functionality mapped into the single-pin BDM interface.
- Real time debug support, with 6 hardware breakpoints (4 PC, 1 address and 1 data) that can be configured into a 1- or 2-level trigger with a programmable response (processor halt or interrupt).
- Capture of compressed processor status and debug data into on-chip trace buffer provides program (and optional slave bus data) trace capabilities.
- On-chip trace buffer provides programmable start/stop recording conditions plus support for continuous or PC-profiling modes.
- Debug resources are accessible via single-pin BDM interface or the privileged WDEBUG instruction from the core.

The debug module architecture supports a number of hardware breakpoint registers that can be configured into single- or double-level triggers based on the PC or operand address ranges with an optional inclusion of specific data values. The triggers can be configured to halt the processor or generate a debug interrupt exception. Additionally, these same breakpoint registers can be used to specify start/stop conditions for recording in the PST trace buffer.

The core includes four PC breakpoint triggers and a set of operand address breakpoint triggers with two independent address registers (to allow specification of a range) and an optional data breakpoint with masking capabilities. Core breakpoint triggers are accessible through the serial BDM interface or written through the supervisor programming model using the WDEBUG instruction.

## Hardware Implementation

A detailed description of the hardware implementation of the on-chip debug modules is outside the scope of this book, and are mostly relevant only to tools developers. Check the Freescale website at: *www.freescale.com* for more information on the implementation of these hardware blocks, message protocols, encodings and physical layer bit timings. A simplified block diagram of the ColdFire V1 core and the debug module is shown in Figure 8.1

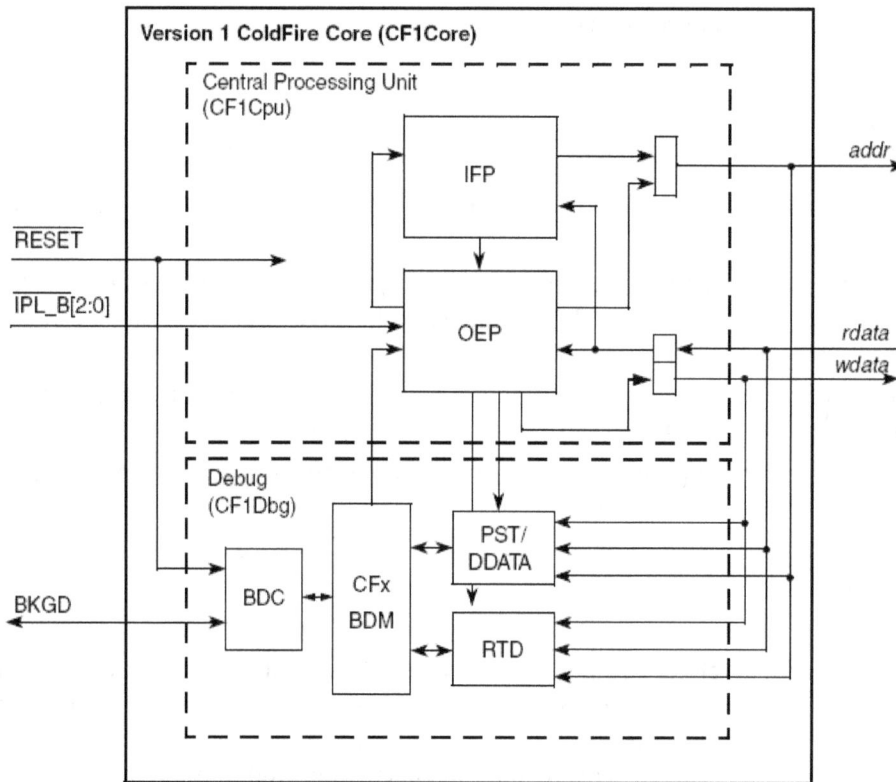

IFP - Instruction Fetch Pipeline
OEP - Operand Execution Pipeline
RTD - Real-Time Debug
BDC - Background Debug Controller
PST/DData - Processor Status/ Debug Data
CFxBDM - ColdFire Background Debug Module

**FIGURE 8.1. ColdFire V1 Debug Architecture**

## HCS08 Debug Module Main Features

- Single pin for mode selection and background communications
- BDC registers are not located in the memory map
- SYNC command to determine target communications rate
- Non-intrusive commands for memory access
- Active background mode commands for CPU register access
- GO and TRACE1 commands
- BACKGROUND command can wake CPU from stop or wait modes
- One hardware address breakpoint built into BDC
- Oscillator runs in stop mode, if BDC enabled
- COP watchdog disabled while in active background debug mode

## HCS08 Background Debug Controller (BDC)

The HCS08 MCU Family implements a single-wire background debug interface that supports in-circuit programming of on-chip nonvolatile memory and sophisticated non-intrusive debug capabilities. Unlike debug interfaces on earlier 8-bit generation. This debugger does not interfere with normal application resources, and does not use any user memory or locations in the memory map and does not share any on-chip peripherals.

BDC commands are divided into two groups:

- Active background mode commands require that the target MCU is in active background mode (the user program is not running). Active background mode commands allow the CPU registers to be read or written, and allow the user to trace one user instruction at a time, or GO to the user program from active background mode.
- Non-intrusive commands can be executed at any time even while the user's program is running. Non-intrusive commands allow a user to read or write MCU memory locations or access status and control registers within the background debug controller.

Typically, a relatively simple interface pod is used to translate commands from a host computer into commands for the custom serial interface to the single-wire background debug system. Depending on the development tool vendor, this interface pod may use a standard RS-232 serial port, a parallel printer port, or some other type of communications such as a universal serial bus (USB) to communicate between the host PC and the pod. The pod typically connects to the target system with ground, the BKGD pin, RESET, and sometimes VDD. An open-drain connection to reset allows the host to force a target system reset, which is useful to regain control of a lost target system or to control startup of a target system before the on-chip nonvolatile memory has been programmed. Sometimes VDD can be used to allow the pod to use power from the target system to avoid the need

**Debug & Development**                                                        **159**

for a separate power supply. However, if the pod is powered separately, it can be connected to a running target system without forcing a target system reset or otherwise disturbing the running application program.

The following sections in this chapter provide details on the BKGD pin, the background debug serial interface controller (BDC), BDM command set summary, a standard 6-pin BDM connector, as well as real-time debug and trace capabilities. The core definition supports revision B+ (DEBUG_B+) of the ColdFire V1 debug architecture.

## Background Debug (BKGD) Pin

The ColdFire V1 and HCS08 processors implement a pseudo-open-drain single-wire background debug interface pin (BKGD) that can be driven by an external controller or by the MCU. Data is transferred most-significant (MSB) bit first at 16 BDC clock cycles per bit (nominal speed).

The primary function of this pin is for bidirectional serial communication of background debug mode commands and data. During reset, this pin selects between starting in active background (halt) mode or starting the application program. This pin also requests a timed sync response pulse to allow a host development tool to determine the correct clock frequency for background debug serial communications.

The background debug controller serial interface requires the external host controller to generate a falling edge on the BKGD pin to indicate the start of each bit time. The external controller provides this falling edge whether data is transmitted or received.

When a development system is connected, it can pull BKGD and RESET low, release RESET to select active background (halt) mode rather than normal operating mode, and then release BKGD. It is not necessary to reset the target MCU to communicate with it through the background debug interface. There is also a mechanism to generate a reset event in response to setting the background debug force reset [BDFR] control bit in the configuration/status register 2 (CSR2) But if the debugger pod is not connected to the standard interface connector the internal pull-up on BKGD selects normal operating mode. In the HCS08, the BDFR control bit is implemented in System Background Debug Force Reset Register (SBDFR).

## Background Debug Serial Interface Controller (BDC)

This communications implements a custom serial protocol used in the HCS08 family. This protocol assumes that the host knows the communication clock rate determined by the target BDC clock rate. The BDC clock rate may be the system bus clock frequency or an alternate frequency source depending on the state of clock switch [CLKSW] control bit in the extended configuration/status register (XCSR). All communication is initiated and controlled by the host which drives a high-to-low edge to signal the beginning of each bit time. Commands and data are sent most significant bit (MSB) first.

If the host is attempting to communicate with a target MCU that has an unknown BDC clock rate, a SYNC command may be sent to the target MCU to request a timed synchronization response signal from which the host can determine the correct communication speed. After establishing communications, the host can read XCSR and write the clock switch (CLKSW) control bit to change the source of the BDC clock for further serial communications if necessary.

The custom serial protocol requires the debug pod to know the target BDC communication clock speed. The clock switch (CLKSW) control bit in the XCSR[31,Äì24] register allows you to select the BDC clock source. The BDC clock source can either be the bus clock or the alternate BDC clock source. When the MCU is reset in normal user mode, CLKSW is cleared and that selects the alternate clock source. This clock source is a fixed frequency independent of the bus frequency so it does change if the user modifies clock generator settings. This is the preferred clock source for general debugging.

When the MCU is reset in active background (halt) mode, CLKSW is set which selects the bus clock as the source of the BDC clock. This CLKSW setting is most commonly used during flash memory programming because the bus clock can usually be configured to operate at the highest allowed bus frequency to ensure the fastest possible flash programming times. Since the host system is in control of changes to clock generator settings, it knows when a different BDC communication speed should be used. The host programmer also knows that no unexpected change in bus frequency could occur to disrupt BDC communications.

Normally, setting CLKSW should not be used for general debugging because there is no way to ensure the application program does not change the clock generator settings. This is especially true in the case of application programs that are not yet fully debugged.

After any reset (or at any other time), the host system can issue a SYNC command to the target MCU to determine the speed of the BDC clock. CLKSW may be written using the serial WRITE_XCSR_BYTE command through the BDC interface. The CLKSW is located in the special XCSR byte register in the BDC module and it is not accessible in the normal memory map of the ColdFire V1 core. This means that no program running on the processor can modify this register (intentionally or unintentionally).

For the HCS08 MCU, the CLKSW control bit is located in the BDC and status and control register (BDCSCR).

## Background Debug Commands

The ColdFire V1 debug command set is based on transmission of one or more 8-bit data packets per operation. Each operation begins with a host-to-target transmission of an 8-bit command code packet. The command code definition broadly maps the operation in 4 formats as shown in Table 8.1, 12.2, 12.3 and 12.5.

### TABLE 8.1 Miscellaneous Commands

| 7 | 6 | 5 | 4 | 3 | 2 | 1 | 0 |
|---|---|---|---|---|---|---|---|
| 0 | 0 | R/$\overline{\text{W}}$ | 0 | MSCMD | | | |
| Optional Command Extension Byte (Data) | | | | | | | |

- Read/$\overline{\text{Write}}$ (R/$\overline{\text{W}}$)

0 = Command is performing a write operation.

1 = Command is performing a read operation.

- Miscellaneous command (MSCMD)

0000 = No operation

0001 = Display the CPU's program counter (PC) plus optional capture in PST trace buffer

0010 = Enable the BDM acknowledge communication mode

0011 = Disable the BDM acknowledge communication mode

0100 = Force a CPU halt (background)

1000 = Resume CPU execution (go)

1101 = Read/write of the debug XCSR most significant byte

1110 = Read/write of the debug CSR2 most significant byte

1111 = Read/write of the debug CSR3 most significant byte

### TABLE 8.2 Memory Commands

| 7 | 6 | 5 | 4 | 3 | 2 | 1 | 0 |
|---|---|---|---|---|---|---|---|
| 0 | 0 | R/W | 1 | SZ | | MCMD | |
| Command Extension Byte (Address, Data) | | | | | | | |

- Memory operand size (SZ). Defines the size of the memory reference.

00 = 8-bit byte

01 = 16-bit word

10 = 32-bit long

- MCMD

Memory command. Defines the type of the memory reference to be performed.

00 = Simple write if R/$\overline{\text{W}}$ = 0; simple read if R/$\overline{\text{W}}$ = 1
01 = Write + status if R/$\overline{\text{W}}$ = 0; read + status if R/$\overline{\text{W}}$ = 1
10 = Fill if R/$\overline{\text{W}}$ = 0; dump if R/$\overline{\text{W}}$ = 1
11 = Fill + status if R/$\overline{\text{W}}$ = 0; dump + status if R/$\overline{\text{W}}$ = 1

TABLE 8.3 Core Register Commands

| 7 | 6 | 5 | 4 | 3 | 2 | 1 | 0 |
|---|---|---|---|---|---|---|---|
| CRG | | R/$\overline{\text{W}}$ | CRN | | | | |
| Command Extension Byte (Data) | | | | | | | |

• Core Register Group (CRG)

01 = CPU's general-purpose registers (An, Dn) or PST trace buffer

10 = DBG's control registers

11 = CPU's control registers (PC, SR, VBR, CPUCR, etc..)

• Core Register Number (CRN)

This field selects one of the core general purpose registers or one of the debug registers

TABLE 8.4 Core Register Group and Register Number Encoding

| CRG | CRN | Register Selected |
|---|---|---|
| 0 1 | 0x00-0x070 | D0-D7 |
| | 0x08-0x0F | A0-A7 |
| | 0x10-0x1B | PST Buffer 0-11 |
| 1 0 | Select one of Debug Registers | |
| 1 1 | 0x00 | Other_A7 |
| | 0x01 | Vector Base Register (VBR) |
| | 0x02 | CPU Configuration Register (CPUCR) |
| | 0x0E | Core Status Register (SR) |
| | 0x0F | Program Counter (PC) |

based on the setting of the CRG bit field provided in the command. Table 8.4 depicts encoding of the CRG and CRN bit field.

TABLE 8.5 Trace Buffer Read Commands

| 7 | 6 | 5 | 4 | 3 | 2 | 1 | 0 |
|---|---|---|---|---|---|---|---|
| 0 | 1 | 0 | 1 | CRN | | | |
| Trace Buffer Data [31:24] | | | | | | | |
| Trace Buffer Data [23:16] | | | | | | | |

**TABLE 8.5 Trace Buffer Read Commands**

| Trace Buffer Data [15:8] |
|---|
| Trace Buffer Data [7:0] |

## Debug Command Set

The table below summarizes the BDM command set and provides a brief explanation of each command.

**TABLE 8.6 Background Debug Command Set Summary**

| Command Mnemonic | Command Classification | Command Brief Description |
|---|---|---|
| SYNC | Always Available | Request a timed reference pulse to determine the target BDC communication speed |
| ACK_DIABLE | Always Available | Disable the communication handshake. This command does not issue an ACK pulse. |
| ACK_ENABLE | Always Available | Enable the communication handshake. Issues an ACK pulse after the command is executed |
| BACKGROUND | Non_Intrusive | Halt the CPU if ENBDM is set. Otherwise, ignore as illegal command |
| DUMP_MEM.sz | Non_intrusive | Dump (read) memory based on operand size (sz). Used with READ_MEM to dump large blocks of memory. An initial READ_MEM is executed to set up the starting address of the block and to retrieve the first result. Subsequent DUMP_MEM commands retrieve sequential operands |
| DUMP_MEM.z_WS | Non_intrusive | Dump (read) memory based on operand size (sz) and report status. Used with READ_MEM{_WS} to dump large blocks of memory. An initial READ_MEM{_WS} is executed to set up the starting address of the block and to retrieve the first result. Subsequent DUMP_MEM{_WS} commands retrieve sequential operands |
| FILL_MEM.sz | Non_intrusive | Fill (write) memory based on operand size (sz). Used with WRITE_MEM to fill large blocks of memory. An initial WRITE_MEM is executed to set up the starting address of the block and to write the first operand. Subsequent FILL_MEM commands write sequential operands |

**TABLE 8.6 Background Debug Command Set Summary**

| Command Mnemonic | Command Classification | Command Brief Description |
|---|---|---|
| FILL_MEM.sz_WS | Non_intrusive | Fill (write) memory based on operand size (sz) and report status. Used with WRITE_MEM_WS to fill large blocks of memory. An initial WRITE_MEM{_WS} is executed to set up the starting address of the block and to write the first operand. Subsequent FILL_MEM{_WS} commands write sequential operands |
| GO | Non_intrusive | Resume program execution |
| NOP | Non_intrusive | No operation |
| READ_CREG | Active_Background | Read one of CPU control registers |
| READ_DREG | Non_intrusive | Read one of debug module's control registers |
| READ_MEM.sz | Non_intrusive | Read the appropriately-sized (sz) memory value from the location specified by the 24-bit address |
| READ_MEM.sz_WS | Non_intrusive | Read the appropriately-sized (sz) memory value from the location specified by the 24-bit address and report status |
| READ_PSTB | Non_intrusive | Read the requested longword location from the PST trace buffer |
| READ_Rn | Active_Background | Read the requested general-purpose register (An, Dn) from the CPU |
| READ_XCSR_BYTE | Always Available | Read the most significant byte of the debug module's XCSR |
| READ_CSR2_BYTE | Always Available | Read the most significant byte of the debug module's CSR2 |
| READ_CSR3_BYTE | Always Available | Read the most significant byte of the debug module's CSR3 |
| SYNC_PC | Non_intrusive | Display the CPU's current PC and capture it in the PST trace buffer |
| WRITE_CREG | Active_Background | Write one of the CPU's control registers |
| WRITE_DREG | Non_intrusive | Write one of the debug module's control registers |
| WRITE_MEM.sz | Non_intrusive | Write the appropriately-sized (sz) memory value to the location specified by the 24-bit address |
| WRITE_MEM.sz_WS | Non_intrusive | Write the appropriately-sized (sz) memory value to the location specified by the 24-bit address and report status |

**TABLE 8.6 Background Debug Command Set Summary**

| Command Mnemonic | Command Classification | Command Brief Description |
|---|---|---|
| WRITE_Rn | Active_Background | Write the requested general-purpose register (An, Dn) of the CPU |
| WRITE_XCSR_BYTE | Always Available | Write the most significant byte of the debug module's XCSR |
| WRITE_CSR2_BYTE | Always Available | Write the most significant byte of the debug module's CSR2 |
| WRITE_CSR3_BYTE | Always Available | Write the most significant byte of the debug module's CSR3 |

## Real-Time Debug and Trace Support

The ColdFire V1 family supports debugging real-time applications. For these types of embedded systems, the processor must continue to operate during debug. The foundation of this area of debug support is that while the processor cannot be halted to allow debugging, the system can generally tolerate the small intrusions with minimal effect on real-time operation.

The classic debug architecture supports real-time trace via the PST/DDATA output signals. For this functionality, the following apply:

- One (or more) PST value is generated for each executed instruction.
- Branch target instruction address information is displayed on all non-PC-relative change-of-flow instructions, where the user selects a programmable number of bytes of target address.

  - Displayed information includes PST marker plus target instruction address as DDATA.

  - Captured address creates the appropriate number of DDATA entries, each with 4 bits of address.
- Optional data trace capabilities are provided for accesses mapped to the slave peripheral bus.

  - Displayed information includes PST marker plus captured operand value as DDATA.

  - Captured operand creates the appropriate number of DDATA entries, each with 4 bits of data.

The resulting PST/DDATA output stream, with the application program memory image, provides an instruction-by-instruction dynamic trace of the execution path.

For the ColdFire V1 core and its single debug signal, support for trace functionality is completely redefined. The ColdFire V1 solution provides an on-chip PST/DDATA trace buffer (known as the PSTB) to record the stream of PST and DDATA values.

Even with the application of a PST trace buffer, problems associated with the PST bandwidth and associated fill rate of the buffer remain. Given that there is one (or more) PST entry per instruction, the PSTB would fill rapidly without some type of data compression.

For the core, the size of the data trace window will define, to a great extent, the amount of debug port bandwidth needed to send the traced data back to the debug tool. Some applications may have widely scattered variables that need to be traced in the same debug session, so it may be undesirable to trace data accesses over the entire span of the memory map that contains only a few scattered variables that you want to observe. Doing so could result in excessive data trace messages that could exceed the bandwidth of the debug port. Fortunately, the PST compression technology was previously developed and included as part of the ColdFire Version 5 core (although very different than the resulting ColdFire V1 implementation).

## Background Debug Connector

In the ColdFire V1 and HCS08 devices, a relatively simple interface pod is used to translate commands from a host computer into commands for the custom serial interface to the single-wire background debug system. Depending on the development tool vendor, this interface pod may use a standard RS-232 serial port, a parallel printer port, or some other type of communications such as a universal serial bus (USB) to communicate between the host PC and the pod. The pod typically connects to the target system with ground, the BKGD pin, RESET, and sometimes VDD.

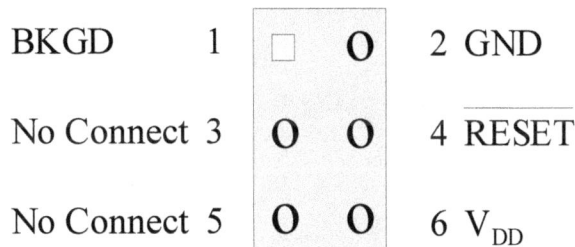

| BKGD | 1 | □ | O | 2 GND |
| No Connect | 3 | O | O | 4 RESET |
| No Connect | 5 | O | O | 6 $V_{DD}$ |

FIGURE 8.2. Recommended BDM Connector

An open-drain connection to reset allows the host to force a target system reset, useful to regain control of a lost target system or to control startup of a target system before the on-chip nonvolatile memory has been programmed. Sometimes VDD can be used to allow the pod to use power from the target system to avoid the need for a separate power supply. However, if the pod is powered separately, it can be connected to a running target system without forcing a target system reset or otherwise disturbing the running application program.

There are many 3rd party tools companies that provide more sophisticated debug capabilities such as large real time trace buffers for those application problems that need it.

# CHAPTER 9    *Mini-FlexBus*

## *Introduction*

Due to the complexity of today's embedded applications, additional memory may be required even when a large amount of on-chip memory is integrated. Software development may also benefit from temporarily using external memory prior to programming code into the on-chip flash. To expand system memory, an external bus interface is implemented to accommodate such requirements. The external bus interface allows system designers to incorporate as much as 2 one Mega byte memory banks using a 20 bit address bus with two chip-selects. For applications that do not require external memory, the external bus interface pins may be used as general purpose I/O (GPI/O) with each pin individually configured as general purpose input or output.

The external bus interface implementation on this device is called Mini-FlexBus. This bus contains the interface signals to access external memory mapped devices. It provides address, data and control signals to allow the application software access external asynchronous and synchronous peripherals such as RAM, ROM, EPROM, and other logic devices that require parallel address and data buses.

The Mini-FlexBus is a subset of the FlexBus module found on other ColdFire microprocessors. The Mini-FlexBus minimizes package pin-outs while maintaining a high level of configurability and functionality.

This bus supports byte (8 bit), word (16 bit), longword (32 bit) data transfers. The programmers model contains 2 chip-select address registers (CSAR1:0), 2 chip-select control registers (CSCR1:0) and 2 chip-select mask registers (CSMR1:0). The chip-select address registers define the base address of the external memory blocks, the chip-select control registers are used to control transfer parameters such as address setup and hold, data size, the number of wait states for the external device being accessed, automatic internal transfer termination enable or disable, and finally, the chip-select mask registers define the port size of the external memory blocks.

For external memory and/or peripherals accesses, the Mini-Flexbus provides an internally generated transfer acknowledge signal (TA) that can be used to complete bus transactions. This signal must always be enabled by setting the auto-acknowledge (AA) enable in the chip-select control register (CSCRn). The TA assertion is determined by a programmable number of wait cycles defined in CSCRn. The number of wait states inserted may be chosen to be between 0 to 63 to satisfy the external memory or peripheral device access time.

**Summary of the Features**

- Two independent, user-programmable chip-select signals (CS[1:0]) that can interface with SRAM, PROM, EPROM, EEPROM, Flash, and other peripherals
- 8 and 16 bit port sizes
- Byte, word, and longword transfers
- Programmable address setup time with respect to the assertion of chip select
- Programmable address hold time with respect to the negation of chip select and transfer direction
- Non-multiplexed operation with 20-bit address and 8-bit data buses
- Multiplexed operation with 20-bit address with either 16-bit or 8-bit data buses

## *Mini-FlexBus Architecture*

To eliminate the need for external components, the Mini-FlexBus contains chip select logic, registers and output signals that can split up the external memory space into different partitions. Typically, each used chip select would be connected to separate external memory device. The pins of unused chip selects may be reconfigured for alternate or GPIO functions.

The address range for a chip select is defined by the values stored in its base address register (CSAR0 & CSAR1) and mask register (CSMR0 & CSMR1). When software performs a load from or store to an address that falls within the programmed range, the corresponding chip-select signal is asserted and the bus cycle is executed with attributes defined in its corresponding chip-select control register (CSCR0 & CSCR1).

When the core drives the address bus, the address is compared to the two chip-select address registers (CSARs), assuming both chip-selects are active. If the address matches one of the CSARs, the appropriate CS signal is asserted, but if the address does not match one of the two registers, the internal bus cycle is terminated with a bus error and *no* chip select is asserted.

Also, if more than one CSAR parameters are configured to cause an address to match multiple CSARs, the bus cycle is terminated with a bus error and *none* of the chip select signals will assert.

For a write access, the address, data, transfer start (FB_ALE), chip select (FB_CS*n*), and all attribute signals change on the rising edge of the bus clock (FB_CLK). Read data is also latched into the device on the rising edge of the clock. Note that while the timing diagrams in this chapter include FB_CLK, this signal is not used by external asynchronous memory devices. In these cases, FB_CLK is only shown as a timing reference.

FB_CLK is needed for transfers between the ColdFire® MCU and synchronous external memory devices, such as synchronous SRAM, where addresses are clocked and latched by the FB_CLK signal.

### Data Alignment and Data Bus Pins Connections

Mini-FlexBus aligns data transfer on one or two byte lanes depending on the memory data port width. The port width on this bus can either be 8-bit or 16-bits. The 16-bit port is only applicable when multiplexed mode is selected. When the data bus width is 16-bits, a long word transfer requires 2 bus cycles to transfer the data, but only requires 1 bus cycle to transfer a 16-bit word. In non-multiplexed mode, the data bus width can only be 8-bit wide

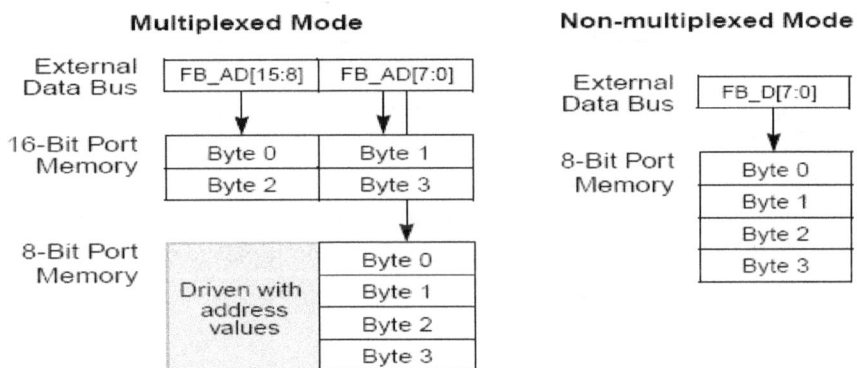

FIGURE 9.1. **16 and 8-bit Data Bus Connections to External Memory**

therefore, 4 bus cycles are required to transfer a 32-bit longword, and 2 bus cycles for a 16-bit word transfer. Figure 9.1 on page 171 shows how the byte lanes connected to the external memory and the sequential transfers of a 32-bit longword transfer for the supported port sizes.

### External Bus Interface Signals

The external bus control signals are shown in Table 9.1 with a brief functional description of each. The Mini-FlexBus can be configured in one of two modes, non-multiplexed and multiplexed mode. In non-multiplexed mode, FB_A[19:0] function as the address bus which is driven externally upon a match on one of the chip-select address registers. In this mode, only 8 data bus pins available FB_D[7:0] that carries the data in and out of the MCU. The data direction is determined by the FB_R/$\overline{W}$ signal depending on whether a read or write cycle is being performed. In multiplexed mode, the interface supports a single 20-bit wide multiplexed address and data bus FB_AD[19:0]. The full 20-bit address is always driven on the first clock of a bus cycle and during the data phase, the FB_AD[15:0] lines are used for the data transfer. The Memory size is determined by the programmed port size for the corresponding chip-select. The Mini-FlexBus continues to drive the

**TABLE 9.1 External Bus Control Signals**

| Signal Name | I/O | Description |
|---|---|---|
| FB_A[19:0][a] | I/O | Address bus signals in a non-multiplexed mode. |
| FB_D[7:0] | I/O | Data bus signals in non-multiplexed mode. In multiplexed bus mode, this bus is not used. |
| $\overline{FB\_CS[1:0]}$ | O | Chip select signals. In multiplexed mode only FB_CS0 is available and FB_CS1 is multiplexed with FB_ALE. |
| $\overline{FB\_OE}$ | O | Output enable signal |
| FB_R/$\overline{W}$ | O | Read/write. 1 = Read, 0 = Write |
| FB_ALE | O | Address latch enable.This signal is multiplexed with FB_CS1 |

a. Address signals [A31:20] are not available externally but these signal are still used internally by the chip select logic to qualify a chip-select match.

address on any FB_AD[15:0] lines not used for data. When multiplexed bus is selected, there are two choices available to the user, 16-bit or 8-bit data bus mode. Figure 9.2 on page 173 depicts the Mini-FlexBus multiplexed and non-multiplexed operating modes and how the address and data bus pins are driven externally.

| Port Size & Phase | | FB_AD | | |
|---|---|---|---|---|
| | | [19:16] | [15:8] | [7:0] |
| 16-bit | Address phase | Address | | |
| 16-bit | Data phase | Address | Data | |
| 8-bit | Address phase | Address | | |
| 8-bit | Data phase | Address | | Data |

FIGURE 9.2. Mini_FlexBus Operating Modes

## Device Memory Map

From power-on or system reset, both chip select logics are disable and the reset vectors are fetched from the internal Flash from address 0x0000_0000 and address 0000_0004. Recall that reset uses two vectors for the initial stack pointer and program counter values. To enable the chip select logic, the application software must configure the appropriate chip-select register set for the proper transfer attributes of the external memory block or blocks.

TABLE 9.2 Memory Map

| Address Range | Block Size | Actual On-Chip Memory | Memory Map |
|---|---|---|---|
| 0x(00)00_0000<br><br>0x(00)3F_FFFF | 4 MB | 128K | Allocated to On-chip Flash Memory |
| 0x(00)40_0000<br><br>0x(00)7F-FFFF | 4 MB | --- | Allocated for Off-chip Memory Expansion |
| 0x(00)80_0000<br><br>0x(00)9F_FFFF | 2 MB | 24K | Allocated to On-chip RAM Memory |

TABLE 9.2 Memory Map

| Address Range | Block Size | Actual On-Chip Memory | Memory Map |
|---|---|---|---|
| 0x(00)A0_0000<br><br>0x(00)BF_FFFF | 2MB | --- | Available to Off-chip Expansion |
| 0x(00)C0_0000<br>0x(00)C0_000F | 16 Bytes | 16 Bytes | Rapid GPIO |
| 0x(00)C0_0010<br><br>0x(00)FF_7FFFF | 3.5MB | --- | Uuimplemented |
| 0x(00)FF_8000<br>0x(00)FF_FFFF | 32K | 32K | On-chip Peripherals |

## Misaligned Accesses

Because operands, unlike opcodes, can reside at any byte boundary, they are allowed to be misaligned. A byte operand is properly aligned at any address, a word operand is misaligned at an odd address, and a longword is misaligned at an address not a multiple of four. Although the processor enforces no alignment restrictions for data operands (including program counter (PC) relative data addressing), additional bus cycles are required for misaligned operands.

Instruction opcodes and extension words must reside on word boundaries. Attempting to prefetch a misaligned instruction word causes an address error exception.

## Bus Errors

There are certain accesses to the Mini-FlexBus that cause the system bus to hang or generate a bus error. It is important to have a good access-error handler to manage these conditions.

One such access is if CSCRn[AA] is cleared, the system hangs. Four other types of accesses cause the access to terminate with a bus error.

• If the Mini-Flexbus module is disabled by the platform peripheral power management, accesses will cause an error termination on the bus and prohibit the access to the external memory or the I/O device.

• Attempted writes to a memory blocks defined as write protected (CSMR*n*[WP] is set) are terminated with an error response and the access is inhibited to the Mini-FlexBus.

• When the driven address by the core does not match either chip select region will terminated with an error response and the access is inhibited to the external bus.

• When the driven address matches both chip select regions will terminate with an error response and the access is inhibited on the external bus.

## *External Signals and Operation*

Table 9.2 above lists the Mini-FlexBus signal names, their reset state and I/O direction.

- **Address bus**

  The external address bus consists of signals A[19:0], which provide up to 1M bytes of external memory space for each chip select. Internal address signals A[31:20] are not available externally, but are used to qualify the address generated by software. Note that the Mini-FlexBus module chip select addresses are limited to two ranges: 0x0040_0000 to 0x007F_FFFF and 0x0080_0000 to 0x009F_FFFF. The used values are typically subsets of these address ranges, and are defined in software by the values written to the BA field of the CSARn and BAM field of the CSMRn registers. See the device memory map in Table 9.2 above.

- **Data Bus and Transfer Operations**

  The data bus may be segmented into two, 8 bit "lanes". For each port size selected, only certain lanes must be attached to the external device. Figure 9.1 on page 172 illustrates the connections for 8 and 16 wide external memories, and the order in which a long word is output when software performs a write to the memory. An external memory with an 8 bit data bus should be connected to the single lane D[7:0] in non-multiplexed bus mode and a 16 bit external memory should be connected to AD[15:0] in multiplexed bus mode.

- **FB_CS[1:0] - Chip Select Signals**

  The Mini-FlexBus two chip select signals allow easy glueless connection to a maximum of two external memory mapped peripherals. Following initial startup, the two chip-selects may be configured by the application software to setup the base address and block sizes to suit the application's use of the external memory map. The base address is defined by the 16 bit BA field in the CSARn registers, while the block size is defined by the 16 bit BAM field in the CSMRn registers. In both registers, the 16 bits refer to the most significant 16 bits of the address emitted by the core. This means the base address must start on a 64K byte boundary, and the block size is a multiple of 64K bytes. The base address is simply the value written to the BA field multiplied by 64K bytes.

---

**Mini-FlexBus** 175

For example, to start the chip select at address 0x0040_0000, BA should equal 0x0040. Normally, the block size accessed by a chip select would be specified by setting bits starting at the least significant bit in the BAM field. This technique results in the block size having a binary weighted range, given by the equation: $2^N * 64K$ bytes, where N is the number of set bits.

For example, to allocate a contiguous 256K byte block of memory to a chip select, BAM should equal 0x0003.

In general, the BAM bits mask the base address when the base address is compared with the address emitted by the core to determine if the chip select should be asserted. Any cleared bit in the BAM bit field causes the corresponding address bit in the base register to be compared with the value emitted by the core; any set bit in the BAM causes the corresponding BA bit to be ignored. The outcome is that non-contiguous right justified set bits in the BAM field will result in multiple images of non-contiguous memory blocks whose total size is determined by the number of set bits in the BAM field, given by the equation above.

- **FB_ALE - Address Latch Enable**

  The address latch enable signal indicates that the device has begun a bus transaction and the address and attributes are valid. This signal is asserted for one bus clock cycle and may be used externally to capture the bus transfer address.

- **$\overline{FB\_OE}$ - Output Enable**

  This signal is asserted to the interfacing memory and/or peripheral to signal a read transfer when a chip select matches the current address decode.

- **FB_R/$\overline{W}$ - Read/Write**

  This signal determines the data transfer direction, it is driven high for data read and driven low for data write operation.

## *External Interface Timing Examples*

The following timing diagrams show the external bus timing for read and write accesses. The signal FB_CLK is shown for reference and generally is not connected to asynchronous external memories. The Mini-FlexBus interface terminates the bus cycle upon the internal assertion of the auto-acknowledge signal. This signal asserts when the AA control bit is set and upon the completion of the programmed number of wait cycles as determined by the wait state (WS) bit field in the CSCRn. The following section explains the Mini-FlexBus read and write cycles.

### Read Bus Cycle

During a read cycle, the ColdFire core receives the data from the selected memory or I/O device using the following steps and as illustrated in Figure 9.3. As stated earlier, the Bus clock for asynchronous memory is not used externally but included in the description for timing reference only.

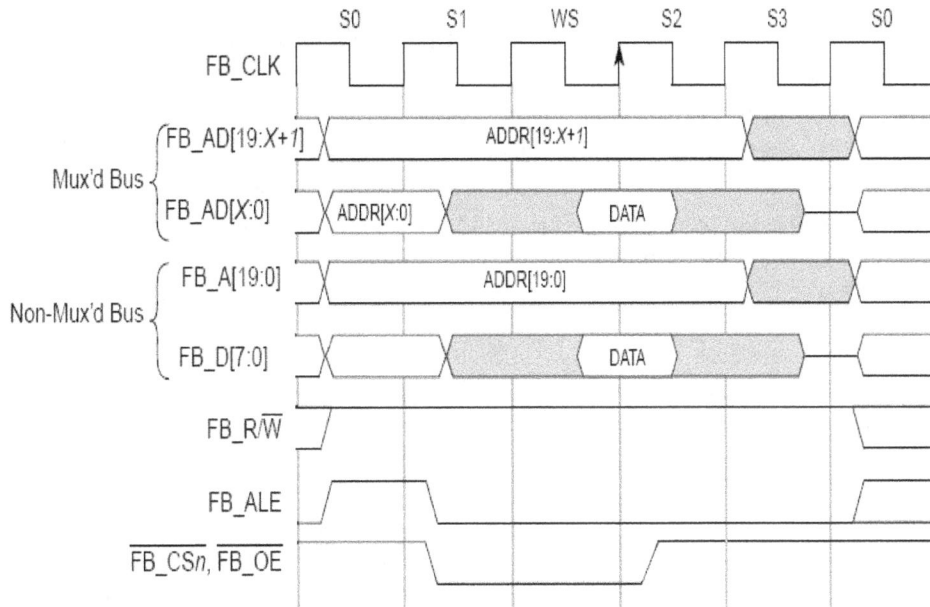

FIGURE 9.3. **Read Bus Cycle Timing Diagram (one wait state)**

- S0- The read bus cycle is initiated on the first clock, the device places a valid address on FB_AD[19:0], asserts the address latch enable (FB_ALE), and drives the read/write (FB_R/W) high for a read operation.

- S1- On the rising edge of the second clock, address latch enable (FB_ALE) is negated on the rising edge of the bus clock (FB_CLK), and chip-select (FB_CS*n)* is asserted. Address bus continues to be driven on the FB_AD pins that are unused for data. Read Data is driven by the external device before the next rising edge of FB_CLK (the rising edge that begins S2).

- S2- The chip select-signal (FB_CS*n)* is negated and the internal system bus transfer is completed. The processor latches data on the rising clock edge entering S2. and the external device may stop driving the data after this edge. However, data can be driven until the end of S3 or any additional address hold cycles.

- S3- All Address, data, and FB_R/W go invalid on the rising edge of FB_CLK at the beginning of S3, terminating the read cycle.

## Write Bus Cycle

The following timing diagram in Figure 9.4 on page 178 depicts a write bus cycle sequence. During this cycle, the selected memory or I/O device drives the data bus and the ColdFire core latches the requested data.

- S0- The write bus cycle is initiated on the first clock, the device places a valid address on FB_AD[19:0], asserts the address latch enable (FB_ALE), and drives the read/write (FB_R/W) low for a write operation.

- S1- On the rising edge of the second clock, address latch enable (FB_ALE) is negated on the rising edge of the bus clock (FB_CLK), and chip-select (FB_CSn) is asserted. Address bus continues to be driven on the FB_AD pins that are unused for data. Write Data is driven on the data bus before the next rising edge of FB_CLK (the rising edge that begins S2).

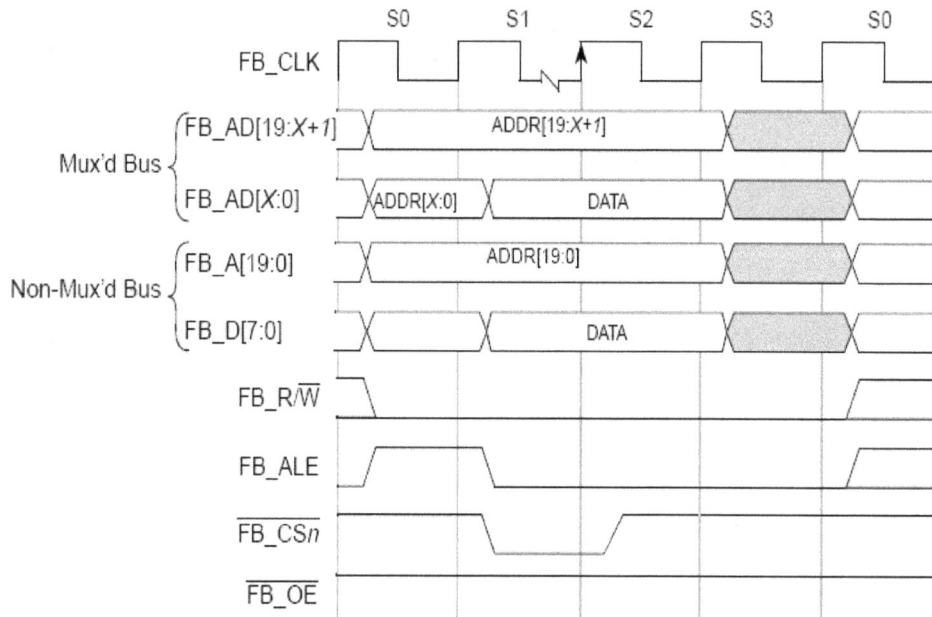

FIGURE 9.4. Write Bus Timing Diagram (no wait states)

- S2- The chip select-signal (FB_CSn) is negated and the internal system bus transfer is completed. The processor continues to drive the data bus on the rising clock edge entering S2 and the external device latches the data on this rising edge. However, data can still be driven until the end of S3 or for any additional address hold cycles.

- S3- All Address, data, and FB_R/W are driven invalid on the rising edge of FB_CLK at the beginning of S3, terminating the read cycle.

### Address Setup and Hold Time

The timing of the assertion and negation of the chip selects, byte selects, and output enable can be programmed on a chip-select basis. Each chip-select can be programmed to assert one to four clock cycles after the address-latch enable (FB_ALE) is asserted. Figure 9.5 on page 179 a read bus cycles with two clocks of address setup.

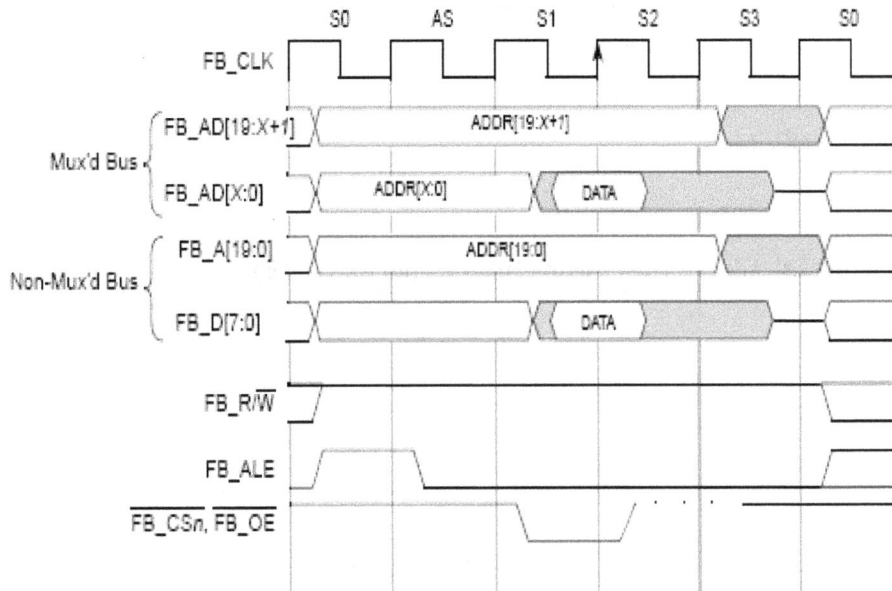

FIGURE 9.5. Address Setup Timing Diagram (no wait states)

In addition to address setup, a programmable address hold option for each chip select exists. Address and attributes can be held one to four clock cycles after chip-select, byte-selects, and output-enable negate. Figure 9.6 on page 180 show a read bus cycles with two clocks of address hold time.

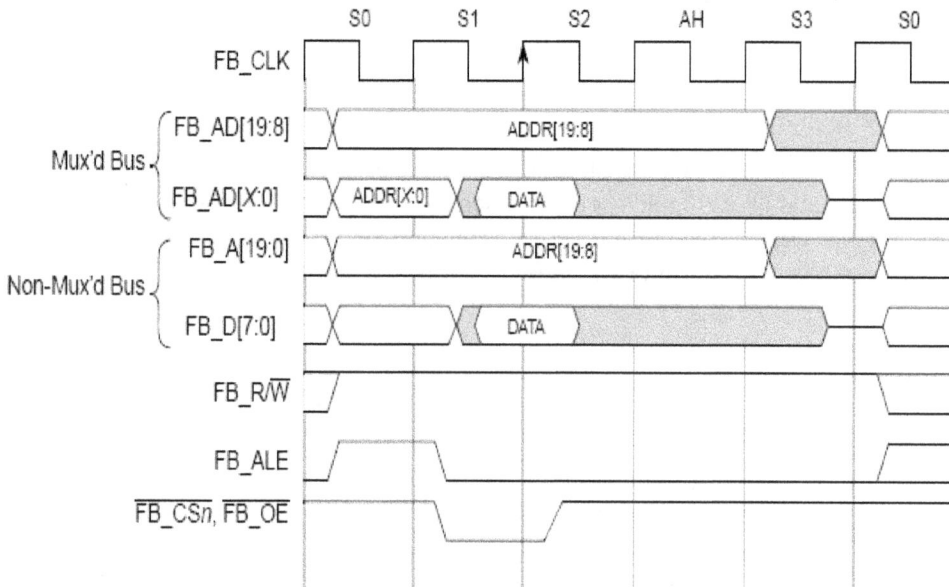

FIGURE 9.6. Address Hold Timing Diagram (no wait states)

## Mini_FlexBus Programmer's Model

The following section provides brief description of the chip-select registers and attempts to give the reader a basic understanding how to configure these registers for a typical external memory requirements connected to the Mini-FlexBus. This bus implements only 2 chip-selects allowing the user to connect up two memory banks to the external bus.

### CSARn - Chip-select Address Register n

This register defines the base address of the memory block connected to external bus. The internal address bus is compared with base address (BA) 16-bit field to determine if a chip select address is matched. The CSARn format is shown in Figure 9.7 on page 181.

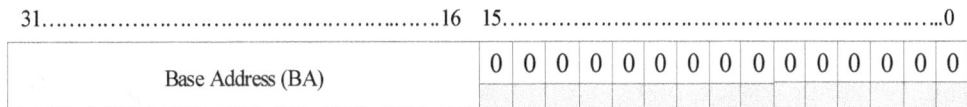

| 31..............................................................16 | 15...............................................................................0 |
|---|---|
| Base Address (BA) | 0 0 0 0 0 0 0 0 0 0 0 0 0 0 0 0 |

FIGURE 9.7. Chip-Select Address Register n

### CSMRn - Chip-Select Mask Register n

This register contains 3 fields to specify the external memory block size, read/write attribute and validity control for this chip-select.

The base address mask (BAM) field which defines the chip-select block size by masking some address bits. A '0' value in the BAM bit field force a compare in the chip-select decode logic and a '1' is ignored by the compare logic.

The second field is the write protect (WP) control to prevent some tasks from writing to this memory block or region when needed. Often, memory or I/O devices may need to be protected from being written and may permit read only operation to secure the system. There are two examples that may require such protection:

1. A Read-only memory block such as ROM, Flash or EEPROM should be configured as write-protected during run time so that any attempt by the application software to perform a write operation will signal a bus error. This bus fault signaling allows the operating system to recover the application quickly from a possible software run-away condition.

2. In multi-tasking environment, it is often necessary to have a block of memory used as a mail box (shared memory) between two processes or tasks. One task may be setup with

write attribute and the other task may have read-only attribute. Any write attempt by the task that has read-only attribute will signal a bus fault to the processor, thus allowing a secure access in inter-task communication.

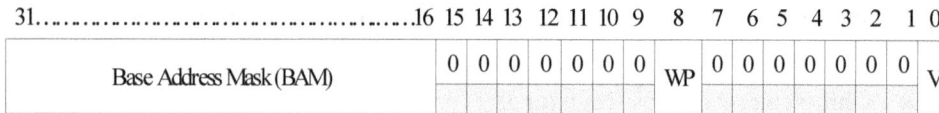

| 31................................................16 15 14 13 12 11 10 9 | 8 | 7 6 5 4 3 2 1 0 |
|---|---|---|

| Base Address Mask (BAM) | 0 | 0 | 0 | 0 | 0 | 0 | 0 | WP | 0 | 0 | 0 | 0 | 0 | 0 | 0 | V |
|---|---|---|---|---|---|---|---|---|---|---|---|---|---|---|---|---|

**FIGURE 9.8. Chip-Select Base Register** *n*

The write protection is controlled by the write-protect (WP) bit shown in the above. When this bit is cleared, read and write accesses are allowed and when set, only read accesses are allowed.

Finally, the last field is the valid (V) control bit which indicates that this chip select is active when set to a logic '1'. All three fields in this register can be written with one long-word transfer operation.

**CSCR*n* - Chip-Select Control Register *n***

This register provides system designers with a method to adjust the bus timing to satisfy a particular memory device requirements. The software can set up external bus timing by configuring the chip select control registers, CSCRn. These registers define the timing between address and chip select assertion, the duration of chip select assertion (wait states), read address hold timing, width of the data bus (port size), and auto-acknowledge of a bus transfer. Figure 9.9 on page 183 depicts the chip-select control register bit field format. A brief description of each field follows:

• ASET- Address Setup Time

This field controls the assertion of the chip-select (CS) signal with respect to assertion of a valid address and attributes. The address and attributes are considered valid at the same time FB_ALE asserts. The CS signal asserts on first, second, third or fourth rising clock edge depending on the ASET control bit field to allow for the additional address setup time.

• RDAH - Read Address Hold or Deselect

This field controls the address and attribute hold time after the termination during a read cycle that hits in the chip-select address space.The hold time applies only at the end of a transfer. Therefore, during a transfer to a port size smaller than the transfer size, the hold time is only added after the last bus cycle. The number of cycles the address and attributes are held may 0, 1, 2, or clock cycles depending on the setting of the RDAH control bit field in the CSCR*n*.

- WRAH - Write Address Hold or Deselect

This field controls the address, data, and attribute hold time after the termination of a write cycle that hits in the chip-select address space. The hold time applies only at the end of a transfer. Therefore, during a transfer to a port size smaller than the transfer size, the hold time is only added after the last bus cycle. This control bit field enables the address and transfer attributes to be held for 1, 2, 3 or 4 clock cycles after the negation of the CS*n* signal.

- WS - Wait States

The number of wait states inserted after FB_CS*n* asserts and before an internal transfer acknowledge is generated. The number of wait states may be set between 0 to 63 cycles.

- MUX - Multiplexed Mode

Selects between multiplexed and non-multiplexed address/data bus mode.
For non-multiplexed bus configuration (MUX =0)., the address information is driven on the address bus (FB_A[19:0]) and data is read/written on the data bus (FB_D[7:0]).
In the multiplexed bus configuration (MUX=1), address information is driven on the address data bus FB_AD[119:0] during the address phase, and low-order address lines FB_AD[7:0] for byte port size or FB_AD[15:0] for word port size during the data phase. The address must be latched using the falling edge of FB_ALE as the latch enable. Data is read/written on FB_AD[7:0] for byte port size and FB_AD[15:0] for word port size.

| 31....................22 | 21 | 20 | 19 | 18 | 17 | 16 | 15....................10 | 9 | 8 | 7 | 6 | 5 4 3 2 1 0 |
|---|---|---|---|---|---|---|---|---|---|---|---|---|
| 0 0 0 0 0 0 0 0 0 0 | ASET | | RDAH | | WRAH | | Wait States (WS) | MUX | AA | PS | | 0 0 0 0 0 0 |

FIGURE 9.9. **Chip-Select Control Register** *n*

- AA - Auto-Acknowledge

This bit determines the assertion of the internal transfer acknowledge for accesses specified by the chip-select address. This bit must always be set when the corresponding chip-select is active. If the chip-select is active and the auto-acknowledge control bit is cleared, the system will hang because this signal is the only method implemented in the Mini-Flex-

Bus to terminate the bus cycle. The auto-acknowledge signal assertion time is determined by the number of the programmed wait states (WS) field.

• PS - Port Size

This control bit field specifies the data port width associated with each chip-select. It determines where data is driven during write cycles and where data is sampled during read cycles.

For 8-bit port width, this field must be set to '$01_2$' to drive or sample the data on data bus pins FB_D[7:0]. When 16-bit port width is selected (PS =1x), the data is sampled or driven on FB_AD[15:0]. The 16-bit port width is supported only in multiplexed mode.

### Mini-FlexBus Initialization Example

The Mini-FlexBus interface comes out of reset with both chip-selects are disabled and it is assumed that the ColdFire V1 MCU will boot from the internal Flash. Figure 9.10 on page 184 depicts one of the ways the Mini-FlexBus interfaces to external Flash and SRAM devices. In this example, the Mini-FlexBus is configured in the non-multiplexed mode with 20 address lines and 8 data lines. The initialization code below configures both chip-select registers with the following parameters:

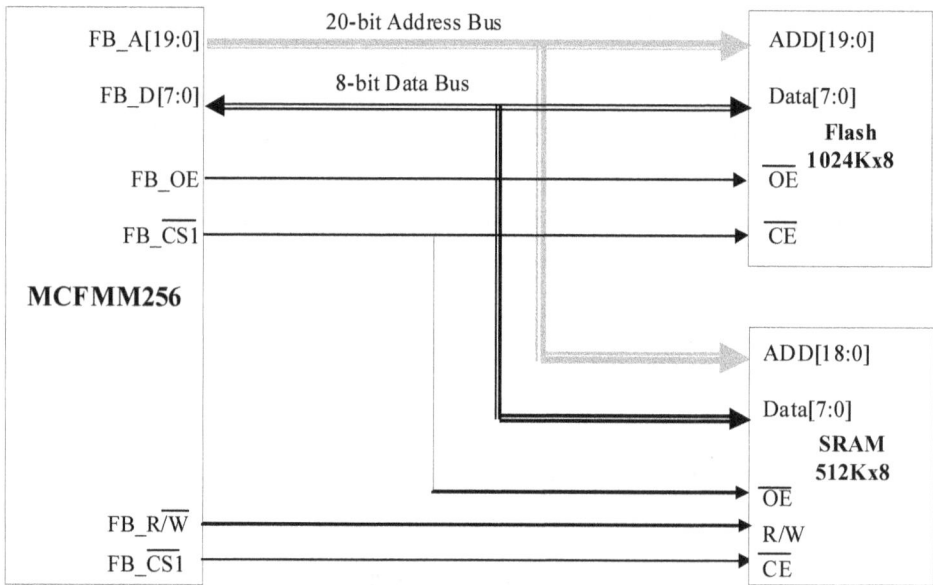

FIGURE 9.10. Mini-FlexBus Interface example

For CS0, the flash base address is setup to start at address 0x40_0000 and the block size is 1 mega bytes. This flash block requires an address setup (ASET) time of 1 cycle, read address hold (RDAH) time of 2 cycle, 2 wait states (WS=2), auto-acknowledge is enabled and memory port width = 8-bits.

For CS1, the SRAM base address is setup to start at address 0x60_0000 and the block size is 512K bytes. The SRAM block requires an address setup (ASET) time of 2 cycles, read address hold (RHAH) time of 1 cycle, 3 wait states (WS=3), auto-acknowledge enabled and memory port width = 8-bits.

```
MCF_MFBPC1 = 0xFF;        //Enable Address Lines A0-A7
MCF_MFBPC2 = 0xFF;        //Enable Address Lines A8-A15
MCF_MFBPC3 = 0xFF;        //Enable Data bus D3-D0 & address A16-A19 lines
MCF_MFBPC4 = 0xFF         //Enable CS[1:0], OE, R/W & Data Lines D7-D4
MCF_CSAR0 = 0x00400000;   //Set Base Address to 0x40_0000 for FB_CS0
MCF_CSMR0 = 0x000F;       //Select memory block size to 1 mega byte
MCF_CSAR0 = 0x00600000;   //Set Base Address to 0x60_0000 for FB_CS1
MCF_CSMR0 = 0x0007;       //Select memory block size to 512K byte
MCF_CSCR0 = 0x00180940;//ASET=1, RDAH=2, WRAH=0, WS=2, PS=8-bits for CS0
MCF_CSCR1 = 0x00240D40;//ASET=2, RDAH=1, WRAH=0, WS=3, PS=8-bits for CS1
```

**Mini-FlexBus Summary**

The Mini-FlexBus can address a maximum memory size of 1 MByte per chip-select. This is independent of the mode or port size used. It is solely based on the twenty address signals available FB_A[19:0]. Twenty address signals can address 2^20 bytes, regardless of whether two bytes are accessed at a time (16-bit port size) or one byte (8-bit port size).
• The FB_ALE and FB_CS1 signals share the same pin. If the multiplexed mode is used, the FB_ALE signal may be required, therefore limiting the number of available chip selects to one.
• Booting externally from the Mini-FlexBus is not supported.
• The chip-select base addresses must be 64-KByte aligned. Valid base addresses are between 0x$40_0000 and 0x7F_FFFF.
• If memory is accessed outside the chip select address range, or if an address match occurs on more than one chip-select, then an access error is generated.
• When internal flash security is enabled, memory accesses through the Mini-FlexBus are blocked.
• Executing code over the Mini-FlexBus is supported.

# CHAPTER 10

*System Integration*

## *Introduction to System Integration*

Both Coldfire and S08 based devices contain a number of modules that are required for correct system operation, or provide some form of system-level protection.

This chapter covers the following functional modules that affect the entire system:

- Multipurpose Clock Generator (MCG)
- Time Of Day Module (TOD)
- Low Voltage Warning (LVW) and Low Voltage Detect (LVD)

In addition the chapter describes status and control registers that affect pin characteristics, clock gating and selection choices, and other power management controls.

## *Multipurpose Clock Generator*

The multipurpose clock generator (MCG) module provides several clock source choices for both Coldfire and S08 devices.

The MCG contains a frequency-locked loop (FLL) and a phase-locked loop (PLL) that can be driven by an internal or an external reference clock. The MCG can select either of the FLL or PLL clocks or either of the internal or external reference clocks as a source for the MCU system clock. The selected clock source is passed through a divider that allows a lower output clock frequency to be derived. The advantage of using the PLL is that it can produce a low jitter clock which makes it suitable to drive serial I/O that requires tight timing. The advantage of the FLL is that it starts up much faster than the PLL, in a few microseconds from power-down modes and makes it suitable for many low power applications that require low interrupt latency. Because the FLL uses a digital filter, its short term jitter is higher than that of the PLL, and might not be suitable for some applications.

If you want to use the USB on devices that have it, the MCG must be configured for PLL engaged external (PEE) mode to achieve an MCGOUT frequency of 48 MHz. The different clocks available in the MCG are listed in Table 10.1.

**TABLE 10.1 MCG Clocks**

| Clock | Description |
|-------|-------------|
| MCGOUT | This clock drives the CPU, debug, RAM, and BDM directly and is divided by two to clock all peripherals (BUSCLK). |
| MCGLCLK | This clock source is derived from the digitally controlled oscillator (DCO) of the MCG. Development tools can select this internal self-clocked source to speed up BDC communications in systems where the bus clock is slow. |
| MCGIRCLK | This is the internal reference clock and can be selected as the TOD clock source. |
| MCGFFCLK | This clock is divided by 2 to generate the fixed frequency clock (FFCLK) after being synchronized to the bus clock. It can be selected as clock source for the TPM modules. The frequency of the FFCLK is determined by the settings of the MCG. |
| MCGPLLSCLK | This clock has a direct connection to the PLL output clock (running at 48 MHz) and thus allows the user to have the flexibility to run the MCGOUT at lower frequencies to conserve power. |

In addition to these clocks, there is a Low Power Oscillator (LPO) and an asynchronous clock for the ADC module (ADACK) that allows it to continue running while the device is in STOP mode.

The LPO clock is generated from an internal low-power oscillator that is completely independent of the MCG module. The LPOCLK can be selected as the clock source to the Time Of Day (TOD) module described later in this chapter or the Computer Operating Properly (COP) system.

Either one of two low power oscillators may be used to clock the MCG module, while one of the oscillators may also clock the TDO module.

### Initializing the MCG

Because the MCG comes out of reset in FEI mode, only the FEE, FBE, or FBI mode can be after reset. Accessing any of the other modes requires first configuring the MCG for one of these three initial modes. Care must be taken to check relevant status bits in the MCGSC register reflecting all configuration changes within each mode.

To change from FEI mode to FEE or FBE modes, follow this procedure:

1. Enable the external clock source by setting the appropriate bits in MCGC2.

2. If the RANGE bit (bit 5) in MCGC2 is set, set DIV32 in MCGC3 to allow access to the proper RDIV values.

3. Write to MCGC1 to select the clock mode.

— If entering FEE mode, set RDIV appropriately, clear the IREFS bit to switch to the external reference, and leave the CLKS bits at %00 so that the output of the FLL is selected as the system clock source.

— If entering FBE, clear the IREFS bit to switch to the external reference and change the CLKS bits to %10 so that the external reference clock is selected as the system clock source. The RDIV bits should also be set appropriately here according to the external reference frequency because although the FLL is bypassed, it remains on in FBE mode.

— The internal reference can optionally be kept running by setting the IRCLKEN bit. This is useful if the application switches back and forth between internal and external modes. For minimum power consumption, leave the internal reference disabled while in an external clock mode.

4. After the proper configuration bits have been set, wait for the affected bits in the MCGSC register to be changed appropriately, reflecting that the MCG has moved into the proper mode.

— If ERCLKEN was set in step 1 or the MCG is in FEE, FBE, PEE, PBE, or BLPE mode, and EREFS was also set in step 1, wait here for the OSCINIT bit to become set indicating that the external clock source has finished its initialization cycles and stabilized. — If in FEE mode, check to make sure the IREFST bit is cleared and the LOCK bit is set before moving on.

— If in FBE mode, check to make sure the IREFST bit is cleared, the LOCK bit is set, and the CLKST bits have changed to %10 indicating the external reference clock has been appropriately selected. Although the FLL is bypassed in FBE mode, it remains on and locks in FBE mode.

5. Write to the MCGC4 register to determine the FLL digitally controlled oscillator (DCO) output (MCGOUT) frequency range. Make sure that the resulting bus clock frequency does not exceed the maximum specified bus clock frequency of the device. Table 10.2 lists the DCO maximum output frequencies as a function of the DMX32 and DRS bits in the

MCGC4 register. When the DMX32 bit is set, the DCO is tuned to support a 32.768kHz reference clock input. When DMX32 is clear, the input frequency range is extended.

**TABLE 10.2 DCO Frequency Ranges**

| DRS | DMX32 | Reference range | FLL factor | DCO range |
|---|---|---|---|---|
| 00 | 0 | 31.25 - 39.0625 kHz | 512 | 16 - 20 MHz |
|  | 1 | 32.768 kHz | 608 | 19.92 MHz |
| 01 | 0 | 31.25 - 39.0625 kHz | 1024 | 32 - 40 MHz |
|  | 1 | 32.768 kHz | 1216 | 39.85 MHz |
| 10 | 0 | 31.25 - 39.0625 kHz | 1536 | 48-60 MHz |
|  | 1 | 32.768 kHz | 1824 | 59.77 MHz |
| 11 | Reserved | | | |

6. Wait for the LOCK bit in MCGSC to become set, indicating that the FLL has locked to the new multiplier value designated by the DRS and DMX32 bits.

Setting DIV32 (bit 4) in MCGC3 is strongly recommended for FLL external modes when using a high frequency range (RANGE = 1) external reference clock. The DIV32 bit is ignored in all other modes

To change from FEI clock mode to FBI clock mode, follow this procedure:

1. Change the CLKS bits in MCGC1 to %01 so that the internal reference clock is selected as the system clock source.

2. Wait for the CLKST bits in the MCGSC register to change to %01, indicating that the internal reference clock has been appropriately selected.

## *Time Of Day Module*

Some devices also have a Time Of Day (TOD) module which may use a dedicated external oscillator as its source clock. The TOD oscillator can also be used as the reference clock into the MCG. The TOD consists of one 8-bit counter, one 6-bit match register, several binary-based and decimal-based prescaler dividers, three clock source options, and one interrupt that can be used for quarter second, one second and match conditions. This module can be used for time-of-day, calendar or any task scheduling functions. It can also serve as a cyclic wake up from low power modes without the need of external components.

Typically, the TOD Control (TODC) register prescaler bits TODPS[2:0] would be configured by initialization software to generate either a quarter second or a one second interrupt. It is possible to enable both interrupts, which may be handled separately in software since each has its own status flag. The prescale bit value for each of the clock sources required to produce a quarter second or one second interrupt is given in Table 10.3.

**TABLE 10.3 TOD Clock Sources & Prescale Values**

| Input | TODPS[2:0] | TOD Clock |
|-------|-----------|-----------|
| 1 kHz LPO | 000 | - |
| 32.768 KHz | 001 | 32.768 KHz |
| 32 KHz | 010 | 32 KHz |
| 38.4 KHz | 011 | 38.4 KHz |
| 4.9152 MHz | 100 | 38.4 kHz |
| 4 MHz | 101 | 32 kHz |
| 8 MHz | 110 | 32kHz |
| 16 MHz | 111 | 32 kHz |

The table also shows the TOD clock output frequency, which will only be present if TODC[TODCLKEN] is set.

The TOD match functionality can also be used to generate interrupts at multi-second time intervals.

**Initializing the TOD**

1. Configure the TODC register

   Configure TOD clock source (TODCLKS bit)

   Configure the proper TOD prescaler (TODPS bits)

2. Write the TOD match register (optional)

   Write the TOD match value to the desired value (TODM register)

3. Enable the TODSC register(optional)

   Enable match functionality

   Enable desired TOD interrupts (QSECIE, SECIE, MTCHIE bits)

4. Enable TODC register. This operation should always be performed last.

   (Optional) Enable TOD clock output (TODCLKEN bit)

   Enable the TOD module (TODEN bit)

## *Low Voltage Detect and Low Voltage Warning*

Voltage detection circuits are provided to protect memory contents and control microcontroller system states during supply voltage variations, and in particular during power on and power off. The circuits include

- a power-on reset (POR) circuit to hold the device in reset when power is initially applied.
- a programmable low voltage detect (LVD) circuit that either generates an interrupt or applies a system wide reset when the power supply voltage drops below one of two specific thresholds.
- a programmable low voltage warning (LVW) circuit that indicates the supply voltage is approaching, but is above the LVD voltage.

When triggered, the LVW sets a flag, and if programmed to do so may generate an interrupt to the core. Like the LVD, the LVW circuit has two selectable trip thresholds. Table 10.4 shows the recommended combinations of LVD and LVW bit values in the SPMSC3 system control register and resultant typical thresholds. The LVDV:LVWV combination 0b10 is not recommended because this results in selecting an LVD trip point that is greater than the LVW trip point, in which case the warning would not be guaranteed to occur.

TABLE 10.4 LVD and LVW Trip Point Values (Typical)

| LVDV:LVWV | LVW Trip Point | LVD Trip Point |
|---|---|---|
| 00 | $V_{LVWL}$ = 2.15 | $V_{LVDL}$ = 1.86 |
| 01 | $V_{LVWH}$ = 2.6 | |
| 10 Not Recommended | $V_{LVWL}$ = 2.15 | $V_{LVDH}$ = 2.33 |
| 11 | $V_{LVWH}$ = 2.6 | |

The choice of thresholds should be based on the tolerance of the power supply plus the rate at which the supply decays. Using the lowest LVW trip point allows the maximum tolerance on the power supply specification, but may reduce the time available to execute code before the LVD triggers and resets the device.

Note that the LVD is disabled upon entering Stop2 or Stop3 modes unless the LVDSE bit is set. If both LVDE and LVDSE are set when the STOP instruction is processed, the device will enter Stop4 mode. The LVD can be left enabled in this mode.

## Clock Selection

The following registers provide control over the clocking of different on-chip modules as well as routing the clock to an external pin.

- SOPT1, SOPT2

  These registers provide controls for the COP timer clock source, its operating mode and timeout values, low power modes, Mini-FlexBus security level, Background Debug, reset pin enablement and BUSCLK routing.

  Note that some of the bits in both registers are write once only, so extra care should be exercised in setting them up.

- SIMCO

  This register defines which internal clock is routed to the CLKOUT pin.

- SCGC1, SCGC2, SCGC3

  These registers enable or disable the clock signals to on-chip modules that are present on the device, and provide some measure of power management by allowing software to turn off the clocks to devices that are not needed. Refer to Freescale documentation-for details.

- SIMIPS

  Most of the bits in this register control SCI features. One bit is provided to select a trigger source for the ADC.

## Power Management

- SPMSC1, SPMSC2, SPMSC3

  These registers contain status and control bits for low voltage detect functions, bandgap voltage reference and the voltage regulator to enable software management of the device power consumption.

## Pin Control

- SOPT3, SOPT4, MFBPC1, MFBPC2, MFBPC3, MFBPC4

  These registers control various combinations of pad drive strengths, slew rates, pull-ups and routing options for the Mini-FlexBus, USB, SCI, I²C and CMT modules.

# I/O Ports and KBI

## Introduction to I/O Ports and KBI

As part of it's control function the microcontroller has to communicate with the outside world, both to receive the necessary inputs for it's control algorithms, and generate outputs to control external devices or communicate information. Analog systems such as ADCs and DACs take care of analog input and output functions, Timers provide high speed and high accuracy digital input and output, and serial communication interfaces connect to other devices. There is still a need for many low speed digital inputs and outputs, for functions such as key inputs and outputs to turn on individual loads. The general purpose parallel Input and Output (I/O) ports provide this functionality. Although these may be fixed as either input or output, typically they are software configurable.

Inputs may be high impedance, or may include a pull-up or pull-down resistor to bias the input state in the absence of an active input. On 3V microcontrollers it may be required to include 5V input compatibility for connecting to external 5V devices.

Outputs are typically actively driven to the supply rails, but may also be open drain to allow connection to a load connected to a higher supply voltage than that of the microcontroller. For larger loads it is generally preferable to drive active low, or sink the load current, as microcontrollers can usually sink more current than they can source. It is also possible to connect several output ports in parallel and sum their sink capabilities for loads beyond the specification of a single output.

Many applications include keys, switches or push buttons for system input, and individual switches may be connected directly to an input pin. If a larger number of keys is required, such as a keypad or keyboard, it is more efficient to arrange them in a scanned matrix. In systems that utilize one of the low power modes it is highly desirable to actively wake up the microprocessor if a key is pressed, and the I/O ports with Key Board Interrupt provide this function.

In order to maintain compatibility between the HCS08 and the ColdFire V1 versions of the

Flexis products, the 8-bit peripherals from the HSC08 product family are used for most common functions, including the 8-bit peripheral bus. Although this increases the Cold-Fire CPU's access time to these peripherals and I/O ports, typically this is not issue. There are, however, cases where the significantly higher throughput of the ColdFire v1 CPU could be used to implement additional I/O functions, such as additional SPI or UART ports, but this requires faster access to the I/O pins that is possible via the 8-bit bus. The Rapid GPIO module was designed specifically to allow the ColdFire V1 CPU direct fast access from the 32-bit processor bus to the I/O pins, bypassing the 8-bit peripheral bus.

With the rich set of peripherals on modern microcontrollers such as the Flexis and Cold-Fire V1 families, most applications do not require all of the available functions. Also, on many microcontrollers, more peripheral functions have been implemented than there are pins to make available externally. To allow the user flexibility in which functions are brought out, the I/O ports typically offer several configuration options for the functions brought out. The general purpose I/O is typically the default function.

Although the base input and output function is the same for all I/O ports, some ports offer additional functionality such as Toggle or pin drive options. Because the assignment of this additional functionality varies by product, it is necessary to refer to the individual product manual to determine which functions are available on which ports.

---

### *Parallel Input/Output*

The I/O is organized into ports with up to 8 pins per port. Some level of functionality and configuration applies to the entire port, although pins may be accessed and configured for data direction individually. Figure 11.1, "Typical I/O Block Diagram" shows the basic pin block diagram. The Data Direction control bit determines is the associated port pin is an input or an output. Following Reset all Data Direction control bits are 0 for input. Writing a the Data Direction control bit turns the corresponding pin to an output. As a pin is switched to an output, the data in the corresponding Data bit is immediately driven onto the pin. Although the port Data bits are also Reset to 0, it is good practice to write the corresponding Data Registers with the required output values prior to switching port pins to outputs

When a port pin is assigned to a shared function, that function will control the I/O pin directly, irrespective of the state of the Data Direction control and Data bits. When the assigned function is digital, a read of the port Data bit will return the value of the Data bit or the pin state, still according to the Data Direction control bit. If the assigned port function is analog, then the port digital input and output buffers are disabled, and a read will return a 0.

For selected ports it is possible to control the following individually for each pin:

---

**Chapter 11**

- Internal Pull-up enable - Enables an internal pull-up for input mode only
- Slew Rate enable - When enabled limits the port output switching slew rate, useful for reducing EMI emissions.
- Drive Strength select - Enables high output drive for both source and sink currents

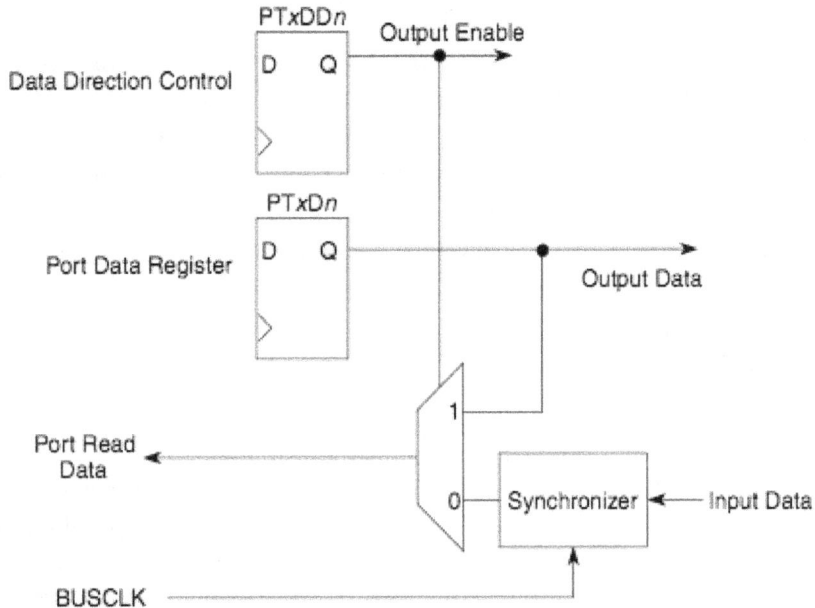

FIGURE 11.1. **Typical I/O Block Diagram**

When using the high output drive, the user must take care that the microcontroller total current source and sink specifications are not exceeded. Also, driving higher loads with fast slew rates is more likely to increased EMC emissions.

**Port Data Set, Clear, and Toggle**

The bit manipulation instructions are very useful for setting, clearing, or toggling individual I/O port data bits. The bit manipulation instructions are implemented as read-modify-write instructions, meaning that the CPU first reads the entire byte of data at the required address, modifies the selected bit, and writes the entire byte back to the same address. This implementation has limitations in some application cases, due to either the number of cycles necessary to execute, or the indeterminate state of the output Data bit for pins that are configured as inputs. This is because the read part of the read-modify-write portion of the instruction reads the input pin, and the write portion writes to the output Data bit. This could be an issue if that pin is later switched to an output without writing the Data bit to a known value. The Port Data Set, Clear and Toggle function, available on specific ports, provides an alternative fast method of changing selected Data bits only, without the use of the bit manipulation instructions. This method also allows the setting, clearing and toggling any combination of bits within a port in a single write.

This function is implemented with three additional registers for each supported port, one each for Set, Clear, and Toggle. The registers function as follows:

- Port Data Set register - Writing a 1 to any bit in this register will set the corresponding port Data bit, writing a 0 will have no effect on the corresponding bit
- Port Data Clear register - Writing a 0 to any bit in this register will clear the corresponding port Data bit, writing a 1 will have no effect on the corresponding bit
- Port Toggle register - Writing a 1 to any bit in this register will toggle the corresponding port Data bit, writing a 0 will have no effect on the corresponding bit

See the following section "Rapid GPIO (RGPIO)" below for more details.

## Rapid GPIO (RGPIO)

The Flexis and ColdFire V1 products implement the peripheral systems from the HCS08 family wherever possible in order to maintain as much compatibility as possible.The design of these peripherals is based on the internal 8-bit bus architecture of the HCS08. The ColdFire V1 CPU utilizes 32-bit internal buses for access to the internal memory system, and the Mini FlexBus if implemented. A peripheral Bridge facilitates the connection between the ColdFire 32-bit bus and the 8-bit peripheral bus, as shown in.Figure 11.2, "ColdFire V1 Bus Structure" . The trade off is that accesses to any location on the 8-bit peripheral bus are slower than accesses on the 32-bit buses. In the vast majority of cases this is not an issue, but in some cases it would be beneficial be able to use the fast access capabilities of the ColdFire V1 CPU to access I/O pins. The Rapid General Purpose I/O (RGPIO) module provides this functionality, by allowing direct single cycle access from the 32-bit ColdFire bus to up to 16 RGPIO pins.

General Purpose I/O is the default function on any port, so the RGPIO has to be enabled

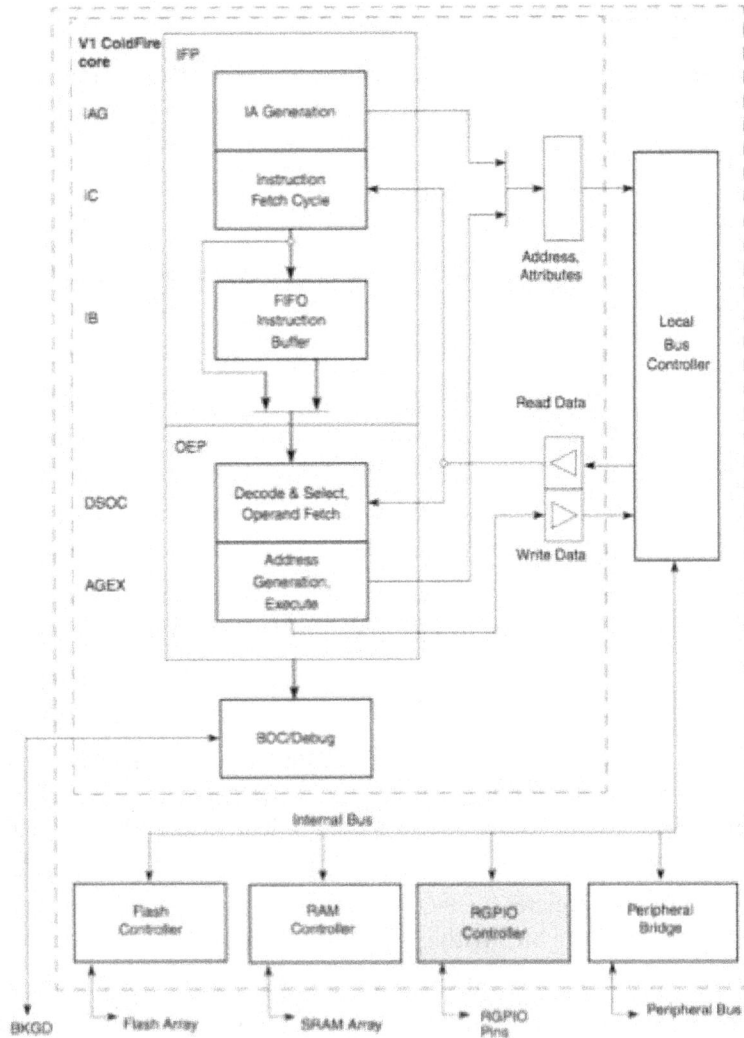

**FIGURE 11.2. ColdFire V1 Bus Structure**

as an alternate function. RGPIO may be enabled on a pin by pin basis in the corresponding bits in the RGPIO Pin Enable (RGPIO_ENB) register. Any RGPIO enabled pins are then configured and controlled by the corresponding bits in the RGPIO registers, not the port registers. Therefore, the data direction for RGPIO is configured in the RGPIO Data Direction Register (RGPIO_DIR), with a 0 for an input, and a 1 for an output. Data is written and read in the RGPIO Data register (RGPIO_DATA)

To facilitate fast control of any combination of RGPIO pins, without the corruption of

**TABLE 1. RGPIO Memory Map**

| Offset Address | Access | Register Name | Register Description |
|---|---|---|---|
| 0x00 | Write | RGPIO_DIR | RGPIO Data Direction Register |
|  | Read |  |  |
| 0x02 | Write | RGPIO_DATA | RGPIO Data Register |
|  | Read |  |  |
| 0x04 | Write | RGPIO_ENB | RGPIO Pin Enable Register |
|  | Read |  |  |
| 0x06 | Write | RGPIO_CLR | RGPIO Write Data Clear Register |
|  | Read | RGPIO_DATA | RGPIO write data register |
| 0x08 | Write | n/a | writes not allowed |
|  | Read | RGPIO_DIR | RGPIO data direction register |
| 0x0A | Write | RGPIO_SET | RGPIO Write Data Set Register |
|  | Read | RGPIO_DATA | RGPIO write data register |
| 0x0C | Write | n/a | writes not allowed |
|  | Read | RGPIO_DIR | RGPIO data direction register |
| 0x0E | Write | RGPIO_TOG | RGPIO Write Data Toggle Register |
|  | Read | RGPIO_DATA | RGPIO write data register |

other RGPIO pins, and without the need for read-modify-write sequences that would significantly slow accesses, the RGPIO provides direct alternative access registers for clearing, setting and toggling pins. These functions are implemented with three additional RGPIO registers as follows:

• RGPIO Write Data Set register (RGPIO_SET) - Writing a 1 to any bit in this register will set the corresponding Data bit, writing a 0 will have no effect

• RGPIO Write Data Clear register (RGPIO_CLR) - Writing a 0 to any bit in this register will clear the corresponding Data bit, writing a 1 will have no effect

• RGPIO Write Toggle register (RGPIO_TOG) - Writing a 1 to any bit in this register will toggle the corresponding Data bit, writing a 0 will have no effect

Note that reading the Set, Clear or Toggle registers returns the value of the RGPIO Data register (RGPIO_DATA), as shown in Table 1, "RGPIO Memory Map,".

## *Keyboard Interrupt (KBI)*

Although the name implies that this function is specifically for keyboard input, it is also suitable for any input that is needed to generate an interrupt, and or wake up the microcontroller from a low power mode. The Keyboard Interrupt (KBI) function is available on specific ports, as defined in the product Data Sheet or Reference Manual.

On the ports that support interrupts, the following is selectable:

- Interrupt enable by pin
- Interrupt detect mode as either Edge sensitive or Edge and Level sensitive per port
- Edge polarity, or Edge and Level sensitivity by pin
- Internal pull-up or pull-down resistors

### Interrupt Detect Sensitivity

Two interrupt detect modes are available:

- Edge only sensitivity, selectable for either positive or negative edges
- Edge and Level sensitivity, selectable for either positive edge and high level, or negative edge and low level

The difference between the two is subtle, and often misunderstood. Edge only sensitivity means that the pin logic has to detect an edge of the selected polarity, after the interrupt has been enabled, and then the asserted level must be present for at least one system clock cycle. If sensitivity for negative edges was selected, then after being enabled, the pin must detect a high level for at least one cycle, followed by a low level for at least one cycle. This will result in the corresponding Interrupt Flag being set, and an interrupt request is generated to the CPU.

For Edge and Level sensitivity either an edge or the asserted level will set the corresponding Interrupt Flag and generate an interrupt request. The main difference is that if the pin is already at the asserted level when the interrupt is enabled, an interrupt will be generated even if no edge is observed at that time.

The Interrupt Flag is cleared by writing a 1 to the Acknowledge bit. In the Edge only mode, any write to the Acknowledge bit will clear the Interrupt Flag. In the Edge and Level mode, the Interrupt Flag will only be cleared by writing a 1 to the Acknowledge bit while all the enabled port pins are at their de-asserted levels.

### Internal Pull-up and Pull-down

The internal pull-up resistor is enabled on a pin by pin basis in the Port Pull Enable Register (PTxPE). In the basic I/O port mode, only a pull-up resistor is available. Internal Pull-down resistors are only available for specific alternate functions, such as Keyboard Interrupts.In the Keyboard Interrupt function, the pull option is also selected using the enable bits in the Port Pull Enable Register (KBI2ES), and the Edge select control bits also select the type of pull resistor.

- 0 selects - Pull-up resistor, Negative edge, Low level
- 1 selects - Pull-down resistor, Positive edge, High level

### KBI and Stop Modes

In Stop3 and Stop4 modes, all I/O is maintained because internal logic circuitry stays powered. Upon wake-up, normal I/O function is available to the user.

In the partial power-down mode Stop2, port states are lost and need to be restored upon wake-up. It is recommended to save CPU register status and the state of I/O registers in RAM before the STOP instruction is executed.

On wake-up, it is first necessary to determine if sufficient power was maintained during Stop2 for a full recovery. This is indicated by the Partial Power Down Flag (PPDF) in the System Power Management Status and Control 2 register (SPMSC2). If sufficient power was maintained, then the PPDF bit will be set, and the I/O register states should be restored from the values saved in RAM. Other peripherals may also require initialization or restoration to their pre-stop condition. Then clear the PPDF bit by writing a 1 to the Partial Power Down Acknowledge bit SPMSC2[PPDACK].

If the PPDF bit is cleared, all I/O must be initialized as if a power-on-reset had occurred.

### Keyboard Interrupt Initialization

When an interrupt pin is first enabled, it is possible to get a false interrupt flag. To prevent a false interrupt request during pin interrupt initialization, do the following:

1. Mask interrupts by clearing the Interrupt Enable bits in the Keyboard Interrupt Status and Control register KBIxSC[KBIE]

2. Select the pin polarity by setting the appropriate Interrupt Edge Select bits in the Interrupt Edge Select register KBIxES[KBEDGn]

3. If using the internal pull-ups/pull-downs, configure the associated enable bits in the Interrupt pin select register KBIxPE

4. Enable the interrupt pins by setting the appropriate Interrupt Pin Select bits in the Interrupt pin select register KBIxPE[KBIPEn]

5. Write to the Interrupt Acknowledge bits in the Keyboard Interrupt Status and Control register KBIxSC[KBACK] to clear any false interrupts

6. Set appropriate Interrupt Enable bits in the Keyboard Interrupt Status and Control register KBIxSC[KBIE] to enable interrupts.

### Keyboard Interrupt Register and Bit Names

When referring to the Freescale technical documentation on the Keyboard Interrupt registers, and control and status bits, the following acronym table will be useful:

TABLE 2. Keyboard Interrupt Acronyms

| Name/Acronym | Type | Meaning |
| --- | --- | --- |
| KBIxxx | Register | KeyBoard Interrupt |
| KBIxSC | Register | Keyboard Interrupt Status and Control |
| KBF | Bit | Interrupt Flag |
| KBACK | Bit | Interrupt Acknowledge |
| KBIE | Bit | Interrupt Enable |
| KBIMOD | Bit | Detection Mode |
| KBIxPE | Register | Interrupt pin select register |
| KBIxES | Register | Interrupt Edge Select register |
| KBEDG | Bit | Interrupt Edge select bit |

# CHAPTER 12

# *Basic Serial Peripherals*

## *Introduction to Basic Serial Peripherals*

This family of devices has a variety of on chip serial peripherals that each have different attributes designed to support data communication with peripheral components such as sensors, port expanders and displays. The serial peripherals are the SCI, SPI and I²C bus controllers.This chapter contains descriptions of these controllers.

## *Serial Communications Interface*

Serial communication is the common and most cost effective ways to transmit digital data by wire over long distances. The SCI implements the industry standard serial communication protocol that has been available on a wide range of microcontrollers since the 1980's. This protocol is compatible with the RS232 interface used on most desktop computers and also used in many industrial, automotive and commercial applications such as point-of-sale equipment. The SCI communicates asynchronously and supports both half and full duplex data transfers and may be programmed to operate at both standard and non-standard baud rates. The SCI uses standard non-return-to-zero (NRZ) format and includes an on-chip baud rate generator. This interface uses only two signals, a transmit data (TxD) and receive data (RxD) pins. If needed, other modem control signals, such as request-to-send (RTS) and Clear-to-send (CTS) may be implemented by software with some of the available general purpose I/O pins.

For example, in ColdFire version1 and HCS08 (QE128) devices implementation with a system clock frequency of 50Mhz[1], the SCI can transmit and receive at rates as high as

---

1. I/O Peripheral are clocked at half system clock speed.

1.5625K baud and as low as 190 baud. An SCI frame is either 8 or 9 bits with one or two stop bits. The most common configuration uses one start bit, 8-bits of character data, and one stop bit. A nine bit configuration uses one start bit, 9-bits of character data, and one stop bit. The ninth bit of character data can be used for an extra stop bit, receiver wake-up function, or a parity bit.The SCI transmitter hardware may optionally generate and transmit parity, while the receiver has parity checking hardware to ensure data integrity of the received data.

The SCI can be interrupt driven with eight flags and interrupt requests. There are 3 separate interrupt requests and vectors, one for the transmitter, one for the receiver, and for error conditions. This provides highly efficient interrupt processing of normal transmit/ receive functions without polling or interrupt checking. All receiver error conditions can be handled by the separate interrupt vector.

## *SCI Architecture*

Most ColdFire® V1 and HCS08 devices contain two or three identical SCI Modules, each consisting of a transmitter and receiver that may be enabled independently. The SCI supports both half and full duplex operation. The transmitter is connected to the serial bus Transmit Data (TxD) pin and the receiver is connected to the Receive data (RxD) Pin. Optionally, request-to-send (RTS) and clear-to-send (CTS) pins may be used to enable transmission-reception flow control to avoid receiver overrun conditions. Figure 12.1 shows the basic block diagram of the SCI module.

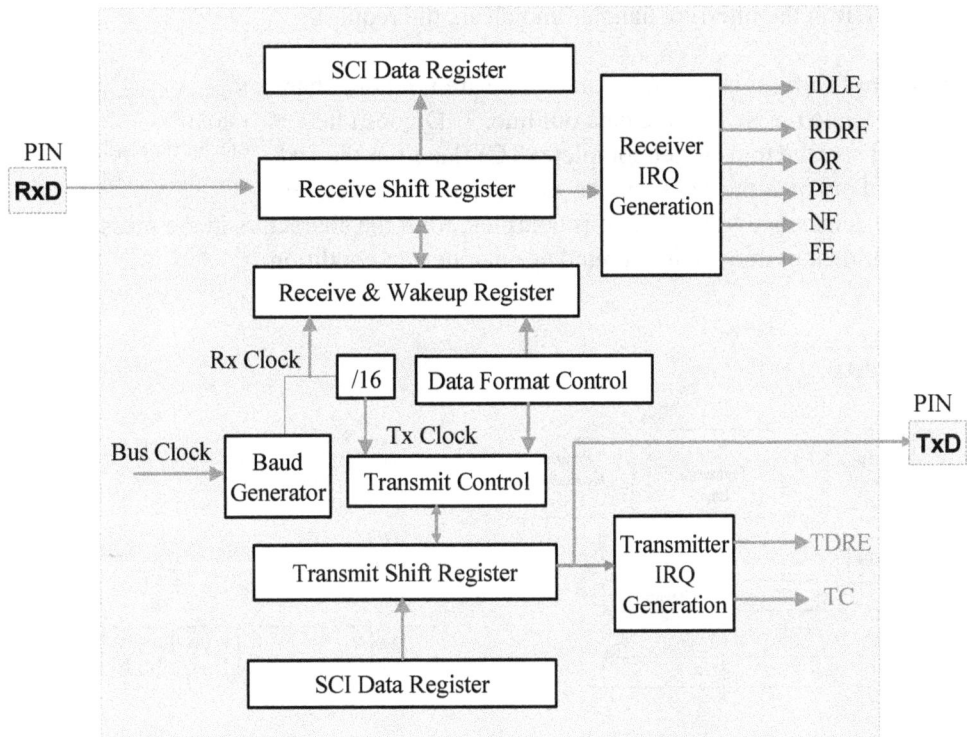

FIGURE 12.1. SCI Functional Block Diagram

## Transmitter Operation

The transmitter is double buffered, consisting of two registers: a serial shifter and an SCI data register (SCDR). Double buffering allows software to write the next character while the previously written character is being shifted out serially. To transmit data, the software writes to the SCDR when the transmitter is ready for the next character. The transmitter automatically appends a start bit, an optional parity bit if enabled, and one stop bit, as shown in the example in Figure 12.2. The number of bits shifted out by the transmitter depends on the serial format. The SCI can be programmed to have a character length between 8 or 9 bit character length depending on the setting of the bit per character (M) bit field specified in the SCI control register 1 (SCIC1). When a character has been completely shifted out, the next data character is transferred from the SCDR to the transmit shift register and at this point the SCDR becomes empty. The SCI indicates that it is ready

**Basic Serial Peripherals**                                                      207

for the next character by asserting transmitter ready status (TDRE) flag in the SCI status register 1 (SCIS1). If the transmitter interrupt is enabled (TIE=1) in the SCI control 2 register (SCIC2), the SCI signals a request to the CPU. The CPU writes the next character into the SCDR in the interrupt handler and clears the request.

If the transmit shift register empties and there are no more character written into the SCDR, the data out line, TxD goes idle, and the SCI sets the transmitter complete (TC) flag after the last bit is shifted out onto the TxD pin. An asserted TC flag may be interpreted as message transmission is complete if all the characters in the message have been transmitted, or may be interpreted as an underrun condition.

Note: An idle state is indicated when there is at least one bit time at logic '1' after the stop bit.

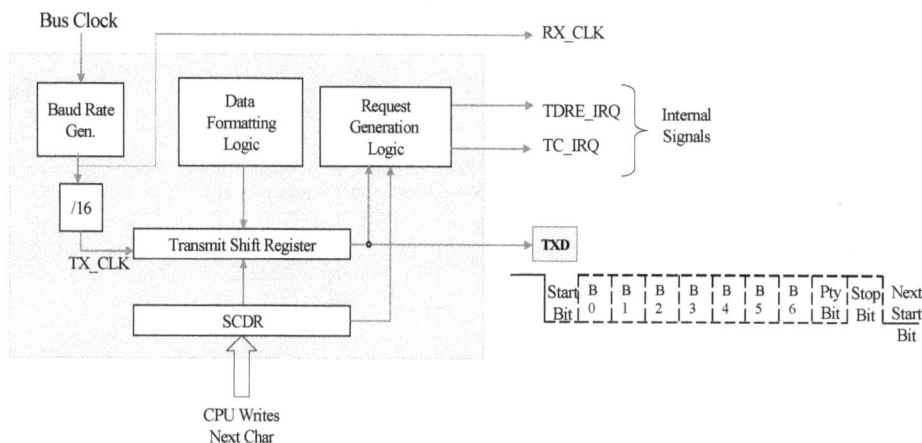

**FIGURE 12.2. Transmitter Simplified Functional Block Diagram**

An underrun is a condition indicated by the SCI transmitter to signal that the CPU was not able to keep up with the transmitter section. This condition may occur if the CPU is busy with other critical tasks or may have possibly occurred due to a software runaway condition. Once the application software determines that an underrun condition has occurred, it may send a break sequence to signal the remote receiver of a synchronization loss during the message transmission. A break sequence may be used to re-synchronize the transmitting and receiving nodes.

The TDRE and the TC flags will automatically clear when the CPU read the SCIS1 followed by a write of a new character into the SCDR. Reading the SCIS1 register is a necessary step to indicate to the software which flag is currently set in order to handle the appropriate signaling event. After the flag is cleared by the interrupt handler, a new character may be written into the SCDR for transmission. If a new character is written to the SCDR without first clearing the TDRE flag, the character will not be transmitted.

The SCI is capable of generating one or more break characters when software sets the send break (SBK) bit in SCCR1. Multiple break characters may be transmitted while SBK is set. Break transmission begins when SBK is set and ends with the transmission in progress at the time that either SBK or transmitter enable (TE) is cleared. To ensure only a single break character is transmitted, software should toggle SBK quickly to one and back to zero. At the end of a break character, at least one bit time of logic level '1' is transmitted to ensure that a subsequent start bit can be detected. If TE remains set, after all pending idle, data and break characters are shifted out, TDRE and TC are set and TxD will be held in the idle state.

To insert a delimiter (idle time) between two messages, or to wake up non-listening receivers between messages, software can clear and set the TE control bit before data in the serial shifter has shifted out. The transmitter finishes the transmission, then sends a preamble of one character of idle time. After the preamble period, the transmitter output may be held in the idle state unless the TDRE has been cleared which indicates the transmitter is loaded with the next character.

The TDRE and the TC interrupt flags are enabled by the transmit interrupt enable (TIE) and transmission complete interrupt enable (TCIE) bits in SCICR1.

The rate at which data bits are transmitted is determined by the baud rate generator which may be programmed to divide the system clock by a minimum of one and a maximum of 8191. The output of the baud rate generator is applied to a fixed divider which divides the system clock by 16 before it is applied to the transmitter clock. The baud rate is controlled by a 13 bit baud rate bit field (SBR) in the SCI control register 1 (SCICR1). Data transmission and reception typically use a common baud rate as in most serial communications interface type systems. For this reason, only one baud rate generator is provided for both the transmitter and the receiver. The receiver clock however, is 16 times the baud rate or the transmitter clock, refer to Figure 12.2 above for the baud clock generation. For the transmitter, only one serial clock cycle is needed to clock each data bit out of the serial shifter, but to guarantee accurate interpretation of serial bus levels, the receiver shifts each data bit at 16 times the transmitter clock to allow for bit sampling. Bit sampling is covered later in this chapter.

### Receiver Operation

As in the transmitter, the receiver is also double-buffered containing a receive serial shifter and a parallel receive data register designated as SCDR. As in the transmitter, the serial shifter cannot be accessed by software. Since double buffering is provided, received data can be held into the SCDR while the next data is shifted in. Refer to Figure 12.3 for the SCI receiver section block diagram, which also show a received frame of 8 data bits, 1 parity bit and 1 stop bit. Once the received data is shifted in, it is automatically transferred to the SCDR. The SCI sets the receive data register full (RDRF) flag which may trigger interrupt request to the CPU if the interrupt is enabled. An interrupt handler may respond and transfer the data from the SCDR to a receive buffer in memory. The application software may need to check the data integrity with each character received. Since the SCI generates one interrupt requests for all receiver errors and uses only one vector to service all, software should poll the SCIS1 register to find out the error type in order to take the appropriate action.

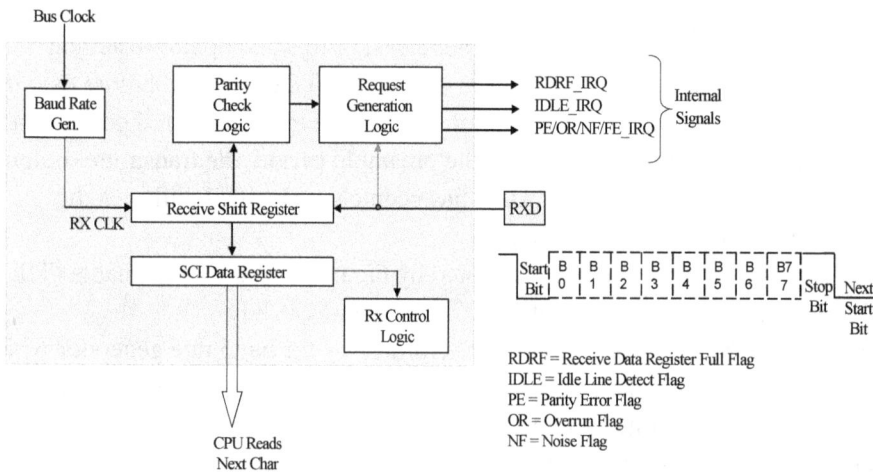

FIGURE 12.3. SCI Receiver Block Diagram

As stated previously, the receiver clock is 16 times the transmitter clock so that the RxD input signal can be sampled at a rate higher than the bit rate. The sampling clock is referred to as the RT clock and is used for synchronization of each data bit. Data is shifted into the receive serial shifter according to the most recent synchronization of the RT clock with the incoming data stream. From this point on, data movement is synchronized with the MCU system clock. To ensure bit stability, a voting logic is used to determine the bit value by taking 3 votes in the middle of the bit time. Two out of three votes determine the bit value as shown in Figure 12.4. The start bit is sampled at points RT1, RT3 and RT5. If a potential valid start bit is detected, the receiver samples 3 more times in the middle of the expected bit at points RT8, RT9, and RT10. If all agree the receiver receives the remaining data bits, otherwise the start is rejected and the SCI begins its search for a start bit. Each bit following a valid start bit is sampled in the middle of the bit time as shown in the in the figure below. If one sample disagrees with the other two samples, a Noise Flag (NF) will

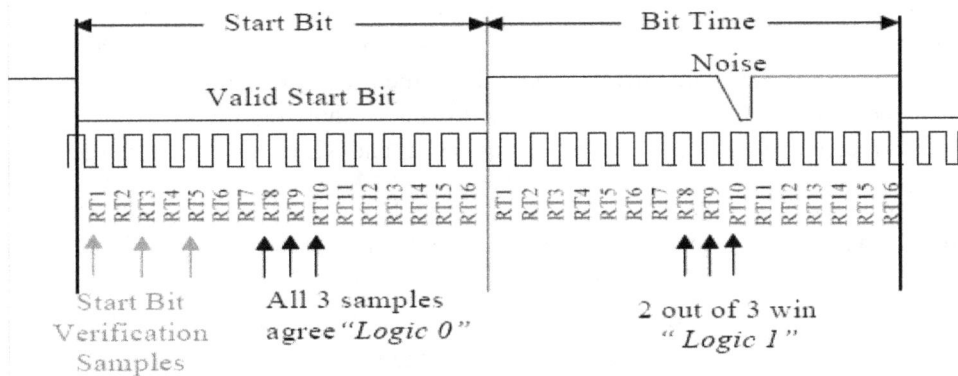

**FIGURE 12.4. Bit Sampling Techniques**

be set in the SCISR. The setting of the NF could trigger an interrupt to the CPU, if enabled.

The number of bits shifted in by the receiver depends on the serial format. The SCI uses 8 bit or 9 bit data lengths depending on the setting of the M bit in the SCCR1. If an 8 bit length is selected, the total frame length will be 10 bits. For a 9 bit length, the total frame length will be 11 bits. Each valid frame must begin with a start bit and must end with at least one stop bit. When the stop bit is received, the data in the serial shifter is transferred to the SCDR. The RDRF must be cleared before the next transfer from the shifter can take place. If the RDRF flag is still set when the shifter receives another character, transfers are inhibited and the overrun error (OR) flag in the SCIS1 will set. The setting of the OR flag indicates that the CPU lost one or more characters and needs to service the receiver at a faster rate. When the OR is set, the data in SCDR is preserved, but the data in the serial shifter is lost. The overrun condition may occur if the CPU is busy with other critical tasks or may possibly occur due to a software runaway condition. Once the application software determines that one or more character have been lost, it should request re-transmission of the message. This will be the proper method to recover message or correct the error.

Other error conditions such as the noise flag (NF), the parity flag (PF), and the framing error flag (FE) are detected by the receiver section as data bits are received. Since these flags are associated with the received character, they are cleared when the RDRF flag is cleared. OR, PF, NF and FE share one interrupt vector. If the application software needs to determine the exact error type generated with the received character, the CPU should poll the SCIxS1 register in the interrupt handler.

## Single Wire Operation

In normal mode of operation, the SCI uses two pins, one for transmission and another for reception. To minimize the number of external pin signals, software may configure the SCI to operate in a single wire mode where the TxD pin is used for both transmission and reception. This mode is enabled by setting the SCI control (SCIC1) register LOOPS and RSRC bits. When single wire operation is enabled, the RxD line is disconnected from the receiver and may be used as a general purpose I/O pin as shown in Figure 12.5. The transmitter and receiver must be enabled by setting the TE and RE bits in the SCICR2.

In single wire mode, multiple transmitters connected to the same line may attempt to transmit data at the same time which may cause contention on the TxD line. To avoid this condition, the SCI has a status bit called receiver active (RAF). The RAF flag indicates message transmission is in progress. Transmitters that wish to transmit a message should first test the RAF flag in their respective SCIS2. If the flag is set, it indicates that another transmitter is currently transmitting its message and this transmitter must wait until the RAF bit is cleared. The detection of an idle line automatically clears this bit, to indicate no one currently is transmitting, allowing another transmitter access to the TxD line.

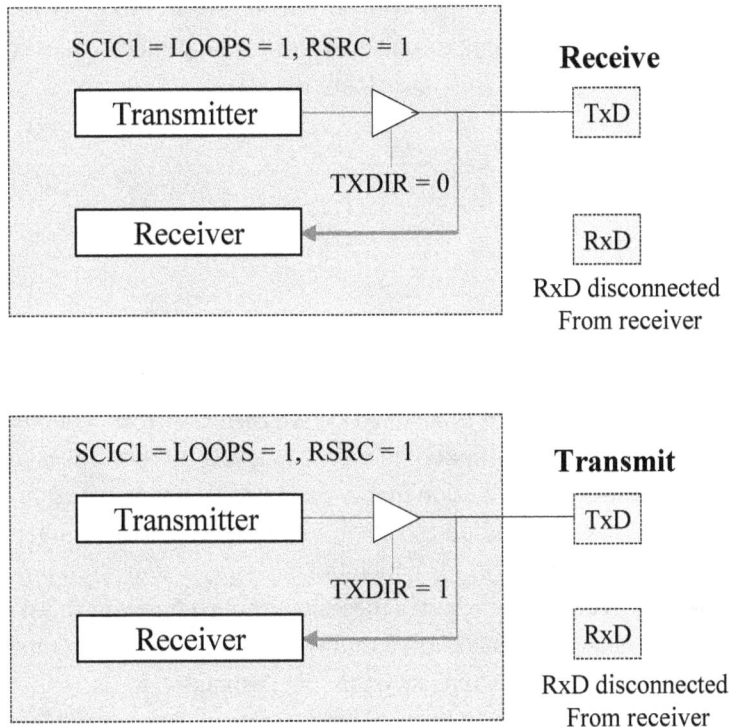

FIGURE 12.5. Single-Wire modes

## Multi-Drop Mode Operation

This mode allows a transmitting device to direct transmission to a single receiver or to a group of receivers by sending an address frame at the start of each message. This means a common messaging protocol implemented in software must be used when all serial interfaces are connected together. For proper multi-drop operation, the receiver is placed in wake-up mode by setting the receive wake-up (RWU) bit in the SCI control register 1 (SCIxC1). When set, the receiver interrupts and status flags and are disabled preventing it from generating interrupt requests to the core. Although software can clear the RWU bit, it is automatically cleared by hardware during wake-up time (when the first character in the next message is received). The WAKE control bit in SCI control register 1 selects one of two wake-up methods. The two methods are: Idle line wake-up and address mark wake-up as shown in Figure 12.6 and Figure 12.7. When WAKE is cleared, idle line wake-up is selected and when set, address-mark wake-up is selected. The idle-line wake-up method allows a receiver to ignore messages until one or more idle characters are received. The receiver clears the RWU bit causing it to wake up when an idle time is detected, waits for the first frame reception, then generates an interrupt request via receive data register full flag (RDRF).

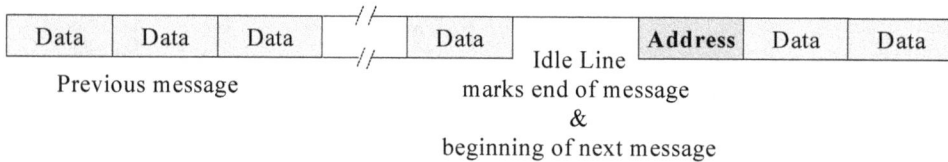

| Data | Data | Data | // | Data | | Address | Data | Data |
|------|------|------|-----|------|--|---------|------|------|

Previous message         Idle Line
marks end of message
&
beginning of next message

**FIGURE 12.6. Idle Line Wake-up Mode**

If the address mark wake-up method is selected (WAKE =1), a logic '1' in the most significant bit (MSB) position of a frame will clear the RWU bit and wakes up all receivers. A logic '1' in the MSB position marks the received character as an address frame that contains address information as shown in Figure 12.7. All receivers evaluate the address character and the receiver for which the message is addressed processes the frames that follow. Any receiver for which a message is not addressed can set its RWU bit to return to the standby state. The RWU bit remains set and the receiver remains in standby until another address frame appears on the RxD signal.

Address mark wake-up method allows back-to-back message transmission and reception without idle time between, while idle line wake-up requires messages to be separated by at least one idle character. The choice of an address mark or an idle line wake-up may be

used to increase the SCI bandwidth. Typically, when messages are too short, an address mark wake-up is used, whereas idle line wake-up is used for long messages.

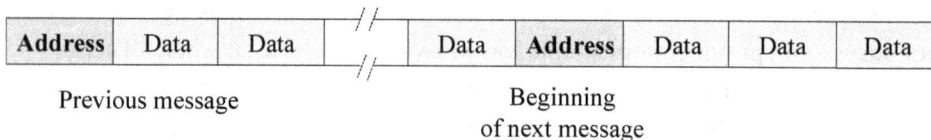

| Address | Data | Data | // | Data | Address | Data | Data | Data |
|---------|------|------|-----|------|---------|------|------|------|

Previous message           Beginning
of next message

**FIGURE 12.7. Address Mark Wake-up Method**

## Interrupts

The SCI can generate multiple interrupt requests and uses 3 separate vectors. Two requests originate in the transmitter and six others originate in the receiver. The transmitter interrupts are generated when the transmitter is empty (TDRE=1) and when transmission is complete (TC=1) both of which share the same vector. A second vector is used for receiver interrupts. The requests from the receiver are generated when the receive data

**TABLE 1. SCI Interrupt Generation**

| Interrupt Source | Flag Bit | Source | Local Mask Bit | Source | Description |
|---|---|---|---|---|---|
| Transmitter | TDRE | SCIxS1 bit 7 | TIE | SCIxC2 bit 7 | Transmit data register is empty |
| | TC | SCIxS1 bit 6 | TCIE | SCIxC2 bit 6 | Transmitter is empty |
| Receiver | RDRF | SCIxS1 bit 5 | RIE | SCIxC2 bit 5 | A character received in SCDR |
| | IDLE | SCIxS1 bit 4 | ILIE | SCIxC2 bit 4 | An Idle condition is detected |
| Receiver | OR | SCIxS1 bit 3 | ORIE | SCIxS3 bit 3 | One or more characters loss indication |
| | NF | SCIxS1 bit 2 | NEIE | SCIxS3 bit 2 | Noise is detected during char reception |
| | FE | SCIxS1 bit 1 | FEIE | SCIxS3 bit 1 | (Frame error) No stop bit was detected in the received frame |
| | PF | SCIxS1 bit 0 | PEIE | SCIxS3 bit 0 | (Parity error) Frame received is corrupted) |

register is full (RDRF=1) and when it detects an idle line condition (IDLE=1). These 2 requests share the same vector. A third vector is used and shared by all receiver errors. Table 1 above lists all the interrupt requests that can be signaled to the core from the SCI module. When interrupted for an error condition, the interrupt handler should read the SCISR1 to determine the error type since there is only one vector assigned for all error requests.

In QE128 devices, there may be 1 or more SCI modules integrated on chip, to differentiate between SCI modules, register names are designated with letter 'x' to indicate which module is being selected for programming. For example, the table above shows some of the SCI registers listed as SCIxS1, SCIxS2 and SCIxC2. The letter 'x' should be replaced with 1 or 2 for a device having two SCI modules, for instance.

- Break Generation

The SCI transmitter section is able to generate a break condition by setting the send break (SBK) bit in the SCIxC2 register. Break characters are a full character time of 10 or 11 bits of logic 0 including a start and a stop bits. The selection between 10 or 11 bit break is determined by the character length selected based on the value of the M bit in SCIxC3 register. A longer break of 13 bit times can also be generated by setting BRK13 bit in the SCIS2 register. The 13 bit break is often used to partially support the LIN bus.

When a break transmission sequence is needed, the application software would normally waits for TDRE to set to indicate the last character of a message has moved to the transmit shifter, then writes a '1' and then write '0' to the SBK bit. This action queues a break character to be sent as soon as the shifter stage is available. Multiple break characters may be transmitted if the SBK bit remains set.

**TABLE 2. Break and IDLE Character Length**

| SBK | M | BRK13 | Break Length | Idle Length |
|-----|---|-------|--------------|-------------|
| 0 | 0 | 0 | N/A | 10 |
| 1 | 0 | 0 | 10 Bits | 10 |
| 1 | 1 | 0 | 11 Bits | 11 |
| 1 | 0 | 1 | 13 Bits | 13 |
| 1 | 1 | 1 | 14 Bits | 14 |

The SCI receiver detects a break condition if a character is received as '0's in all data bits. The reception of all '0' in the received data causes the framing error (FE) flag to set. The framing flag status bit may be interpreted in one of two ways: a good break character or a framing error. If all data bits of the character are received as '0's and the framing error status bit FE =1, then a break character is received, but if the data bits of the received character are non-zero and the FE is set, then the received character is corrupted. A framing error is typically detected by the receiver when a character is received without the Stop bit.

The start and stop bits are generated by transmitter section to synchronize data transmission with the receiver. Every character begins with a transition from high-to-low called the start bit, and sometimes referred to as mark-to-space. The transition from high-to-low resynchronizes the transmitter with the receiver on every character. In order for the receiver to be able to detect the next start bit of the next character, the transmitter appends a stop bit as a delimiter at the end of each character. The receiver will signal a framing error by setting the FE flag if the expected stop bit as the last bit of the received character is not detected. This condition indicates the received character is corrupted due to loss of synchronization between the receiver and the transmitter.

The length of the break character is affected by the BRK13 and M bits as shown in Table 2. Also, the idle character length is affected by the SCIxC1 M and BRK13 control bits as shown in the table above.

### SCI Module Initialization Procedure

1. Select the baud rate as required by the application. Use the following formula:

$$\text{SCI Baud Rate} = \text{Bus Clock}/(16 \times BR^2) \qquad \textbf{(EQ 1)}$$

For example, to select a baud rate of 9600 assuming system clock frequency of 50MHz, the SBR should be set to 163.

2. Write the SBR value into the SCI baud high and low (SCIxBDH and SCIxBL) registers to start the baud rate generator. In the same write cycle you can also select the following configurations:

3. Write SCIxC1 to select the following configurations:

   - Enable parity if needed by setting the PE bit and parity type, odd or even controlled by the PT bit.
   - Select normal or short idle line detection controlled by the ILT bit.
   - Character length of 8 or 9 bits controlled by the M bit.
   - If needed, select wake-up method. The wake-up of either idle line or address mark is controlled by the WAKE bit.

4. Write the SCIxC2 register to select the following configurations:

   - If desired, enable receiver and transmitter interrupt by setting the following interrupt enable bits: TIE, TCIE, RIE, and ILIE.
   - Enable transmitter and Receiver by setting TE and RE bits.
   - If desired, place the receiver in the stand-by mode by setting the RWU bit.

   Note: As soon as the transmitter is enabled, an interrupt request is generated since TDRE & TC flags are

---

2. The 13 SBR[12:0] bits in the SCI baud rate high/low (SCIxBDH/L) registers are referred to collectively as BR, and they set the modulo divide rate for the SCI baud rate generator.

After writing this register, preamble (idle) character(s) will now be shifted out of the transmit shift register.

5. Write SCIxC3 to select the following configurations:
   - If the SCI is configured for 9-bit data (M = 1), T8 may be thought of as a

ninth transmit data bit to the left of the MSB of the character data in the SCIxD register. When transmitting 9-bit data, the entire 9-bits are transferred to the SCI shift register after SCIxD is written. T8 should only be written if it needs to change from its previous value before SCIxD is written. If T8 does not need to change for the next character (such as when it is used to generate address mark or a parity), it need not be written each time SCIxD is written.

If SCI module needs to be configured in the single-wire mode or half-duplex operation as controlled by the SCIxC1 (LOOPS = RSRC = 1) bits, the TXDIR bit should be written to determines the direction of data at the TxD pin. A value of '0' configures the TxD pin as input for reception, and a value of '1' configures it as output for transmission. If needed, set the transmit data inversion (TXINV[3]) bit to select inversion of the output data levels on the TxD line.
   - If desired, enable all receiver error interrupts by setting the following bits in the SCIxC3: ORIE, NEIE, FEIE, and PEIE.

**.ISR Transmit sequence for each byte:**
1. To determine which transmitter flag is set (TDRE or TC), read the SCIxS1 register. Reading SCIxS1 is also required to clear the transmitter interrupt flags.
2. If the TDRE flag is set, write the next character to be transmitted to SCIxD. For 9 bit character, the T8 bit should be written first when needed.
3. Repeat Step 1 and 2 for each subsequent character transmission.

When the last character in is written into the SCIxD, both TDRE and the TC flags will set after all data bits are shifted out of the serial shifter. To clear both flags, read SCIxS1 and perform one of the following steps:
   - Write a new character in the SCIxD to transmit new data
   - Queue a preamble by clearing transmit enable (TE) bit
   - Queue a break character by setting the send break (SBK) bit in the SCIxC2 register.

---

3. Setting TXINV inverts the TxD output for all cases: data bits, start and stop bits, break, and idle.

4. Return from interrupt

**ISR Receive Sequence for each byte:**

1. Read the SCIxS1 to determine which interrupt flags (RDRF or IDLE). This will be one of two steps needed in order to clear the flags.

2. If RDRF is set, read the SCIxD register. Now the receiver flags will be cleared.

3. Store received character into memory.

4. If IDLE flag is set, this condition could be used to indicate message reception is complete, and it is safe now to close receive message buffer in memory.

5. Return from interrupt

## *Serial Peripheral Interface*

The Serial Peripheral Interface (SPI) provides a method of exchanging serial data between QE128 devices and other microcontrollers and I/O peripherals. Connected devices can range in complexity from simple shift registers to multi-mega bit serial Flash memory devices. This interface is practically implemented on every MCU manufactured and supplied by Freescale Semiconductor to provide a way for 8, 16 and 32-bit architectures to be able to communicate with each other when having multiple devices implemented in the target application.

The SPI is a full-duplex, high speed synchronous bus designed to communicate with other on-board SPI peripherals. The maximum number of SPI bus is 4-wire interface signals. These signals are: Master-out-slave-in (MOSI) to transmit data, Master-in-slave-out (MISO) to receive data, serial clock (SPSCK) to clock data in and data out. The fourth signal is the chip-select ($\overline{SS}$) to select the SPI slave device.

The SPI block consists of transmitter and receiver sections, a baud rate generator, and a control unit. The transmitter and receiver sections use the same clock derived from the SPI baud rate generator. During transfers, SPI data is exchanged between the master and slave simultaneously using synchronous bit clock provided from the SPI master. The bus flexibility provides a wide variety of transfer rate, clocking, and interrupt-driven communication options.

## Main Features

- Two SPI Modules (SPI1 and SPI2)
- Operates in master or salve mode
- Fault Mode Detection and Interrupt Generation
- Transmit and receive interrupt generation
- Multi-master support with mode fault detection
- Supports frame size of 8-bit with either LS or MS bit first transmission ability
- Flexible baud rates and clock generation
- Bidirectional mode operation

## SPI Operation

The SPI is fully compatible with other SPI systems found on many Freescale's 8, 16 and 32 bit microcontrollers. Data transfer between two SPI peripherals can be performed in full-duplexed with minimum number of four-wires or optionally half duplexed with only three-wires. A common practical example of two SPI peripherals connection is shown in Figure 12.8 below.

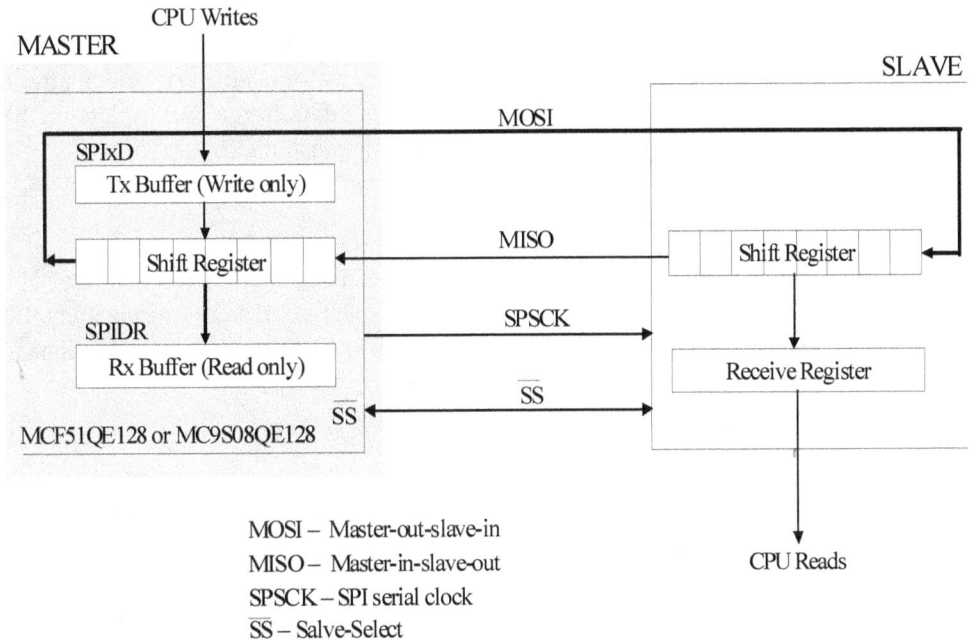

MOSI – Master-out-slave-in
MISO – Master-in-slave-out
SPSCK – SPI serial clock
$\overline{SS}$ – Salve-Select

FIGURE 12.8. **SPI Master-Slave Interface Example**

Data transfer is initiated by the master with an 8-bit data frame write into the SPI data register (SPIx[4]D). Notice that this register is double buffered, allowing the SPI to write a new data while the current frame being shifted out on the MOSI pin. The double buffering provides additional throughput than the former SPI found on the MC68HC11, for instance. Once the data is written into the transmit buffer, the SPI hardware transfers the data into the shift register as soon as the previous data transfer is completed. As can be seen from the figure above, the receiver section is also double buffered. When the data transfer is complete, the received data will automatically be transferred into the receive register.

The SPI transmitter is able to generate an interrupt request to the CPU whenever the Tx buffer becomes empty. Likewise, it is able to generate a second request to the CPU when the Rx buffer is full, assuming transmit and receive interrupts are enabled.

---

4. Where $x$ is either SPI1 or SPI2

During a transfer, the master shifts data out on the master-out-slave-in (MOSI) pin to the slave, while simultaneously shifting data in on the master-in-slave-out (MISO) pin from the slave. The transfer effectively exchanges the data that was in the SPI shift registers of the two SPI systems.

The master selects the slave device through its $\overline{SS}$ output signal, this signal is asserted one half serial clock before data transfer begins and it is negated one half serial clock after the entire frame transfer is complete. Once the transfer of the data frame is complete the $\overline{SS}$ signal is negated until the next transfer is initiated by the master SPI.

## SPI Interface Signals

There are 4 basic pins associated with each SPI module. The function and behavior of these signals depends on their configuration in their respective SPI control registers. If an SPI module is not used by the target application, MOSI, MISO, SPSCK and $\overline{SS}$ signals revert to general purpose input/output pins.

- Master-out-slave-in (MOSI) and Master-in-slave-out (MISO)

These two signals are used to exchange data between the master and the slave SPI device. With each clock pulse, a data bit is driven on the MOSI by the master and a data bit is received on the MISO from the slave.

When configured as a master, this pin is the serial data output (MOSI), and when configured as slave, this pin is the serial data input (MISO). The master/slave configuration is controlled by the master (MSTR) bit is the SPI control register 1 (SPIxC1).

If SPC0 is set to enable single wire bidirectional mode and master mode is selected, this pin becomes the bidirectional data I/O pin or master-out-master-in (MOMI). More information is provided later on the bidirectional mode in this chapter.

## SPI Clock (SPSCK)

The SPI clock is driven by the master to exchange data with the slave device. Data transfer is synchronized with the internally-generated serial clock. There are two control bits associated with the SPI clock, clock phase (CPHA) and clock polarity (CPOL). These two bit

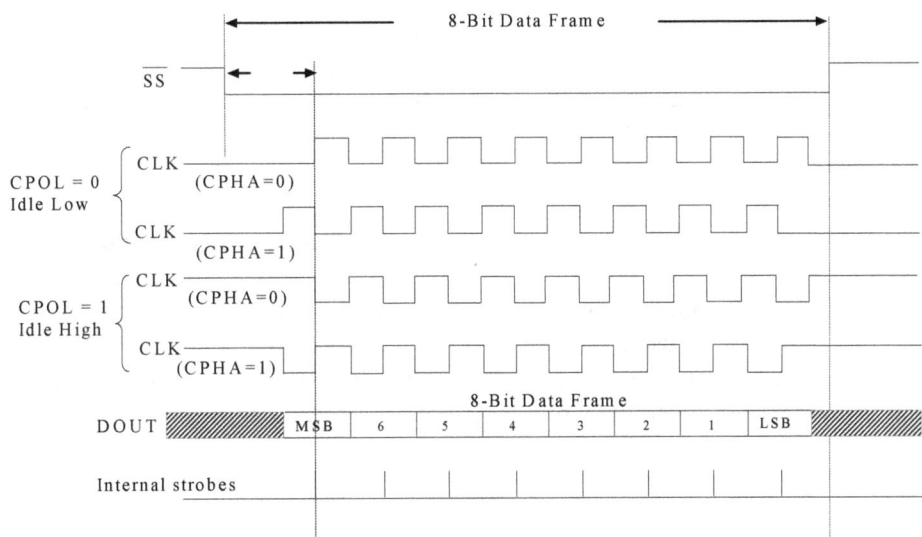

FIGURE 12.9. **SPI Clock Phase and Polarity Timing**

are located in the SPI control register 1 (SPIxC1) and are used to select the required clock type to satisfy the slave device. The CPOL selects the inactive state of the serial clock (SPSCK), while the CPHA selects which clock edge causes the data to change and on which edge to capture the incoming data bits. These two control bits can be programmed to produce one of 4 suitable clock signals for the selected slave. Refer to the SPI timing diagram in Figure 12.9.

When configuring the SPI for various peripherals, it is important to carefully examine the peripheral's electrical characteristics to determine the correct transfer format, clock phase and clock polarity. In some cases, it may be necessary to change the SPI configuration between transfers to accommodate a mix set of peripherals connected on the same bus. In addition to the transfer format and clock configuration options, each SPI peripheral's data sheet must be examined for the peripheral device timing and slave select requirements. While automatic assertion y the SPI master of the $\overline{SS}$ signal reduces some software over-head and intervention, it will probably not be useful with vast majority of peripherals, especially when longer than 8 bits of data transfer is required by the slave. In these cases, the automatic $\overline{SS}$ signal should be controlled entirely by the software. In other words, the $\overline{SS}$ signal should be configured as a general purpose output pin to allow the software full control of the pin assertion and negation at the beginning of multi-byte transfer and at the end of transfer, respectively.

ColdFire and HCS08 (QE128) devices only support transfers in multiple of 8 bits and the SPI bus may not be able to accommodate some other SPI peripherals in straightforward manner. Therefore, it may be necessary to utilize a combination of hardware and software techniques to be able to interface to these peripherals.

It is also necessary to consider other SPI peripherals requiring data transfers that are not an integer multiple of 8 bit frames. Obviously, in these cases additional software intervention is needed for proper communication.

In summary, the selection of the data transfer format, clock phase and clock polarity depends on the attached device requirements. However, as a general guideline, slave peripherals driving their first data bit on the MISO line when slave select line is asserted are generally candidates for using the transfer format defined when the CPHA control bit is zero. On the other hand, peripherals requiring a clock edge before placing their first data bit on the MISO line will most likely require the transfer format defined when the CPHA control bit is one.

### Slave Select Signal $\overline{SS}$

The $\overline{SS}$ pin is used as an output signal from the master to select a slave SPI device. This signal is automatically asserted when the master is ready to transmit data and automatically negated when data transmission is complete. The $\overline{SS}$ signal can be programmed in one of the following multiple options.

6. Automatic output signal when in master mode
7. Input signal when in the slave mode
8. General purpose output (not used by the SPI bus)
9. General purpose input (not used by the SPI bus)

When having a system with multiple SPIs implemented and function in both master and slave modes, the $\overline{SS}$ pin direction will be controlled by the slave select output enable (SSOE) bit in SPI control register 1. In the master mode, if the mode fault function is disabled, the $\overline{SS}$ pin is not used and can be configured as GPI/O, otherwise this signal is an input to allow another master to select *this* SPI master as a slave. Refer to Table 3 for additional information on $\overline{SS}$ pin options.

**TABLE 3. Slave Select Pin Function**

| MODFEN | SSOE | Master Mode | Slave Mode |
|--------|------|-------------|------------|
| 0 | 0 | GP I/O (not used by SPI function) | Slave Select Input |
| 0 | 1 | GP I/O (not used by SPI function) | Slave Select Input |
| 1 | 0 | $\overline{SS}$ Input for Mode Fault | Slave Select Input |
| 1 | 1 | Automatic $\overline{SS}$ Output | Slave Select Input |

## *Functional Description*

### Transmit Operation

To transmit data to a slave device, the master SPI writes the Tx buffer after checking the SPI transmit empty flag to ensure it is set. This flag is set by default out of reset and the SPI can assume the transmitter is empty after it is enabled. Subsequent writes to the SPI Tx data buffer must not occur until the SPTEF is set. Writing data while this flag is cleared will result in data loss and the data frame will not be transmitted. As previously noted, the transmitter is double buffered allowing the application software or the interrupt handler to write a new frame into the Tx data buffer while the current frame transmission is in progress. Once the data frame in the shifter is fully transmitted, the new frame in the Tx data buffer will automatically be transferred to the shifter. The data transfer into the shifter from the Tx buffer sets the SPTEF and generates interrupt request if the SPI transmitter empty interrupt enable is set in the SPI control register 2. In the interrupt service routine, the handler must clear the SPTEF before returning to the main program. To clear the flag, the interrupt handler must read the SPIxS to determine who is requesting and then write the Tx data register with another frame, if SPTEF flag is set.

The SPI interface is able to generate 3 interrupt conditions, but these interrupts share one vector. For this reason, the handler needs to read the SPIxS first to determine which one of the three SPI events is requesting in order to take the appropriate action.

### Receive Operation

Data transmission and reception is always initiated by the master SPI. The clock supplied from the master causes the data to be shifted out of the master and into the slave, and out of the slave into the master. As soon as a data frame is shifted in, the receiver moves the data from the shifter to the Rx buffer and sets the SPRF. If the receiver interrupt is enabled, the SPI triggers an interrupt request to the CPU. Again, to clear the interrupt flag, the handler must read the SPIxS to determine the SPI interrupt request type followed by a read of the Rx buffer, if the SPRF flag is set.

Note that the Tx buffer is write only register, while the Rx buffer is a read only. These two registers share the same address with a single memory mapped I/O address and are referred to as SPIxD. In other words, the CPU writes the SPIxD to transmit data and reads the SPIxD to receive data.

## Modes of Operation

The SPI interface supports two basic operating modes to exchange data between the master and the slave. The two modes are: Normal and Bidirectional. A detailed description of the two modes are provided below:

- Normal Mode

The normal mode, most familiar to HCS08 or HC11 developers, utilizes the master-out-slave-in (MOSI) and master-in-slave-out and (MISO) signals to exchange data between the two SPIs in full-duplex. Since the SPI is a synchronous interface, a clock is needed to transfer the data. This clock is referred to as the SPI serial clock (SPSCK) and must be provided from the master. The SPSCK is used to clock data out from the master into the slave on one clock edge, while the slave device output its data on the same clock but on the opposite edge.

- Bidirectional Mode

Often, the application may require additional signals (pins) than the target MCU can provide. If some of the on-chip peripherals are able to function with less pins and with minimal performance penalty, the additional pins needed by the target application may be spared. For example, the serial interfaces integrated on this architecture can be configured to transmit and receive with less signal if they are configured in a certain way. For this reason, the SPI and the SCI can optionally be configured in what is called "Bidirectional Mode" and "Single-Wire Mode", respectively. If we take a closer look at the MCF51QE128 and its counterpart the MC9S08QE128, we find two SCIs and two SPIs modules on these 2 devices. If the application utilizes the bidirectional and single-wire modes on these peripherals, we can spare up to 4 pins, which can be configured either as GP I/O or be assigned to alternate function.

Furthermore, if only two SPI peripherals connected, one as a master and the other connected *always* as a slave, the master $\overline{\text{SS}}$ pin can also be spared by having the slave select signal connected to ground permanently, thus eliminating the need for the additional slave select ($\overline{\text{SS}}$) signal.

The bidirectional mode is selected when the SPC0 bit is set in SPI control register 2. In this mode, the SPI uses only one serial data pin for the interface with external device(s). The MSTR bit in the SPIxC1 register decides which pin to use for the data transfer. The MOSI pin becomes the serial data I/O (MOMI) pin for the master mode, and the MISO pin becomes serial data I/O (SISO) pin for the slave mode. The MISO pin in master mode and MOSI pin in slave mode are not used by the SPI. The direction of each serial I/O pin depends on the BIDIROE bit. If the pin is configured as an output, serial data from the shift register is driven out on the pin. The same pin also is the serial input to the shift register.

The bidirectional mode is selected by setting the SPI control (SPC0) bit and the direction of the pin is controlled by the bidirectional mode output enable (BIDIROE) bit. These two bits are provided in the SPIxC2. Refer to Table 4 for additional information.

TABLE 4. SPI Pin Control 0 and Bidirectional Mode Output Enable Encoding

| SPC0 | BIDIROE | Master Mode | Slave Mode |
|---|---|---|---|
| 0 | x | Normal Mode MOSI & MISO pins used | Normal Mode MISO & MOSI pins used |
| 1 | 0 | Master-in (MI) | Slave-out (SO) |
| 1 | 1 | Master-out (MO) | Slave-in (SI) |

If the mode fault mode is enabled in bidirectional master mode, both data pins MISO and MOSI can be occupied by the SPI, though MOSI is normally used for transmissions in bidirectional mode and MISO is not used by the SPI. If a mode fault occurs, the SPI is automatically switched to slave mode. In this case MISO becomes occupied by the SPI and MOSI is not used. This must be considered, if the MISO pin is used for another purpose. To illustrate the bidirectional mode, refer to Figure 12.10.

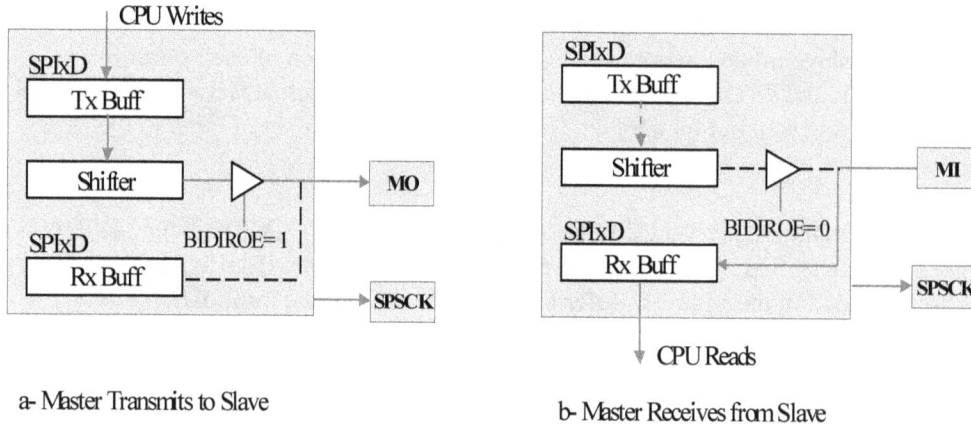

a- Master Transmits to Slave          b- Master Receives from Slave

**FIGURE 12.10. Master in Bidirectional Mode**

### Mode Fault Function

The mode fault is mainly intended for a system requiring multiple master support. A mode fault will occur only if MODFEN bit is set and SSOE bit is cleared. The mode fault detection feature can be used in a system where more than one SPI device might become a master at the same time. This condition is detected when a master's $\overline{SS}$ input pin is driven low, indicating that some other SPI device is trying to address *this* master as a slave. This could indicate a harmful output driver conflict, so the mode fault logic is designed to disable all SPI output drivers when such an error is detected.

When a mode fault condition is detected, the MODF flag is set and the MSTR control bit is automatically cleared which changes the SPI configuration from master to slave mode. The output drivers on the SPSCK and MOSI pins become disabled to prevent contention on the bus.

### Connecting Multiple Slaves

The SPI interface may be used to connect multiple slaves as shown in Figure 12.11 below. In this application example, the master SPI can use some of the available general purpose I/O pins to generate multiple slave select (SS) signals. The assertion and the negation of these signals must be controlled in software. Also, the configuration of the master $\overline{SS}$ pin should not be used as automatic slave select output. Instead, this pin should be configured as general purpose output and therefore must be controlled in the same way as the other slave select signals.

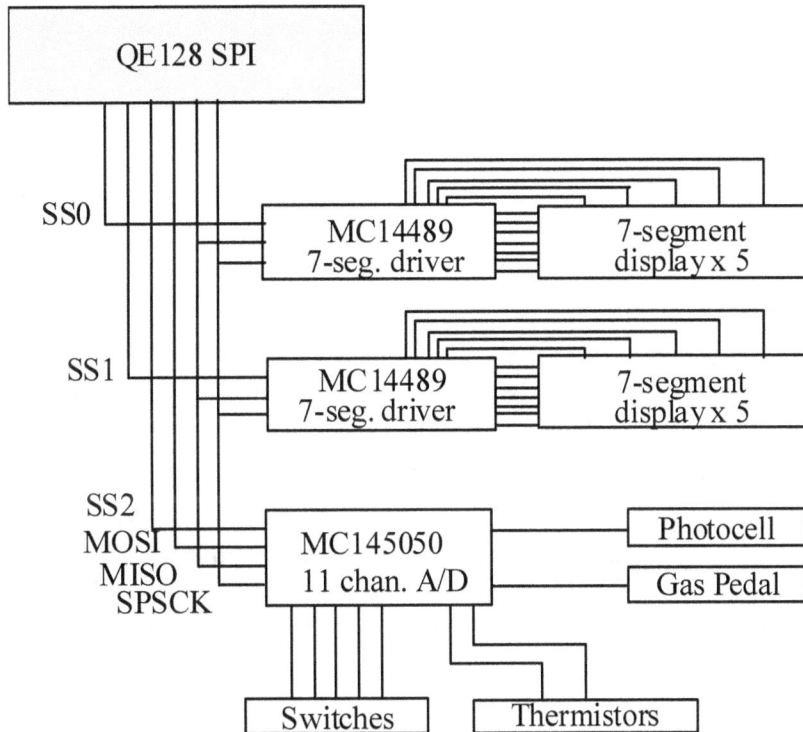

FIGURE 12.11. Connecting Multiple SPI Slaves Application Example

## SPI Baud Rate Selection

The maximum SPI clock frequency is one-half the frequency of the internal bus clock (fsys/2). Baud rate is selected by writing two values into the SPI baud rate prescaler divisor (SPPR) and the SPI baud rate divisor (SPR) in the SPIxBR register.

TABLE 5. Baud Rate Selection with Bus Clock of 25 MHz

| SPPR | SPR | Divider | SPI_CLK |
|------|-----|---------|---------|
| 0 | 0 | 2 | 12.5 MHz |
| 1 | 1 | 8 | 3.125 MHz |
| 2 | 3 | 48 | .52 MHz |

**TABLE 5. Baud Rate Selection with Bus Clock of 25 MHz**

| SPPR | SPR | Divider | SPI_CLK |
|------|-----|---------|---------|
| 3 | 4 | 128 | 195 KHz |
| 4 | 5 | 320 | 78.125 KHz |
| 5 | 6 | 768 | 32.5 KHz |
| 7 | 7 | 2048 | 12.2 KHz |

Table 5 depicts some examples of clock rate selection. The equation below may be used to get a different number of serial clock frequency flavors.

$$SPSCK = Bus\ clock/(SPPR)*(SPR)$$
(EQ 2)

## SPI Interrupts

The SPI is able to generate 3 interrupts via a single vector to the interrupt controller with Tx buffer empty and Rx buffer full being the most common requests. There are three flag bits, two interrupt mask bits, and one interrupt vector associated with the SPI system.

• Transmitter Empty Interrupt

The SPI transmitter interrupt is enable by setting the (SPTIE) in the SPIxC1 register. The transmitter interrupt is signaled to the interrupt controller when the data in the SPIxD register is transferred to the shift register. This action sets the SPITEF flag bit.

The interrupt handler should clear the SPITEF flag by first reading the SPIxS followed by data write to the SPIxD register.

• Receiver Interrupts

The SPI receiver interrupts are enabled by setting (SPIE) mask bit in the SPIxC1 register. This bit enables requests when the receiver is full or the mode fault condition is detected. When one of the flag bits are set, (SPRF or MODF), a hardware interrupt request is sent to the interrupt controller. The SPI interrupt service routine (ISR) should check the flag bits to determine which event type caused the interrupt.

To clear the Rx buffer full (SPRF) flags, the interrupt handler should read the SPIxS followed by a read of the SPIxD. Whereas, the MODF flag is cleared by reading the SPIxS followed by a write to the SPI control register 1.

Recall that the MODF flag can only be set in the master SPI if the mode fault is enabled.

## SPI Initialization Example

The following initialization steps setup the SPI module to satisfy certain requirements of a particular slave. We will assume the slave SPI requires 8 bits of data frames with most-significant bit shifted first, 2 MHz serial clock, data shifted on first edge with low idle clock polarity, and SPI interrupts are enabled.

1. Write the SPIxBR register to select 2MHz serial clock
2. Set SSOE = 1 in SPIxC1 and MODFEN = 1 in SPIxC2 to enable automatic $\overline{SS}$ output.
3. Write the SPI control register 1 (SPIxC1) to select the following options:
   - Set the SPE and MSTR bits to enable and configure the SPI in master mode
   - Set SPTIE and SPIE bits to enable transmitter and receiver interrupts
   - Clear and set the CPOL and CPHA bits so that data shifted on the first clock edge.
   - Set MSBFE bit to enable shifting most significant bit out first

Since the SPI transmitter empty (SPTEF) flag is set by default, an interrupt will be requested immediately as soon as step 3 is executed.

The interrupt handler executes the following steps:

1. reads the SPIxS to find out which interrupt flag is set
2. writes the SPIxD if SPITEF flag is set. (write operation causes this flag to clear, if was set)
3. reads the SPIxD if the SPRF is set. (read operation causes SPRF to clear, if was set)
4. write the SPIxC1 if MODF[5] flag is set. (write operation causes the MODF to clear, if was set)

When transmission and reception is completed and no additional transfer is needed, the software can stop the SPI operation by clearing the SPI enable (SPE) bit in the SPIxC1.

### Peripheral Clock Gating

QE128 devices include a clock gating system to manage the bus clock sources to the individual peripherals. Using this system, the user can enable or disable the bus clock to each of the peripherals at the clock source, eliminating unnecessary clocks to peripherals which are not in use and thereby reducing the overall run and wait mode currents. Out of reset, all peripheral clocks are enabled. For lowest possible run or wait currents, user software should disable the clock source to any peripheral not in use. The actual clock will be enabled or disabled immediately following the write to the Clock Gating Control registers 1 and2 (SCGC1 and SCGC2). Any peripheral with a gated clock can not be used unless its clock is enabled. Writing to the registers of a peripheral with a disabled clock has no effect.

---

5. The MODF condition will only occur in the master SPI if the $\overline{SS}$ pin is configured as input.

The application software should disable the peripheral before disabling the clocks to the peripheral. When clocks are re-enabled, the peripheral registers need to be reinitialized by the user software.

In the stop modes, the bus clock is disabled for all gated peripherals, regardless of the settings in the SCGC1 and SCGC2 registers.

## *Inter-Integrated Circuit Bus Module*

The Inter-Integrated Circuit Bus, or $I^2C$ bus was developed in the early 1980's by Philips Semiconductors. Its original purpose was to provide an easy way to connect a microcontroller to peripheral chips in a television.

Today, the $I^2C$ bus is used in many more application fields than just consumer audio and video equipment. The bus is generally accepted in the industry as a de-facto standard. The $I^2C$ bus has been adopted by several leading chip manufacturers to provide an interface to a wide range of peripheral components such as analog to digital converters, temperature sensors, real time clock chips, LCD displays and I/O port expanders.

Moreover, $I^2C$ has been used as the basis for a number of other industry standard serial communication protocols.

For example, $I^2C$ is the protocol adopted in the VESA DDC standard for communicating setup information and allowing user control of PC monitors. It is also the basis of the System Management Bus (SMBus) used in PC motherboards and smart chargers to provide status and management of power supply systems. Serial Presence-Detect (SPD) operation on SDRAM modules uses the $I^2C$ protocol to interrogate the embedded EPROM containing performance attributes of SDRAM memories used in personal computers.

The $I^2C$ bus is a multi-master bus. This means multiple devices that are all capable of initiating a data transfer can be connected to it. The $I^2C$ specification states that the device that successful initiates a data transfer on the bus is considered the **Bus Master** and for the duration of data transfer, all other ICs are **Bus Slaves**. The $I^2C$ specification incorporates arbitration and clock stretching protocols to deal with situations where multiple devices request simultaneous bus access. The behavior of the bus under these conditions is described later in this chapter.

The $I^2C$ physical interface on ColdFire® microcontrollers is designed to operate at up to 100 Kbps with maximum bus loading. The device is capable of operating at higher baud rates, up to a maximum of the internal bus clock divided by 20, with reduced bus loading. The maximum communication length and the number of devices that can be connected are limited by a maximum bus capacitance of 400 pF.

**Summary of the I²C Module Features**

- Compatibility with I²C bus standard version 2.1
- Support for 3.3-V tolerant devices
- Multi-master operation
- Software programmable for one of 64 different serial clock frequencies
- Software selectable acknowledge bit
- Interrupt-driven
- Byte data transfer
- Loss of arbitration interrupt with automatic mode switching from master to slave
- Address identification interrupt
- START and STOP signal generation/detection
- Repeated START signal generation
- Acknowledge bit generation/detection
- Bus busy detection
- General call recognition
- 10-bit address extension
- Support System Management Bus Specification (SMBus), version2

## I²C *Controller Architecture*

The I²C bus physical connections are 2 active wires and a ground connection. The active wires, called SDA and SCL, are both bi-directional. SDA is the Serial Data line, and SCL is the Serial Clock line.

For I²C compliance, all devices connected to these two signals must have open drain or open collector outputs attached to external pull-up resistors.

Every device attached to the bus has its own unique address, no matter what its function. Devices can act as receivers, transmitters or both depending on the functionality. A device such as an LCD driver may be only a receiver, while a programmable memory would be both a transmitter and a receiver.

After a hardware reset, the I²C module can be programmed to be a master or programmed to respond to a slave address.

An important point to note is that the I²C protocol does not use the concept of chip selects as in SPI synchronous transfers, for example, and other similar serial protocols. Instead, the I²C bus protocol defines that the first byte output from a bus master represents a 7 bit address plus a read/write direction flag that is targeted at a specific slave. The one exception is a "general call" address which can address all devices. The I²C specification revision 2.1 contains a table that describes two groups of addresses that have reserved meanings. These are values in the range 0x0000000 to 0x0000FFF and 0xFFFF000 to 0xFFFFFFF.

While the I²C committee coordinates allocation of I²C addresses, in a deeply embedded system using custom I²C devices, it may be possible to arbitrarily choose address ranges.

In a typical application, after the system is reset, software may configure the ColdFire® device as a bus master and expect the I²C bus to be idle. Then on a software request, the master initiates a transfer by issuing an address to select a slave connected to its external SDA and SCL pins. Depending on the state of the read/write bit, the slave will either accept the next byte from the ColdFire® device or return a byte.

The receiver of any byte is usually obliged to generate an acknowledge to the transmitter, to indicate successful arrival of the byte, so the slave will normally acknowledge reception of the address from the master. If for some reason, the slave is unable to do so, the master will recognize this situation and will either abort further transfers, or try again by retransmitting the address.

A nice feature of the I²C protocol is that it allows slow receivers to hold off the transmitter's processing of the acknowledge signal. So if the receiver is busy handling some higher priority event, at can still respond successfully to the address when it gets an opportunity to do so.

In multi-master systems, all masters may attempt to initiate a transfer by outputting addresses at the same time. I²C specifies that no master may output an address unless the bus is idle. This I²C module has a status bit IICS[BUSY] that indicates the bus is busy.

Arbitration of the address takes place on the SDA (data line) while the SCL (clock line) is a logic 1. Because all SDA signals are wired-OR, a master that transmits a logic 1 will lose arbitration while another master is concurrently transmitting a logic 0, because the level on the bus doesn't correspond to the level it transmitted.

Arbitration can continue past the address phase. If the masters are each trying to address the same device, arbitration continues with comparison of the data bits if they are master-transmitter, or acknowledge bits if they are master-receiver. Because address and data information on the I²C bus is determined by the winning master, no information is lost during the arbitration process. A master that loses the arbitration can generate clock pulses until the end of the byte in which it loses the arbitration.

## I²C *Protocol Implementation*

The normal data transfer protocol begins with a START bit, a 7 bit address, a read/write bit, and an acknowledge bit. Then the addressed transmitter will output a byte, followed by an acknowledge bit from the receiver, and this sequence of data followed by an acknowledge bit continues until the master decides to halt further transfers.

If the master is a receiver, it notifies the slave transmitter that no more data is required by not acknowledging the last byte received. If the master is a transmitter, then no special protocol is required.

To complete the transfer protocol, the master issues a STOP bit, or what's termed a repeated START bit (which looks just like a normal START condition). If the master issues a STOP, the bus goes idle, which allows other masters to attempt to gain control. If the master issues a repeated START, it retains control of the bus, and can then address the same or another device and start another transfer sequence.

Figure 12.12 shows an I²C trace taken with an Intronix logic analyzer. The figure shows the decoded serial protocol using the tool's integrated I²C interpreter. The trace represents part of a query that a ColdFire® device has made to an SPD enabled SDRAM module on an evaluation board. After an initial start bit labeled 'S', the ColdFire® device outputs a write address of 0x50, and the SPD device responds with an acknowledge bit labeled 'A'. The ColdFire® device then writes a value of 0x00, and receives another acknowledge bit. The ColdFire® device then outputs a repeated start bit, labeled 'Sr', to indicate the next byte it transmits is another address. In this example, it outputs a read address of 0x50, the SPD device again acknowledges the address and sends back the value of 0x80.

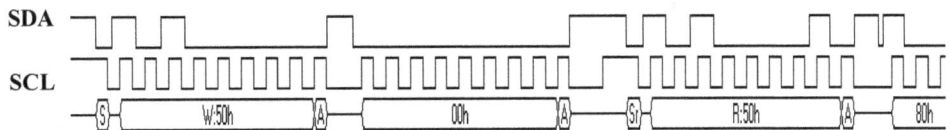

FIGURE 12.12. **I²C Trace from SDRAM SPD Data transfer**

## Start Condition and Address Transmission

When no other device is bus master (both SCL and SDA lines are at logic high), a device can initiate communication by sending a START signal as shown in Figure 12.13. A START signal is defined as a high-to-low transition of SDA while SCL is high. This signal denotes the beginning of a data transfer and awakens all slaves.

The master sends the slave address in the first 7 bits of the first byte after the START signal and then sends the R/W bit which tells the slave the direction of data transfer for further bytes. A bit value of 0 means write transfer and a 1 means read transfer.

Each slave must have a unique address. An I²C master must not transmit its own slave address; it cannot be master and slave at the same time.

The slave whose address matches that sent by the master pulls SDA low at the ninth serial clock to return an acknowledge bit.

## Acknowledge

The transmitter releases the SDA line high during the acknowledge clock pulse as shown in Figure 12.13. The receiver pulls down the SDA line during the acknowledge clock pulse so that it remains low during the high period of the clock pulse.

- If it does not wish to acknowledge the master, the slave receiver must leave SDA high. The master can then generate a STOP signal to abort the data transfer or generate a START signal (actually a repeated start) to start a new sequence. If the master receiver does not acknowledge the slave transmitter after a byte transmission, it means end of data to the slave. The slave must release SDA to allow the master to generate a STOP or a repeated START, which are described later in this section.

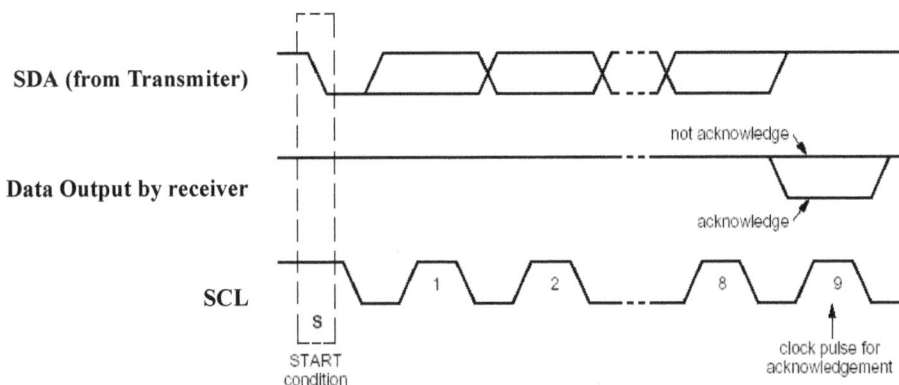

FIGURE 12.13. START, R/W and Acknowledge States

## Data Transfer

When a slave is successfully addressed, multiple data transfers with the MSB first, can proceed, as shown in the Intronix logic analyzer trace in Figure 12.14. Data transfer direction is specified by the bit that follows the 7 initial address bits. The I²C interpreter of the Intronix logic analyzer displays an initial read address of 0x50. Data can be changed only while SCL is low and must be held stable while SCL is high, as shown in the trace. SCL is pulsed once for each data bit. The receiving device must acknowledge each byte by pulling SDA low at the ninth clock, which means a data byte transfer takes nine clock pulses.

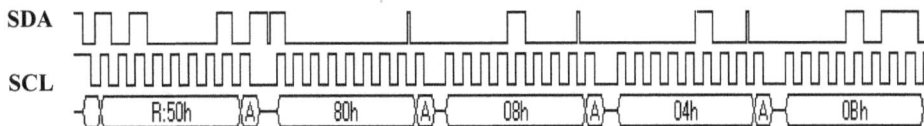

FIGURE 12.14. Multiple Read transfers

## Stop Condition

The master can terminate communication with a slave by generating a STOP signal. This causes the bus to go to idle. A STOP signal is defined as a low-to-high transition of SDA while SCL is a logical 1, as shown in Figure 12.15. The master must have control of SDA

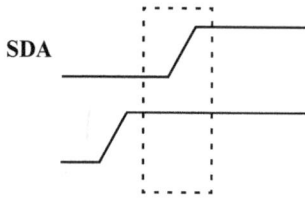

**FIGURE 12.15. STOP Condition**

and SCL to ensure it is able to generate the STOP condition. To guarantee this situation exists, after the master clocks in a slave's acknowledge bit (logic 0 on SDA), the slave must release the bus to give the master the opportunity to generate a STOP condition. Note that the master does not have to generate a STOP signal. Instead, it may generate another START signal, referred to as a repeated START and an address, to prevent the bus from becoming idle.

**Repeated Start**

A repeated START signal is a START signal generated without first generating a STOP signal to terminate the communication. The master would typically use a repeated START to communicate with another slave or with the same slave in the reverse or even same direction without releasing the bus. The repeated START prevents other masters from attempting to gain control of the bus.

Various combinations of read/write formats are then possible, as shown on the simplified formats of Figure 12.16.

- Example 1 is the case of a master transmitter transmitting to a slave receiver. The transfer direction is not changed.

- Example 2 shows the master reading the slave immediately after the first byte. At the moment of the first acknowledge, the master transmitter becomes a master receiver and the slave receiver becomes a slave transmitter.

- In Example 3, the START condition and slave address are both repeated using the repeated START signal. This is to communicate with the same slave in a different mode without releasing the bus. After first addressing the slave, the master reads data from the slave, and signals end of data by not acknowledging. Then the master addresses the slave again without relinquishing the bus, but this time clears the R/W bit to indicate a write operation will follow. The slave acknowledges the first byte. The master may choose to ignore the slave's acknowledgement of the second byte because the master requires no further data. We know this because the master has terminated the transfer with a STOP condition.

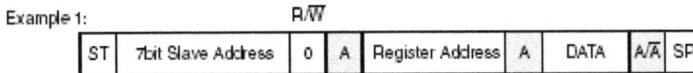

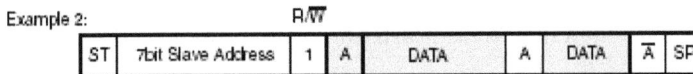

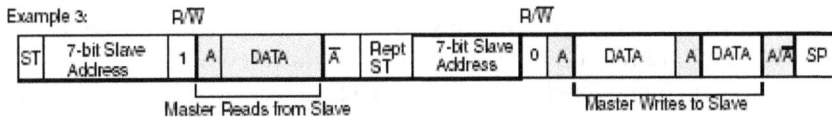

FIGURE 12.16. **Read-Write**

## Multi-Master Arbitration

I²C is a true multi-master bus that allows more than one master to be connected to it. If two or more master devices simultaneously request control of the bus, a clock synchronization procedure determines the bus clock, using integrated hardware timers in each master device. Because wired-AND logic is implemented on the SCL line, a high-to-low transition on the SCL line affects all the devices connected on the bus. The devices start counting their low period and once a device's clock has gone low, it holds the SCL line low until the clock high state is reached. However, the change of low to high in this device's clock may not change the state of the SCL line if another device clock is still within its low period. Therefore, the bus level of SCL is held low by the device with the longest low period. Devices with shorter low periods enter a high wait state during this time, as shown in Figure 12.17. When all devices concerned have counted out their low period, the SCL bus line is released and goes high. At this point, the device clocks and the SCL line are synchronized, and the devices start counting their high periods. The first device to complete its high period pulls the SCL line low again.

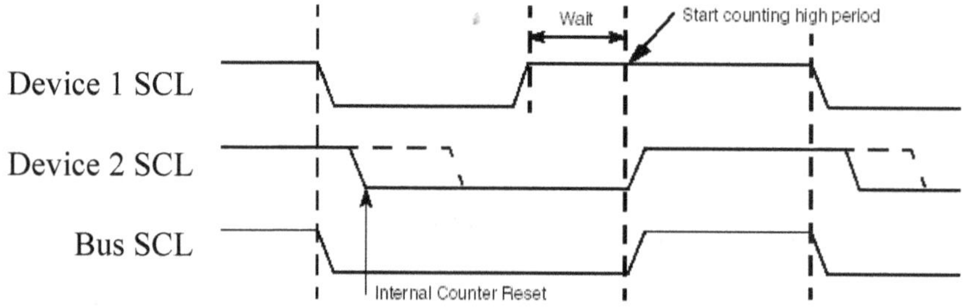

FIGURE 12.17. SCK Synchronization in a Multi-Master bus configuration

The relative priority of the contending masters is determined by a data arbitration procedure. A bus master loses arbitration if it transmits a logic 1 while another master transmits a logic 0. The losing masters immediately switch over to slave receive mode and stop driving SDA output, as shown in Figure 12.18. In this case the transition from master to slave mode does not generate a STOP condition. In this I²C module, the loss of arbitration is indicated by a set IICS[ARBL] bit.

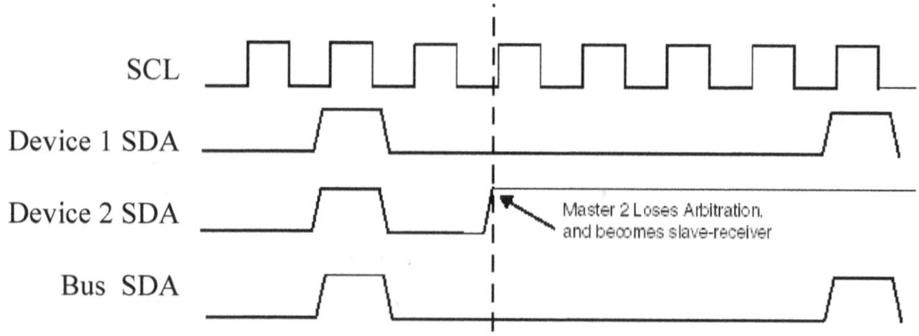

FIGURE 12.18. Arbitration in a Multi-Master bus configuration

### Using Clock Stretching to Handshake Data Transfers

The clock synchronization mechanism can be used as a handshake in data transfers. Slave devices can hold SCL low after completing one byte transfer. In such a case, the clock mechanism halts the bus clock and forces the master clock into wait states until the slave releases SCL. Slaves may also slow down the transfer bit rate. After the master has driven SCL low, the slave can drive SCL low for the required period and then release it. If the slave SCL low period is longer than the master SCL low period, the resulting SCL bus signal low period is stretched.

There are a number of minor drawbacks involved when implementing this. If the SCL becomes stuck due to an electrical failure of a circuit, the master can halt permanently. Of course this can be handled by timeout counters. In any event, if the bus fails in this manner I²C communication is not working anyway.

Another potential drawback of clock stretching is reduced overall bus bandwidth and longer latencies before other masters may attempt to access the bus.

This technique does not interfere with the arbitration mechanism described earlier because holding SCL signal low will lead to only stalling other devices that might want to claim the bus. So the technique has no function limitations, only a loss of bandwidth and increased latency in arbitration in a multi master system.

## I²C Programming Example for Master Mode

Software should execute the following steps to configure the I²C module for interrupt operation as a bus master

- Write IICF to set the IIC baud rate (example provided in this chapter)
- Write IICC1 to enable IIC and interrupts
- Initialize RAM variables (IICEN = 1 and IICIE = 1) for transmit data
- Initialize RAM variables used by routine shown in Figure 12.19
- Write IICC1 to enable TX
- Write IICC1 to enable MST (master mode)
- Write IICD

Figure 12.19 shows the flow chart for a typical interrupt handler that caters for transmission and reception in both master and slave modes.

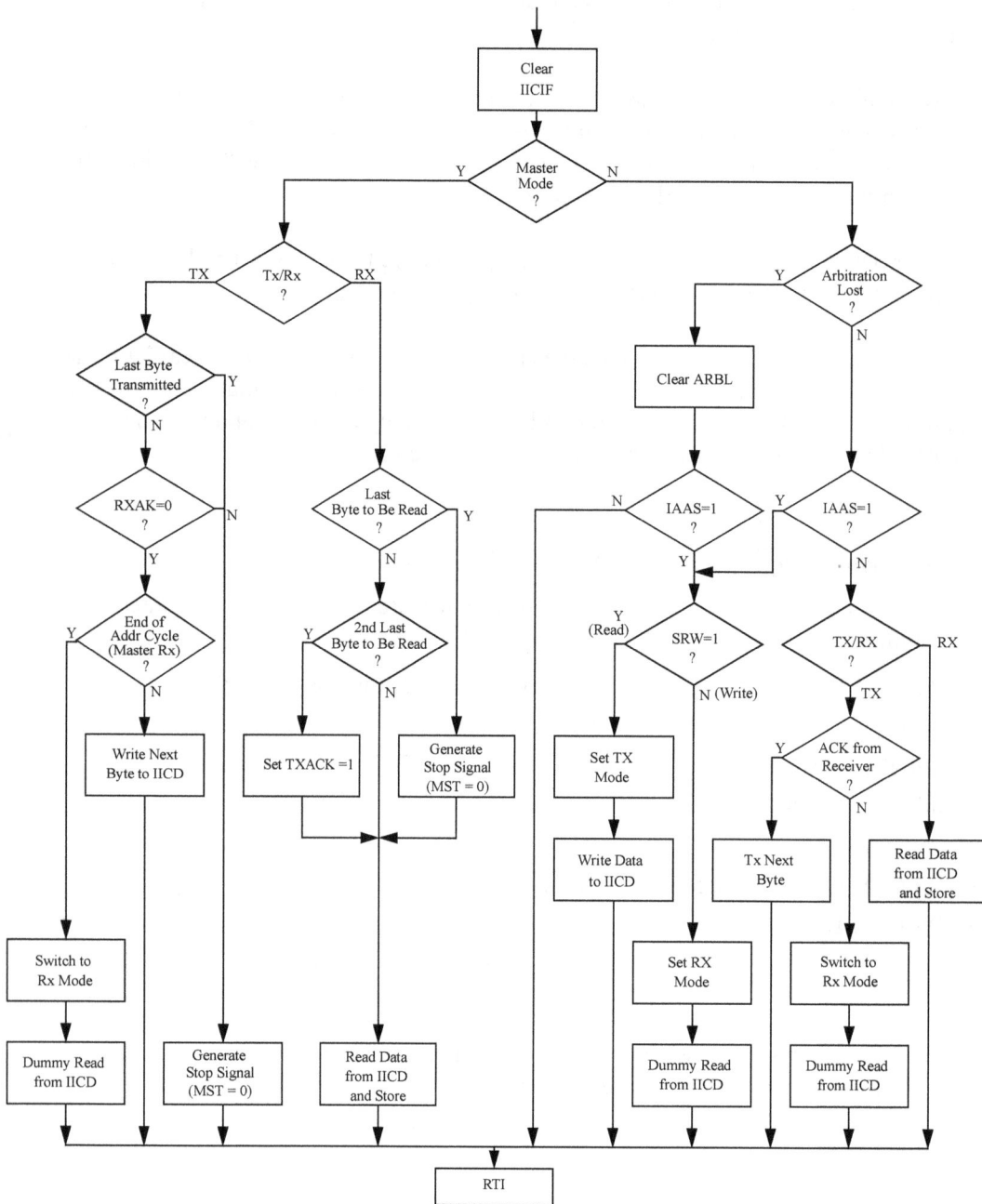

**FIGURE 12.19. Typical I²C Interrupt Routine**

## Lost Arbitration

The I²C bus is a true multi-master bus that allows more than one master to be connected simultaneously. If two or more masters try to control the bus at the same time, the relative priority of the contending masters is determined by a data arbitration procedure. The I²C module asserts this interrupt when it loses the data arbitration process and the IICS[ARBL] bit is set.

Arbitration is lost in the following circumstances:

- SDA sampled as a low when the master drives a high during an address or data transmit cycle.
- SDA sampled as a low when the master drives a high during the acknowledge bit of a data receive cycle.
- A START cycle is attempted when the bus is busy.
- A repeated START cycle is requested in slave mode.
- A STOP condition is detected when the master did not request it.

Software must clear the ARBL bit by writing a 1 to it.

## SMBus

The SMBus is supported by the addition of four registers to the I²C controller:

- IICSMB - Control and Status Register
- IICA2 - Address Register 2
- IICSLT1, IICSLT2 - SCL LowTime Out Registers 1 and 2

The IICSMB contains various control bits and status bits.

The control bit FACK enables software to issue a Fast NACK in the event that the Packet Error Checking (PEC) value is incorrect or an invalid command has been received.

The ALERTEN bit enables the SMBus alert response address, while the SIICAEN bit enables the SMBus device default address, contained in the IICA2 register.

The TCKSEL bit selects the source of the SCL timeout counter, either from the SIM directly or from a prescaled value of the SIM. To use this feature, software must load a non-zero 16 bit value to the two 8 bit registers IICSLT1 and IICSLT2.

Two timeout flags are provided to indicate that the SCL clock is either stuck low (SLTF flag) or high (SHTF flag).

# Control Area Network (MSCAN)

## Introduction to the Controller Area Network

The controller area network (CAN) is a serial communications protocol targeted for automotive and industrial applications that require a high level of data integrity. Some application examples include heavy duty trucks, industrial controls and automotive electronics. The CAN bus is a multi-master protocol which uses non-destructive collision resolution to ensure the highest priority message is transmitted on to the bus first. The CAN bus supports bit rates of up to 1Mbits/second. Not all ColdFire and HCS08 microcontrollers have the MSCAN module, but on some derivatives there may be one or two, supporting CAN specification version 2.0A and 2.0B. Each module has 5 entry receive and 3 entry transmit message buffers.

The bus protocol has built in error detection and error signaling features, along with automatic retransmission of corrupted messages. The CAN bus also distinguishes between temporary errors and permanent node failures and the ability to prevents a faulty node from causing long term disruptions of network traffic. In this chapter we will first cover the CAN bus protocol and later explain the MSCAN architectural details.

### Layered Architecture

The CAN specification is intended to achieve compatibility between any two CAN implementations according to the ISO/OSI reference model. Even though there are many implementations that are architecturally different, the bus messaging protocol must be compatible in order to be CAN compliant. The CAN specification ISO reference model, defines a data link layer and a physical layer. The data link layer is composed of two sub-layers:

- The logical link control sub-layer, or LLC, is concerned with message filtering, overload notification, and recovery management.

- The medium access control sub-layer, or MAC, is responsible for message framing, arbitration, acknowledgment, error detection, and signaling. The MAC sub-layer provides a self checking mechanism, called fault confinement, that distinguishes between temporary and permanent errors.

The CAN specification does not define the physical layer and there is no specified transceiver, allowing the user to define a custom physical layer based on an application's requirements. Acceptable physical media could include a twisted pair (shielded or unshielded), a single wire, fiber optic cable, or transformer coupled to power supplies. The physical layer deals with signal transmission, bit timing, bit encoding, and synchronization. Typical CAN bus transmission rates range from 5K to 1M bits/sec. Slower transmission rates allow for connections of up to one kilometer, whereas higher transmission rates are targeted for shorter connections below 40 meters. Most implementations use NRZ bit formatting over twisted pair bus.

## CAN Messaging Protocol

The CAN bus protocol supports a number of message frames, with transmit and receive data frames being the most common. A multiplexed bus is a scheme of connecting multiple CAN nodes on a single interface which is shared by all nodes. In this section, we will assume either the standard (2.0A) or extended (2.0B) ID frame formats are used.

### Data Frame

A data frame, shown in Figure 13.1, is composed of seven fields. Data frames always begin with a start of frame symbol (SOF). All CAN transmitting nodes use the SOF symbol to start their transmission sequence. Since the CAN is a multi-master bus, all transmitting nodes synchronize themselves with a start of frame symbol and begin to transmit their messages at the same time. After the SOF symbol, transmitting CAN nodes begin arbitrating for the bus. Arbitration for the bus is carried out during the message ID transmission time. Note that Figure 13.1 shows an extended format ID consisting of 29 bits. The standard ID consists of 11 bits. The node having the highest priority (lowest ID number) wins arbitration. All other nodes that attempted to gain access of the bus withdraw and automatically become receivers. Each bit transmitted is echoed and monitored by the transmitting node to verify its access to the bus. The implication of this is the CAN receiver must be enabled in every transmitting node for the arbitration process to operate. A transmitting node loses arbitration when a recessive bit is transmitted and echoed as a dominant. In CAN terminology, recessive typically means logic 1 and dominant, logic 0.

When the arbitration process is completed, the bus winner continues to transmit its mes-

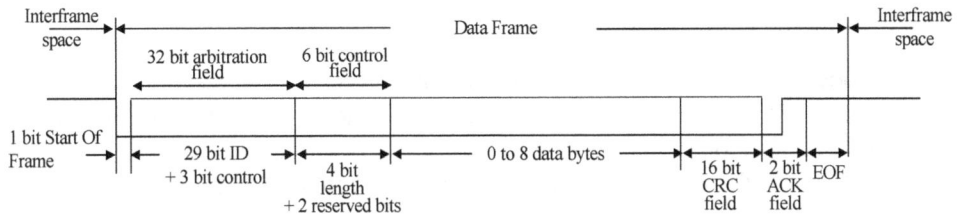

FIGURE 13.1. **Extended ID Data Frame Format**

sage frame.

To ensure data integrity, the CAN module hardware calculates a cyclical redundancy check character (CRC) during message transmission and it is checked by all receiving nodes. The CRC is a 15 bit field followed by one recessive delimiter bit.

The last field transmitted in a data frame is a 2 bit acknowledge (ACK) field indicating that a message was received successfully without any errors. The acknowledge field is asserted by the receivers as one dominant bit followed by one recessive bit. If a transmitter does not detect an ACK it re-transmits the data frame.

Frames are separated by an end of frame (EOF) symbol to provide a synchronizing method among all transmitting nodes. The EOF symbol consists of seven recessive bits driven on the CAN bus following the acknowledgment field.

**Remote Frame**

A remote frame is similar to a data frame. It is distinguished from a data frame by the remote transmission request (RTR) bit, and has a lower priority than a data frame. This frame is used by a node to request a data frame from another node. The transmitting node message buffer (MB) that contained the remote frame automatically becomes a receive MB for the requested data. When a remote frame is received in by the MSCAN, its ID is compared to all transmit message buffer IDs containing a code field 1010. If there is a match, the matching MB will be transmitted as a response. Note that a code field of 1010 indicates that the data frame will only be transmitted as a response to a remote frame. A received remote frame is not stored in a receive MB and it is only used to trigger a transmission of the requested data.

If a data frame and a remote frame with the same message ID are initiated at the same time, the data frame will prevail. The RTR bit which identifies a remote from a data frame is used during the bus arbitration process. Since the RTR bit is dominant in a data frame and recessive in a remote frame, the node transmitting a remote frame loses arbitration during the RTR bit transmission time.

## Overload Frame

Receiving nodes requiring additional processing time between successive data or remote frames will transmit an overload frame. CRC delimiter, acknowledgment delimiter, end-of-frame and intermission fields, all have fixed-form bits and must be sensed as expected by the CAN bus. If a fixed-form is destroyed, it is interpreted as an overload frame by the transmitting nodes. An overload frame consists of two fields: an overload flag of six dominant bits followed by a delimiter of eight recessive bits. For additional details refer to the Bosch Controller Area Network document, Version 2.0.

## Error Detection

To achieve a high degree of data integrity, the CAN bus provides supports error signaling and self checking. Once a node wins arbitration, all others become receivers and listen to the transmitted frame, even if not destined for them. This ensures that only error free frames are processed and all others are signaled with an error. Error signaling and recovery is an important property of the CAN bus because it is reported by any node detecting an error condition. To be CAN compliant, the bus must be able to recognize up to 5 global errors.These errors are:

1. **Bit Error**: Each transmitted bit is monitored by the node driving the bus. If the monitored bit has the opposite polarity of the transmitted bit, the CAN bus interprets this condition as a bit error. During transmission of the ID, no bit error is signaled if the transmitted bit is of opposite polarity to the monitored bit on the bus. This merely indicates that the node has lost arbitration and must withdraw from the transmission process.

2. **Stuff Error**: To guarantee a sufficient number of transition during message transition, the transmitter always stuffs (inserts) a bit of the opposite polarity when transmitting more than five consecutive bit of same polarity. The stuff bit is used for transmitter-receiver synchronization purposes. This bit is removed after it is checked by the receiver. If the receiver does not detect the stuff bit, an error flag is signaled.

3. **Cyclical Redundancy Check (CRC)**: Serial data transmission over a long distance is sometime prone to data loss in a noisy environment. To ensure data loss detection, a CRC is calculated by the transmitter and transmitted with the frame. The receiver calculates a CRC from the received frame, and compares it with the received CRC. If the two CRCs are not equal, a CRC error is reported by one or more receiving nodes by transmitting an error frame. The signaling of an error frame causes the transmitter to immediately abort message transmission. Error signaling on the CAN bus is recoverable and considered a temporary error.

4. **Form Error**: This error occurs whenever the fixed-form bit field contains one or more illegal bits. The CRC delimiter, acknowledgment delimiter, end-of-frame and intermission fields, all have fixed-form bits and must be sensed as expected by the CAN bus otherwise a form error is signaled and may be interpreted as an overload frame.

5. **Acknowledgment Error**: Each time a frame is transmitted, an acknowledgment must be signaled to the transmitter to indicate successful message reception. The acknowledge error is a 2 bit field composed of one dominant followed by one recessive bit. If the transmitter does not receive an acknowledgment symbol an error flag is signaled.

Besides signaling the error with an error frame on the bus, each CAN node may signal an interrupt request by setting the corresponding error flag. The MSCAN module reports errors by setting the appropriate error flag in the error status register (ESR).

## *Error Signaling*

If any of the five errors discussed above is detected during message transmission and reception, an error flag is signaled. For bit, stuff, form and acknowledgment errors, an error flag is signaled at the next bit time. For a CRC error it is signaled one bit time after the acknowledgment field. The error flag signaling causes the transmitter to immediately abort message transmission. There are two forms of error flags: active and passive.

### Error Active and Error Passive Nodes

An error-active node transmits an error flag consisting of six consecutive dominant bits to signal an error condition. Bit stuffing is applied to all fields from the start of a frame to the end of a frame. A sequence of six consecutive dominant bits on the bus violates bit stuffing rules, and may cause other nodes to react by transmitting another error flag of six more consecutive dominant bits. The total length of an active error flag can vary between a minimum of six to a maximum of twelve dominant bits as shown in Figure 13.2.

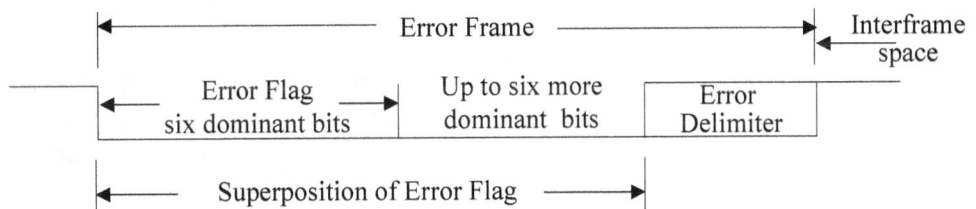

FIGURE 13.2. **Active Error Frame**

An error passive node detecting an error attempts to transmit a passive error flag consisting of six consecutive recessive bits. A passive error node waits for at least three or more consecutive recessive bits before transmitting an error flag of six consecutive recessive bits, as shown in Figure 13.3.

An error flag consists of two distinct fields. The first is given by the superposition of error flags contributed from other nodes and the second is an error delimiter consisting of eight recessive bits.

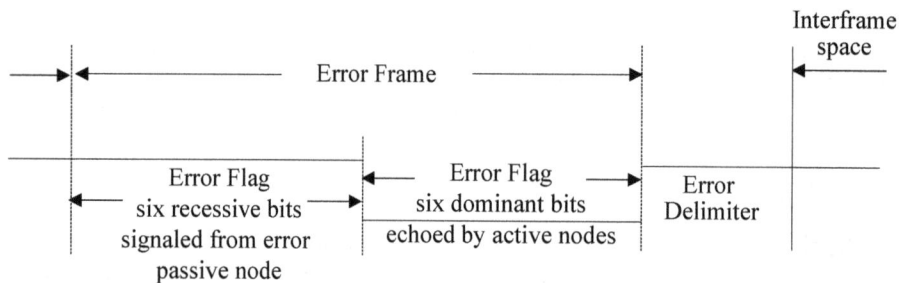

FIGURE 13.3. **Error Passive Frame**

## *Fault Confinement*

The specification of the CAN bus guarantees that a message is accepted by all nodes or rejected by all nodes, ensuring only uncorrupted messages are processed. Corrupted messages are signaled with an error flag which causes the transmitting node to immediately abort transmission. Message abortion is temporary on the CAN bus, and after a maximum of 29 bit times, retransmission of the previously corrupted message is automatically initiated. To avoid long term disruptions of the network, a faulty node may no longer be allowed to participate in the transmission process depending on its status.

A node on the network can be in one of three possible states: error active, error passive or bus off. In order for the CAN hardware to differentiate between states, each node has an 8 bit receive error (CANRXERR) counter and 8 bit transmit error (CANTXERR) counter. These two 8 bit error counter are read-only registers and are incremented or decremented depending on unsuccessful or successful message reception, respectively. An error active node state is indicated by the values of its receive and transmit error counters. When both counter values are in the range of 0 to127 counts, the node is said to be an error active node. In this state, the node will transmit an error active flag during error frames. If one of the error counters exceeds 127 counts, the node becomes an error passive node. In this state, the node will transmit error passive flag during error frames. Refer to the error active frame format in Figure 13.2 and error passive frame format in Figure 13.3. Error counters are incremented by either 1 or 8 counts upon detecting one of the five global errors, whereas a successful message transmission decrements the counter value by 1 count.

The state of the transmitter and the receiver are indicated to the CPU by the CAN status change (CSCIF) interrupt flag in the CANRFLG register. The CPU is able to check the state of the of the MSCAN node by reading the receiver status (RSTAT) and transmitter status (TSTAT) bits in the CANRFLG register. Each of the two bit field indicate the current TX/RX error counter value and the current node state as shown in Table 1 below:

**TABLE 1. Receive and Transmit Error Counters and Node States**

| RSTAT/TSTAT | Receive Error Counter | Transmit Error Counter | Node State |
|:---:|:---:|:---:|:---:|
| 0 0 | 0 <= error counter <= 96 | 0 <= error counter <= 96 | Active |
| 0 1 | 96 < error counter <= 127 | 96 < error counter <= 127 | Active |
| 1 0 | >127 but <255 | >127 but <255 | Passive |
| 1 1 | > 255 | > 255 | Bus-off |

Also, the application software is able to read the exact receive and transmit error counters by reading the CANRXERR and the CANTXERR, respectively.

A bus-off state is entered when either of the node error counters reaches the count value of > 255. In this state, the node switches off its output drivers and no longer participates in the network as a transmitter, but continues to monitor the bus. If the bus-off recovery mode (BORM) bit in the MSCAN control (CANCTL1) register 1 is negated, automatic recovery from the bus off state is enabled according to the CAN specification 2.0B. An enabled bus-off recovery node will return to error active state following the detection of 128 occurrences of 11 recessive bits.

Receive error conditions and node state changes are reflected in the MSCAN receive flag (CANRFLG) register and may be reported to the CPU with interrupt requests. Node state changes and error conditions signaling are enabled by setting the appropriate interrupt mask bits in the MSCAN receive interrupt enable (CANRIER) register.

## *MSCAN Architecture*

The MSCAN module is based on the Control Area Network architecture and is implemented with five receive message buffers and three transmit message buffer. The contents of a message buffers are either sent to or fetched from the CAN bus through two separate serial message buffers as shown in Figure 13.4.

As can be seen in the figure below, the MSCAN module implements only the digital layer which must connect to the physical layer with external transceiver. The transceiver is capable of driving the high current needed for the CAN bus and usually has current protection against a defective bus.

On the receiver side, the CAN message is assembled into the serial receive message buffer, and if the entire message is received with no errors, the message ID is compared to the acceptance filter after passing through the acceptance mask and is placed in the receive buffer if the message ID matches. Once received, the receiver buffer full flag (RXF) is set in the CAN receive flag (CANRFLG) register and an interrupt may be signaled to the CPU if the interrupt is enable. In the interrupt handler, the CPU reads message buffer and clears the interrupt flag.

On the transmitter side, the transmit message buffer having the highest priority is loaded into the transmit serial shifter. Once successfully transmitted, the MSCAN signals the CPU with an interrupt request by setting the transmit buffer empty (TXEx) flag in the MSCAN transmitter flag (CANTFLG) register. In the interrupt handler, the CPU clears the request and may fill the empty buffer with another message for transmission.

**Receive Operation**

MSCAN received messages are stored in a five stage input First-in-first-out (FIFO). There are four background receive (RxBG) and one receive foreground (RxFG) buffer. The five buffers are mapped into a single memory area consisting of 16 memory-mapped I/O locations. The background receive buffers (RxBG) are only accessible to the MSCAN, but the foreground receive buffer (RxFG) is addressable by the CPU. This scheme simplifies the interrupt handler software because only one address area is used for the receive process.

The receive buffers consist of 16 bytes to store the CAN control bits, the identifier (standard or extended), the data contents, and a time stamp. The receiver full flag (RXF) indicates the status of the foreground receive buffer. This flag is set when the buffer contains a correctly received message with a matching identifier.

On reception, each message is checked by the MSCAN to see whether it passes the filter. After successful reception of a valid message, the MSCAN shifts the message into the receiver FIFO, sets the RXF flag, generates a receive interrupt to the CPU, and the message is simultaneously written into the active RxBG. The interrupt handler must read the received message from the RxFG and then clear the RXF flag to acknowledge the interrupt in order to release the foreground buffer. If a new message is received, which could follow immediately after the inter-frame sequence (IFS) field of the CAN frame, it will be written into the next available RxBG. However, if the received message is invalid due to wrong identifier, transmission errors, etc, the message is ignored and will not be shifted into the receive FIFO.

When the MSCAN module is transmitting, it will receives its own transmitted messages into the background receive buffer (RxBG) but does not shift it into the receiver FIFO, generate a receive interrupt, or acknowledge its own messages on the CAN bus. The exception to this rule is when the MSCAN is configured in the loopback mode. In the loopback mode, the MSCAN treats its own messages exactly like all other incoming messages from the CAN bus. During ID transmission, all nodes arbitrate for the CAN bus, and in case of arbitration loss due to a higher priority node winning the bus, the MSCAN automatically becomes a receiver. Bus arbitration loss is normal and is not treated as a fault condition.

If the receive FIFO buffers are filled with correctly received messages and another message is correctly received from the CAN bus with an accepted identifier, the latter message is discarded and an error interrupt with overrun indication is generated, if the overrun interrupt is enabled (OVRIE) in the CANRIER. The MSCAN remains able to transmit messages while the receiver FIFO is full, but all incoming messages are discarded. As soon as a receive buffer in the FIFO is available again, new valid messages will be accepted.

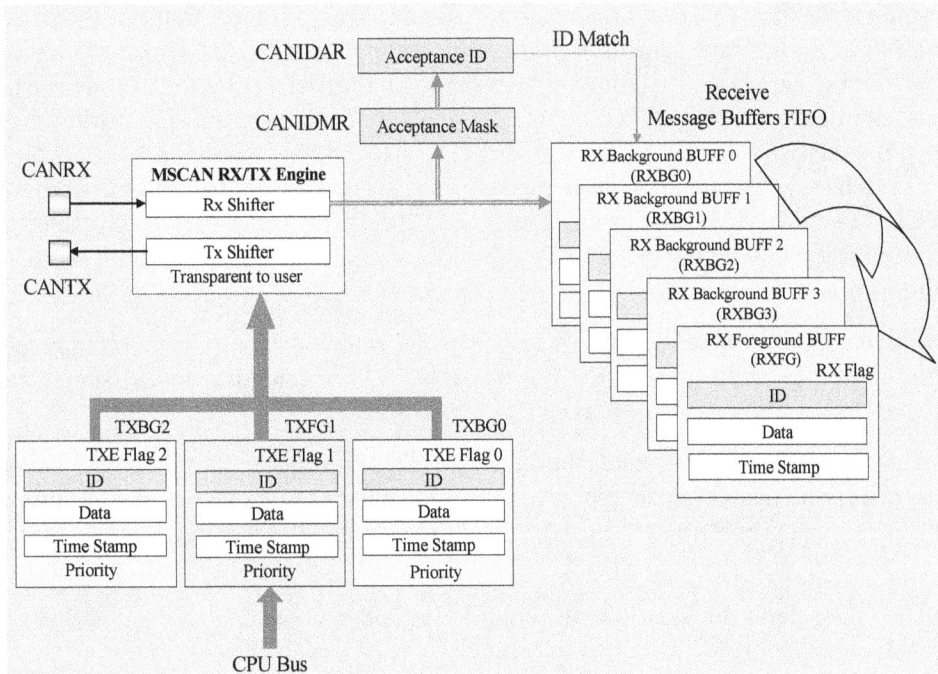

FIGURE 13.4. MSCAN Architecture

## Transmitter Operation

The MSCAN transmitter has three prioritized message buffers. The triple transmit buffer scheme optimizes real-time performance by allowing multiple messages to be set up in advance. Of the three, only one transmit buffer (TX) is accessible to the CPU at a time. To select the next available TX buffer, the CPU reads the MSCAN transmitter interrupt flag (CANTFLG) register and then write the value into the MSCAN transmit buffer selection (CANTBSEL) register.

Before message transmission, the application software writes the required message information and assign a priority to each transmit message buffer. The message with the highest buffer priority will be transferred to the TX serial buffer in preparation for transmission. The message buffer priority is programmed into the transmit buffer priority (TBPR) register.

To transmit a message, the CPU must identify an available transmit buffer, which is indicated by a set transmitter buffer empty (TXEx) in the CANTFLG register. If a transmit buffer is available, the CPU must set a pointer to this buffer by writing to the transmit buffer select (CANTBSEL) register. This makes the respective buffer accessible within the MSCAN transmit foreground buffer (CANTXFG) address space. There are three transmit empty flags (TXE2:0) in the CANTFLG register, and three transmit buffer select control (TX2:0) bits, one pair for each message buffer. The MSCAN selects the lowest buffer number based on the setting of the three TX buffer select control bits. At reset, all transmit buffer empty flags (TXE2:0) are set, indicating the three transmit buffers are empty. In other words, transmit buffers TX0, TX1 and TX2 are available, but only one TX buffer can be selected at a time. In this case, the value read from CANTFLG will be 0b0000_0111. When writing this value back to CANTBSEL, TX buffer TX0 is selected as the transmit foreground buffer (CANTXFG) because the lowest numbered bit set to '1' is at bit position '0'.

Reading back this value out of CANTBSEL results in 0b0000_0110, because only the lowest numbered bit position set to '1' is presented. This mechanism eases the application software the selection of the next available TX buffer.

The algorithmic feature associated with the CANTBSEL register simplifies the transmit buffer selection. In addition, this scheme makes the handler software simpler because only one address area is applicable for the transmit process, and the required address space is minimized.

The CPU then stores the identifier, the control bit field, and the data content into the selected transmit buffer.

The MSCAN then schedules the message for transmission and signals successful transmission of the buffer by setting the associated TXE flag. A transmit interrupt can be generated when TXEx is set and can be used to drive the application software to re-load the buffer.

If more than one buffer is scheduled for transmission when the CAN bus becomes available for arbitration, the MSCAN uses the local priority setting of the three buffers to determine the prioritization. For this purpose, every transmit buffer has an 8-bit local priority field (PRIO). The application software programs this field when the message is set up. The local priority reflects the priority of this particular message relative to the set of messages being transmitted from this node. The lowest binary value of the PRIO field is defined to be the highest priority. The internal scheduling process takes place whenever the MSCAN arbitrates for the CAN bus.

### Aborting Message Transmission

Sometime, it may be necessary to abort a lower priority message transmission in favor of scheduling a higher priority one. Because messages that are already in transmission cannot be aborted, the application software can request abortion of the message by setting the corresponding abort request (ABTRQ) bit in the MSCAN transmitter abort request (CANTARQ) register. The MSCAN then grants the request, if possible, by:

1. Setting the corresponding abort acknowledge flag (ABTAK) in the MSCAN message abort acknowledge (CANTAAK) register.
2. Setting the associated TXE flag to release the buffer.
3. Generating a transmit interrupt. The transmit interrupt handler software can determine from the setting of the ABTAK flag whether the message was aborted (ABTAK = 1) or sent (ABTAK = 0). Finally, the buffer is flagged as ready for transmission by clearing the associated TXE flag. Once the message is transmitted successfully, the MSCAN module signals the CPU via an interrupt request, if enabled.

### Message Buffer Structure

Transmit message buffers are used to store message frames to be transmitted while the receive message buffers are used to store incoming messages from the CAN bus. Each message buffer (MB) is composed of 16 bytes and behave as read/write memory space in the MSCAN module.

There are 4 ID (IDR0-3) registers, 8 data segment (DSR0-7) registers, a priority (TBPR)[1]

| Read/write | Register name | Bit 7 | Bit 6 | Bit 5 | Bit 4 | Bit 3 | Bit 2 | Bit 1 | Bit 0 |
|---|---|---|---|---|---|---|---|---|---|
| R/W | **IDR0** | ID28 | ID27 | ID26 | ID25 | ID24 | ID23 | ID22 | ID21 |
| R/W | **IDR1** | ID20 | ID19 | ID18 | SRR | IDE | ID17 | ID16 | ID15 |
| R/W | **IDR2** | ID14 | ID13 | ID12 | ID11 | ID10 | ID9 | ID8 | ID7 |
| R/W | **IDR3** | ID6 | ID5 | ID4 | ID3 | ID2 | ID1 | ID0 | RTR |
| R/W | **DSR0** | DB7 | DB6 | DB5 | DB4 | DB3 | DB2 | DB1 | DB0 |
| R/W | **DSR1** | DB7 | DB6 | DB5 | DB4 | DB3 | DB2 | DB1 | DB0 |
| R/W | **DSR2** | DB7 | DB6 | DB5 | DB4 | DB3 | DB2 | DB1 | DB0 |
| R/W | **DSR3** | DB7 | DB6 | DB5 | DB4 | DB3 | DB2 | DB1 | DB0 |
| R/W | **DSR4** | DB7 | DB6 | DB5 | DB4 | DB3 | DB2 | DB1 | DB0 |
| R/W | **DSR5** | DB7 | DB6 | DB5 | DB4 | DB3 | DB2 | DB1 | DB0 |
| R/W | **DSR6** | DB7 | DB6 | DB5 | DB4 | DB3 | DB2 | DB1 | DB0 |
| R/W | **DSR7** | DB7 | DB6 | DB5 | DB4 | DB3 | DB2 | DB1 | DB0 |
| R/W | **DLR** | | | | | DLC3 | DLC2 | DLC1 | DLC0 |
| R/W | **TBPR**[1] | PRIO7 | PRIO6 | PRIO5 | PRIO4 | PRIO3 | PRIO2 | PRIO1 | PRIO0 |
| R | **TSRH**[2] | TSR15 | TSR14 | TSR13 | TSR12 | TSR11 | TSR10 | TSR9 | TSR8 |
| R | **TSRL**[2] | TSR7 | TSR6 | TSR5 | TSR4 | TSR3 | TSR2 | TSR1 | TSR0 |

**FIGURE 13.5. Transmit/Receive Message Buffer Organization**

register and 2 time stamp high/low (TSRH/L)[2] registers. The message buffer organization is shown in Figure 13.5 above. A brief description of the message buffer register fields is provided in the following paragraphs:

### Identification Registers (IDR0-IDR3).

The identification registers, support the extended ID format consisting of 32 bits; ID[28:0], substitute remote request (SRR) bit, ID extended (IDE) bit and remote transmission request (RTR) bit. When extended ID is selected, all 32 bits are used in the message buffer, but if the standard ID is selected, only 13 bits are needed. The standard ID format bits are; ID[10:0], RTR, and IDE as shown in Figure 13.6.

---

1. The priority register is applicable to transmit buffers only
2. Time stamp are read-only registers

The CAN bus protocol supports two frame formats, standard and extended defined by the identifier extension bit (IDE). The standard format is an 11 bit ID (IDE=0) and is written to or read from MB ID[28:18], whereas the extended is a 29 bit ID (IDE=1) and is written to or read from MB ID[28:0].

The substitute remote request (SRR) bit is used only in the extended ID format and software must set it to '1' in the TX message buffer. When it is received, the SRR bit is stored, as is, in the RX buffer. A received value of '0' indicates arbitration is lost.

Prior to transmitting a message, the software writes the ID in the Tx MB to identify the transmitted message destination, type and priority. When listeners receive the message, the ID field is compared to the acceptance filter and if the message has matching ID, is automatically filtered into the receive FIFO.

| Read/ write | Register name | Bit 7 | Bit 6 | Bit 5 | Bit 4 | Bit 3 | Bit 2 | Bit 1 | Bit 0 |
|---|---|---|---|---|---|---|---|---|---|
| R/W | **IDR0** | ID10 | ID9 | ID8 | ID7 | ID6 | ID5 | ID4 | ID3 |
| R/W | **IDR1** | ID2 | ID1 | ID0 | RTR | IDE | | | |
| R/W | **IDR2** | | | | | | | | |
| R/W | **IDR3** | | | | | | | | |

FIGURE 13.6. **Standard ID Format**

- Data Segment Registers (DSR7:0)

These registers are written by the CPU when a message buffer is set ready for transmission, or contain data written form the CAN bus when a valid message is received. The maximum number of data stored are 8 bytes, but a transmitted or received frame could range from 0 to 8 bytes.

- Data Length Register (DLR)

This 4 bit field contains the data byte count to be transmitted or the data byte count received. The CPU writes the length field to indicate the number of bytes will be transmitted in this frame, while the CAN bus updates this field to indicate how many bytes are received in the frame. Since this field is 4 bits, a higher value than 8 count will still be treated as 8 bytes.

- Transmit Buffer Priority Register (TBPR)

This register is used to assign message priority local to the CAN module and is only applicable to transmit buffers. The highest priority transmit buffer (lowest binary number) will be processed for transmission first.

- Time Stamp Registers High/Low (TSRH/L)

Received and transmitted messages may be stamped in the active message buffer as soon as the message is acknowledged. The time stamp may be used by the CAN bus for transmit/receive message synchronization among all nodes connected to the CAN bus. The time stamp is obtained from a 16-bit free running counter integrated in the MSCAN module. The free running counter uses the bit rate clock to increment the count. If stamping is enabled, the MSCAN captures the 16 bit time stamp in the TSRH/L in the active message buffer upon successful message transmission/reception. These two registers are read-only and can not be written by the host CPU.

### Message Filtering

The MSCAN integrates two 32-bit identifier acceptance filters to define the acceptable patterns of the standard or extended identifier ID[10:0] or ID[28:0]. The filter mechanism can be managed and controlled by the identifier acceptance mode control (IDAM1:0) bits of the MSCAN identifier acceptance filter (CAIDAC) register to increase the number of filters from 2 to 4 or even 8. Increasing the number of filters will of course decrease the number of the ID bits to compare against the acceptance filter.

To simplify the handler, a filter hit is indicated to the application software by three identifier hit flag (IDHIT[2:0]) bits in the CANIDAC register to identify the filter section that caused the acceptance. The IDHIT bits simplify the application software's task to identify the cause of the receiver interrupt. The IDHIT indication reduces much of the CPU filtering overhead. If more than one hit occurs (two or more filters match), the lower hit has priority. Table 2 and Table 3 depict the identifier acceptance mode and filter hit indication.

**TABLE 2. ID Acceptance Filter Mode**

| IDAM1 | IDAM0 | Identifier Acceptance Mode |
|-------|-------|----------------------------|
| 0 | 0 | Two 32-bit Acceptance Filters |
| 0 | 1 | Four 16-bit Acceptance Filters |
| 1 | 0 | Eight 8-bit Acceptance Filters |
| 1 | 1 | Filter Closed |

**TABLE 3. ID Acceptance Filter Hit Indication**

| IDHIT2 | IDHIT1 | IDHIT0 | ID Acceptance Hit |
|--------|--------|--------|-------------------|
| 0 | 0 | 0 | Filter 0 Hit |
| 0 | 0 | 1 | Filter 1 Hit |
| 0 | 1 | 0 | Filter 2 Hit |
| 0 | 1 | 1 | Filter 3 Hit |
| 1 | 0 | 0 | Filter 4 Hit |

**Chapter 13**

TABLE 3. **ID Acceptance Filter Hit Indication**

| IDHIT2 | IDHIT1 | IDHIT0 | ID Acceptance Hit |
|--------|--------|--------|-------------------|
| 1 | 0 | 1 | Filter 5 Hit |
| 1 | 1 | 0 | Filter 6 Hit |
| 1 | 1 | 1 | Filter 7 Hit |

As shown in the table above, the filter is programmable to operate in one four different modes. An explanation of each mode is detailed in the following paragraphs:

1. Two identifier acceptance filters, each to be applied to the full 29 bits of the extended identifier, RTR, IDE and SRR of the CAN 2.0B frame, or applied only to the 11 bits of the standard identifier plus the RTR and IDE bits of the CAN 2.0A/B message.

   Figure 13.7 shows two 32-bit filter bank configuration. Identifier acceptance registers 0 -3 (CANIDAR0-CANIDAR3) and identifier mask registers 0-3 (CANIDMR0-CANIDMR3) pairs make up the first bank to produce a filter 0 hit.

**CAN 2.0B Extended ID (IDE=1)**

| ID28 | IDR0 | ID21 | ID20 | IDR1 | ID15 | ID14 | IDR2 | ID7 | ID6 | IDR3 | RTR |
|------|------|------|------|------|------|------|------|-----|-----|------|-----|
| ID10 | IDR0 | ID3 | ID2 | IDR1 | IDE | | | | | | |

**CAN 2.0A/B Standard ID (IDE=0)**

| AM7 | CANIDMR0 | AM0 | AM7 | CANIDMR1 | AM0 | AM7 | CANIDMR2 | AM0 | AM7 | CANIDMR3 | AM0 |
|-----|----------|-----|-----|----------|-----|-----|----------|-----|-----|----------|-----|

Filter 0

| AC7 | CANIDAR0 | AC0 | AC7 | CANIDAR1 | AC0 | AC7 | CANIDAR2 | AC0 | AC7 | CANIDAR3 | AC0 |
|-----|----------|-----|-----|----------|-----|-----|----------|-----|-----|----------|-----|

ID Accepted (Filter Hit 0)

**CAN 2.0B Extended ID (IDE=1)**

| ID28 | IDR0 | ID21 | ID20 | IDR1 | ID15 | ID14 | IDR2 | ID7 | ID6 | IDR3 | RTR |
|------|------|------|------|------|------|------|------|-----|-----|------|-----|
| ID10 | IDR0 | ID3 | ID2 | IDR1 | IDE | | | | | | |

**CAN 2.0A/B Standard ID (IDE=0)**

| AM7 | CANIDMR4 | AM0 | AM7 | CANIDMR5 | AM0 | AM7 | CANIDMR6 | AM0 | AM7 | CANIDMR7 | AM0 |
|-----|----------|-----|-----|----------|-----|-----|----------|-----|-----|----------|-----|

Filter 1

| AC7 | CANIDAR4 | AC0 | AC7 | CANIDAR5 | AC0 | AC7 | CANIDAR6 | AC0 | AC7 | CANIDAR7 | AC0 |
|-----|----------|-----|-----|----------|-----|-----|----------|-----|-----|----------|-----|

ID Accepted (Filter Hit 1)

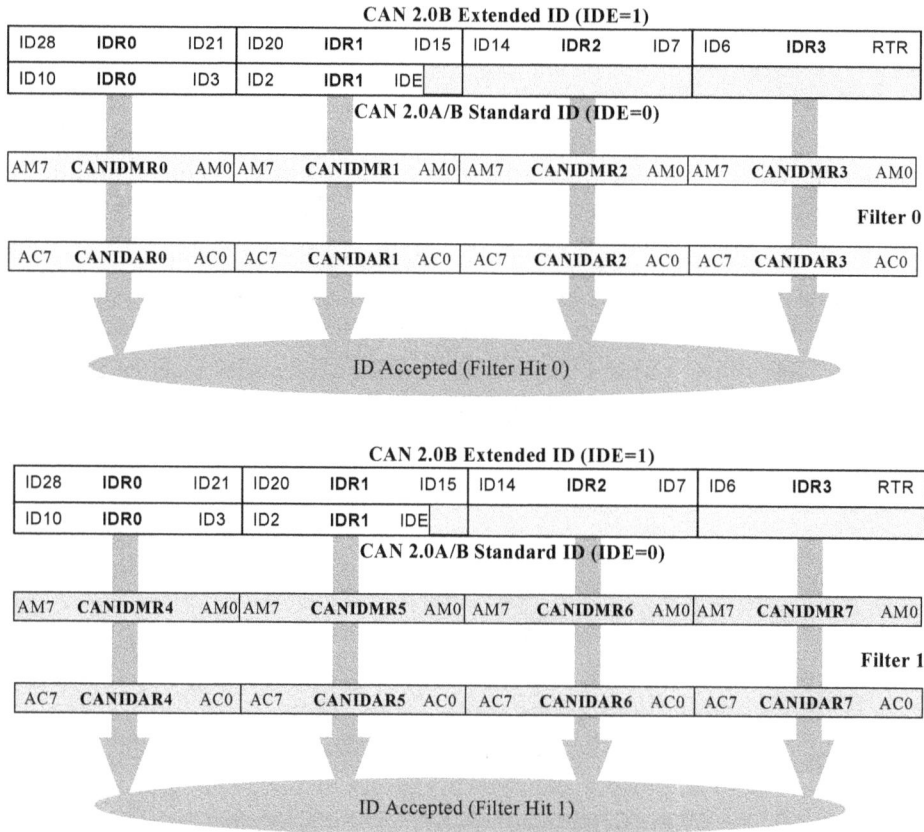

FIGURE 13.7. MSCAN Two 32 bit Message Filters 0 and 1

Similarly, the second filter bank is produced by identifier acceptance registers 4-7 (CANIDAR4-CANIDAR7) and identifier mask registers 4-7 (CANIDMR4-CANIDMR7) pairs to produces a filter 1 hit.

2. Four identifier acceptance filters, each to be applied to the 14 most significant bits of the extended identifier plus the SRR and IDE bits of CAN 2.0B messages. A second option is to apply it to the 11 bits of the standard identifier, the RTR and IDE bits of CAN 2.0A/B messages. Figure 13.8 shows how the first 32-bit filter bank

CANIDAR0-3 CANIDMR0-3) produces filter 0 and 1 hits. Similarly, the second filter bank CANIDAR4-7, CANIDMR4-7 are used to produce filter 2 and 3 hits.

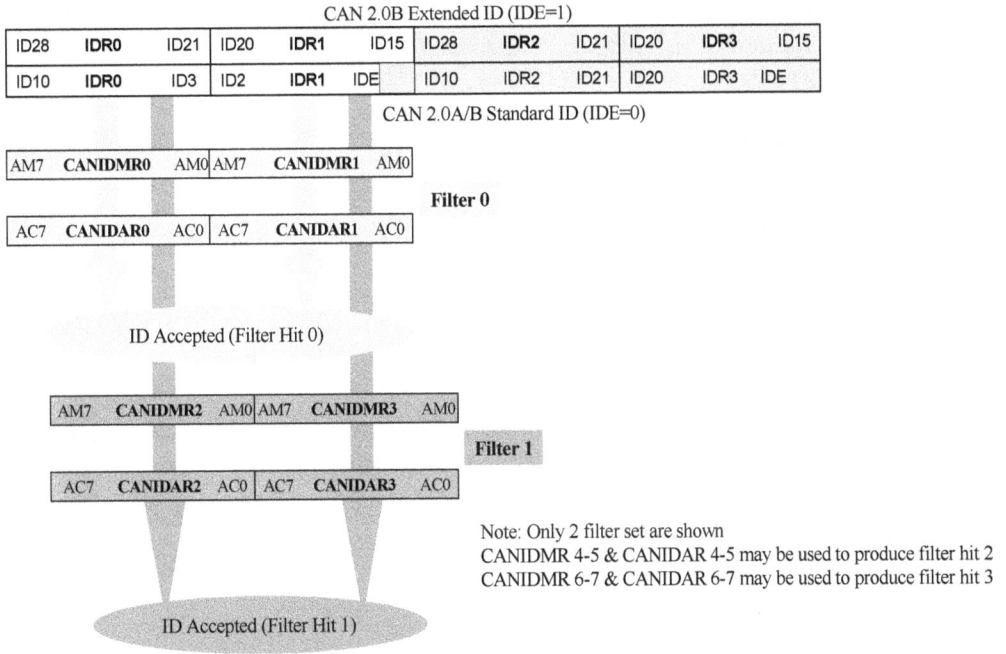

**FIGURE 13.8. MSCAN 16 Bit Filters (0 &1)**

3. Eight identifier acceptance filters, each to be applied to the first 8 bits of the identifier. This mode implements eight independent filters for the first 8 bits of a CAN 2.0A/B compliant standard identifier or a CAN 2.0B compliant extended identifier. Figure 13.9 shows how the first 32-bit filter bank (CANIDAR0-CANIDAR3) and CANIDMR0–CANIDMR3) produces filter 0 to 3 hits. Similarly, the second filter bank (CANIDAR4-CANIDAR7 and CANIDMR4-CANIDMR7) produces filter 4 to 7 hits.

CAN 2.0B Extended ID (IDE=1)

| ID28 | **IDR0** | ID21 | ID20 | **IDR1** | ID15 | ID28 | **IDR2** | ID21 | ID20 | **IDR3** | ID15 |
|------|------|------|------|------|------|------|------|------|------|------|------|
| ID10 | **IDR0** | ID3 | ID2 | **IDR1** | IDE | ID10 | **IDR2** | ID21 | ID20 | **IDR3** | IDE |

CAN 2.0A/B Standard ID (IDE=0)

| AM7 | **CANIDMR0** | AM0 |
|-----|----------|-----|

Filter 0

| AC7 | **CANIDAR0** | AC0 |
|-----|----------|-----|

ID Accepted (Filter Hit 0)

| AM7 | **CANIDMR1** | AM0 |
|-----|----------|-----|

Filter 1

| AC7 | **CANIDAR1** | AC0 |
|-----|----------|-----|

Note: Only 4 filter set are shown
CANIDMR4 & CANIDAR4 may be used to produce filter hit 4
CANIDMR5 & CANIDAR5 may be used to produce filter hit 5
CANIDMR6 & CANIDAR6 may be used to produce filter hit 6
CANIDMR7 & CANIDAR7 may be used to produce filter hit 7

ID Accepted (Filter Hit 1)

| AM7 | **CANIDMR2** | AM0 |
|-----|----------|-----|

Filter 2

| AC7 | **CANIDAR2** | AC0 |
|-----|----------|-----|

ID Accepted (Filter Hit 2)

| AM7 | **CANIDMR3** | AM0 |
|-----|----------|-----|

Filter 3

| AC7 | **CANIDAR3** | AC0 |
|-----|----------|-----|

ID Accepted (Filter Hit 3)

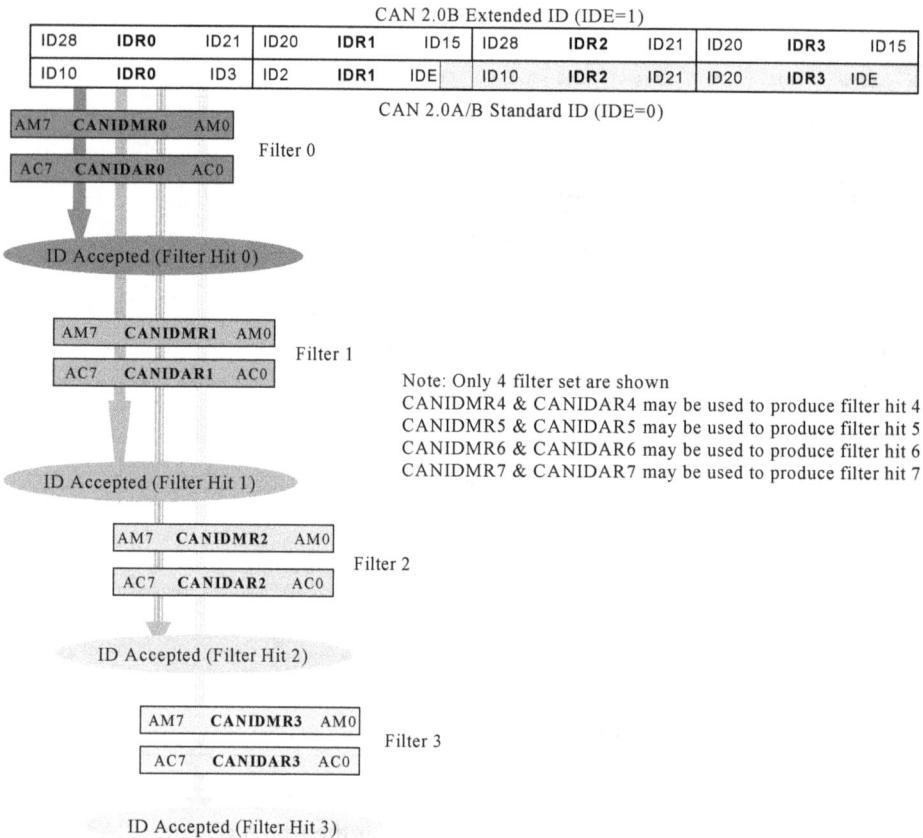

**FIGURE 13.9. MSCAN 8 Bit Filters (0-3)**

4. Closed filter. No CAN messages are copied into the foreground buffer RxFG, and the RXF flag is never set.

To allow multiple messages to be accepted by the filters, eight identifier mask registers (CANIDMR0-7) are provided. For extended ID, CANIDMR0-3 registers are one filter mask set and CANIDMR4-7 are the second filter mask set. For standard ID, CANIDMR0-1, CANIDMR2-3, CANIDMR4-5 and CANIDMR6-7, are respectively filter mask set, 0, 1, 2 and 3. Likewise when 8 bit ID filter is used, then each of the CANIDMR registers behave as one of the 8 mask filter sets. For more information refer to Figure 13.9.

Mask registers are used to determine which bits of the acceptance registers would be used to filter the incoming messages. If the mask register contains a '1' in certain bits it means that the corresponding bits on the acceptance register won't be used for the data filtering. If the bit contains a '0' it means that the value of the corresponding bit of the acceptance register must match with the corresponding bit of the incoming message's identifier value. Any bit or group of bits can be marked as 'don't care' in the MSCAN identifier mask (CANIDMR) registers. In other words, a set bit ignores the compare, while a cleared bit forces the compare. On reception, the incoming message ID must pass through the filter to be accepted. Accepted messages will enter the receive background buffer (RxBG) and rejected ones are discarded. The following example shows how typical acceptance code and mask is set up for incoming message filtering:

**Example 13.1. Acceptance Code and Mask Definitions**

/* Acceptance Code definition */

```
#define ACC_CODE_ID100 0x2000
#define ACC_CODE_ID100_HIGH      ((ACC_CODE_ID100&0xFF00)>>8)
#define ACC_CODE_ID100_LOW       (ACC_CODE_ID100&0x00FF)
```

/* Mask Code Definitions */
```
#define MASK_CODE_ST_ID 0x0007
#define MASK_CODE_ST_ID_HIGH     ((MASK_CODE_ST_ID&0xFF00)>>8)
#define MASK_CODE_ST_ID_LOW      (MASK_CODE_ST_ID&0xFF)
```

For standard identifiers the lowest three bits of the mask register should be set to 1.

Out of reset, all three transmit message buffers are empty and the TXE[2:0] flag bits in the CANTFLG are set by default to indicate empty status. Once the MSCAN is enabled by the software by setting the CAN enable control (CANE) bit in the MSCAN control (CANCTL1) register 1, the transmitter will generate an interrupt request to the CPU if the transmitter empty interrupt enable (TXEIE) bits are set in the CANTIER register.

Most CAN messages are data frames, and to a lesser extent remote frames. The bus protocol distinguishes between the two frame types by allocating a bit in the MB, referred to as remote transmission request (RTR). A value of 1 in this bit position indicates the current MB has a remote frame instead of a data frame.

A remote frame has no data associated with it, whereas a data frame may contain a minimum of zero bytes and a maximum of eight data bytes. The number of data bytes contained in a message is defined by the 4 bit length control field. On the transmitter side, the software writes the length field into the MB to indicate the length in bytes of the message to be transmitted. On the receiver side, the CAN hardware writes the length field to indicate the number of bytes received in the message. The value of the length field is normally set between binary 0000 and 1000 for message lengths between zero (no data) and a maximum of eight bytes. Any value greater than eight will default to a length of eight bytes.

## MSCAN Clock Source

The MSCAN module clock can be configured to use the crystal oscillator clock or may use the bus clock. Figure 13.10 shows clock selection for the ColdFire system as well as the CAN bus.

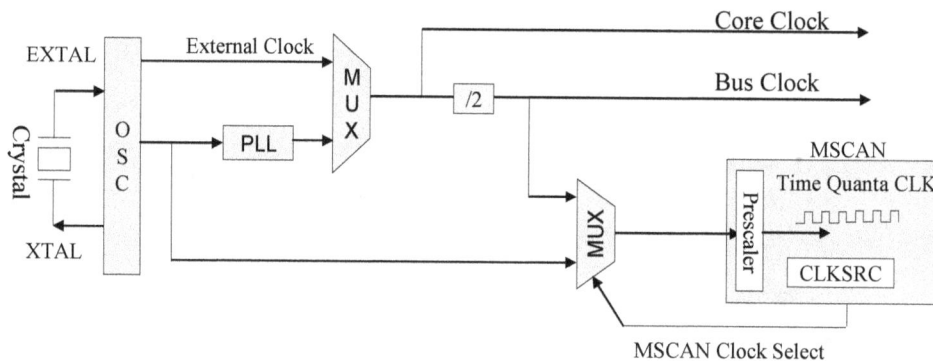

FIGURE 13.10. MSCAN Clock Selection Logic

The clock selection choice is controlled by the clock source (CLKSRC) bit in the MSCAN control register 1 (CANCTRL1). Selecting the crystal oscillator clock eliminates any timing errors (jitter) that may occur if the PLL is operating in frequency **dither** mode. Once the source is selected, the CAN clock passes through an 6 bit prescaler to generate the serial bus clock, referred to as Time Quanta Clock (tq). The MSCAN prescaler can be set to divide the selected frequency from 1 to 64.

## Bit Timing

The MSCAN module provides a flexible clock scheme to meet the nominal bit rate timing required by the CAN bus. The nominal bit rate is defined as "number of bits per second without resynchronization by an ideal transmitter". In other words, the bit time (referred to as nominal bit time) determines transmission baud rate as per the following formula:

$$\text{The Nominal Bit Time} = 1/\text{Nominal Bit Rate} \qquad \text{(EQ 1)}$$

The CAN bus specification requires any bus loading or round trip delay be compensated to minimize synchronization issues. For this reason, a nominal bit time (NBT) is composed of four non-overlapping time segments. Each of the time segments are defined in terms of time quanta (tq). tq is the period of s_clk shown in the figure above. In the MSCAN implementation, the propagation segment (PROP_SEG) and phase_segment 1 (PHASE_SEG1) are represented by TSEG1, and phase_segment 2 (PHASE_SEG2) is represented by TSEG2. A brief explanation of the four time segments are stated in Table 13.1 shown below. Bus values are read at the sample point at the end of TSEG1 to determine the bit

**TABLE 13.1 Nominal Bit Time Composition**

| Segment | Segment Length | Network Function |
|---------|----------------|------------------|
| SYNC_SEG | Always 1 tq | Synchronizes all CAN nodes on the bus. An edge is expected to lie within this segment. |
| PROP_SEG | Represented by TSEG 1, Programmable from 4 to 16 tq | Compensates for bus loading and round trip delays |
| PHASE_SEG1 | Represented by TSEG 1, Programmable from 4 to 16 tq | Compensates for oscillator drifts and positive phase difference between transmitting and receiving nodes |
| PHASE_SEG2 | Represented by TSEG 2, Programmable from 2 to 8 tq | Compensates for oscillator drifts and negative phase difference between transmitting and receiving nodes |

level, as shown in nominal bit timing and sample point in Figure 13.11. Due to bus loading, clock oscillator drifts, and round-trip delays, the sampling point may be skewed and could be interpreted incorrectly. To ensure that the sample point is always taken at the correct time, TSEG1 and TSEG2 are automatically lengthened or shortened depending on the input bit edge phase error. An input edge is considered synchronous if it occurs inside SYNC_SEG. If the edge occurs before SYNC_SEG, the phase error is negative, otherwise it is positive.

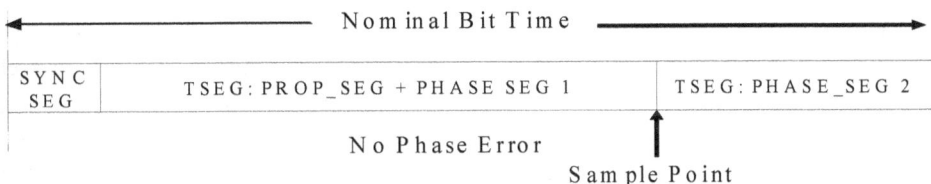

**FIGURE 13.11. Nominal Bit Timing and Sample Points**

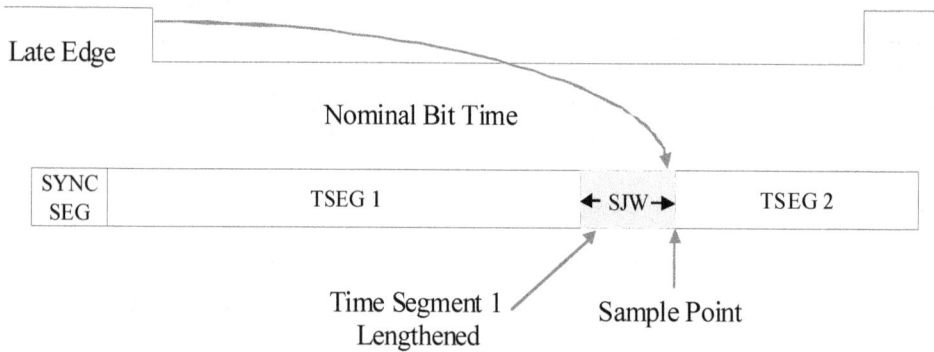

**FIGURE 13.12. Positive Phase Error (TSEG1 Lengthened)**

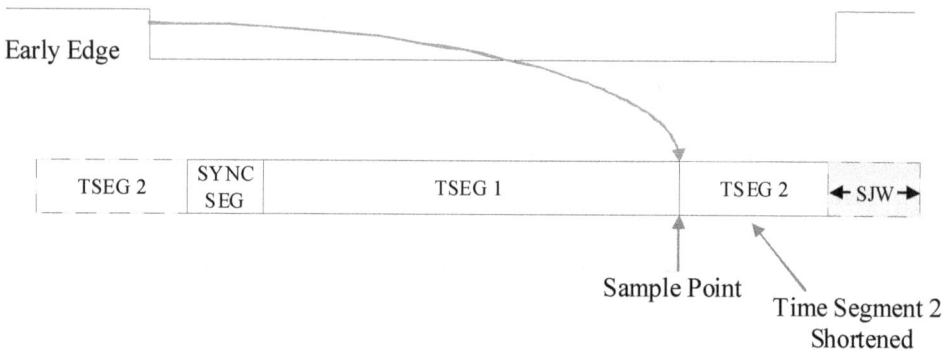

**FIGURE 13.13. Negative Phase Error (T_SEG2 Shortened)**

Figure 13.12 shows the case where the bit edge occurs late, causing TSEG1 to be lengthen so that the sample point is correctly distanced from the edge. In Figure 13.13, the bit edge occurs early, before SYNC_SEG, causing TSEG2 for the previous bit to be shortened, and SYNC_SEG to be omitted so that the sample point is again correctly distanced. Adjustments to the lengths of TSEG1 and TSEG2 are temporary and are restored to their nominally programmed value for the next bit.

The amount by which TSEG1 is lengthened and TSEG2 is shortened is determined by the resynchronization jump width (RJW), which is used to compensate for phase differences between transmitting and receiving nodes. TSEG1 has programmable value from 2 to 16 tq, TSEG2 have a programmable value from 2 to 8 tq, whereas SJW has programmable value from 1 to 4 tq. Software can select the required number of time quanta for each time segment in the MSCAN control register (CANCTL0). The following formula is used to calculate the nominal bit time in time quanta:

$$\text{Bit Time} = \text{SYNC\_SEG} + \text{TSEG1} + \text{TSEG2} \qquad \textbf{(EQ 2)}$$

$$\text{Sample Point} = 1 + \text{TSEG1} \qquad \textbf{(EQ 3)}$$

SJW is usually set to a value equal to or less than TSEG2

## CAN Modes

The MSCAN module has four functional modes: normal mode, freeze mode, listen-only mode and loop-back mode. Additionally, it has one low power mode.

### Normal Mode

This is the operational mode in which the MSCAN module receives and transmits message frames and error frames with all the CAN Protocol functions are enabled. The CANE bit must be set in the MSCAN control register 1 (CANCTL1) to enter this mode.

### Listen Only Mode

In this mode, the MSCAN operates as an error passive node and freezes its receive and transmit error counters. Since some of the ColdFire family members contain 2 MSCAN modules, the listen-only mode provides a method of increasing the effective number of receive buffers and filters seen on a single CAN bus. This is achieved by connecting together the receiver inputs from multiple MSCAN modules. In this mode, one of the MSCAN modules will be set up normally while the other module or modules should be set up in listen-only mode. Received messages with a matching ID are filtered into the buffers[1], but will not transmit any acknowledgments of received messages.

The MSCAN module enters this mode when the listen-only mode (LISTEN) bit in the Control Register (CANCTRL) is set to a logic 1.

### Loop Back Mode

This mode is entered by setting the loop back (LOOPB) bit in the MSCAN Control Register 1 (CANCTRL1). In this mode, MSCAN performs an internal loop back that can be used for self testing. The bit stream output of the transmitter is internally fed back to the

---

1. Messages not acknowledged by other nodes will not be received by a listen-only Mode (LISTEN) node.

receiver input. The RXCAN input pad is ignored and the TXCAN output goes to the recessive state (logic 1). MSCAN behaves as it normally does when transmitting, and treats its own transmitted messages as if they were received from a remote node. As messages are transmitted or received, the MSCAN will generate interrupts to the CPU as it does in normal mode. In loop back mode, the pad normally assigned to the transmitter output may be used as GPIO or the alternate function if available.

### Wait Mode

The WAIT instruction may be executed to put the CPU and selected I/O devices in a low power consumption stand-by mode. Each on-chip I/O peripheral has a wait mode control bit to place it into the wait mode when set, or simply ignore low power mode entry. For example, the MSCAN has CAN stop wait control (CSWAI) bit in the MSCAN control register 0 (CANCTL0). When set, the MSCAN clocks are stopped when the CPU execute the wait instruction. But if the CSWAI bit is cleared, the MSCAN continues to operate in normal mode and generate interrupts while the CPU is in wait mode. The MSCAN can also operate in any of the low-power modes depending on the values of the SLPRQ/ SLPAK and CSWAI bits.

### Stop Mode

In this mode, the MSCAN clocks are shut down when the CPU execute stop1 or stop 2 instruction. In stop3 mode, power down or sleep modes are determined by the SLPRQ/SLPAK values set prior to entering stop3.

### Sleep Mode

Often, the application software may need to shut down the MSCAN and possibly other on-chip modules to conserve power when there are no activities in the system. The CPU can request the MSCAN to enter sleep mode by asserting the sleep request (SLPRQ) bit in the CANCTL0 register. If the MSCAN is neither transmitting nor receiving, it immediately goes into sleep mode. The MSCAN will continue to operate normally and not enter the low power mode if there are one or more message buffers scheduled for transmission. Likewise, if a message is currently being received, the MSCAN continues to receive the message before it enter the sleep mode.

As the bus becomes idle, the MSCAN will enter the sleep mode.

Once the MSCAN enter sleep mode, it indicates that condition to the CPU by setting the sleep acknowledge (SLPAK) status bit in the CANCTL1 register.

## Power Down Mode

This low power mode is entered when the CPU executes either wait or stop instruction while the CSWAI control bit is set in bit in the MSCAN control register 0 (CANCTL0). On entry, the module immediately shuts down the clocks to the CAN protocol interface and message buffer management. Any message transmission or message reception in progress will be aborted. Before entering the power down mode, It is strongly recommended to enter the sleep mode first. This guarantees that all messages in progress will be transmitted/received correctly. In power down mode, the MSCAN control registers are no longer accessible to the CPU.

The MSCAN can recover from stop or wait via the wake-up interrupt. This interrupt can only occur if the MSCAN was in sleep mode (SLPRQ = 1 and SLPAK = 1) before entering power down mode, the wake-up option is enabled (WUPE = 1), and the wake-up interrupt is enabled (WUPIE = 1).

## Initialization Mode

Sometime it may become necessary to stop the MSCAN immediately and reset it in order to be able to reconfigure it in case of synchronization or other unusual errors. Upon entry, transmission and reception is aborted and the MSCAN CANCTL0, CANRFLG, CANRIER, CANTFLG, CANTIER, CANTARQ, CANTAAK, and CANTBSEL registers will reset to their default values. In addition, the MSCAN enables the configuration of the CANBTR0, CANBTR1 bit timing registers; CANIDAC; and the CANIDAR, CANIDMR message filters.

## Interrupts

The MSCAN supports up to four unique interrupt vectors. Three interrupt requests are originated in the transmitter and share the same vector, one interrupt request from the receiver and has its own vector, two error requests share one vector and finally, wake-up interrupt request which has its own vector. Table 14 depicts a summary of all requests and their associate interrupt enable and status.

**TABLE 14. MSCAN Interrupt Sources**

| Interrupt Source | Local Interrupt Enable | Status Register [Flags] |
|---|---|---|
| Transmit Interrupt | CANTIER [WUPIE] | CANTFLG (TXE[2:0] |
| Receive Interrupt | CANRIER [RXIFE] | CANRFLG [RXF] |
| Wake-up Interrupt (WUPIF) | CANRIER [WUPIE] | CANRFLG [WUPIF] |
| CAN Status Change & Overrun | CANRIER [CSCIE & OVRIE] | CANRFLG [OVRIF & CSCIF] |

## MSCAN Transmit Sequence

This section describes the steps needed for message transmission and reception. The steps assume that the MSCAN control registers have been initialized to meet the application requirements and that the PAR_UART in the GPI/O module have been initialized to route the MSCAN signals to their assigned pads.

To transmit a message, software should execute the following steps:

1. Request initialization mode by setting the INITRQ in the CANCTL0 register.
2. Wait for initialization acknowledge by polling the INITAK bit in CANCTL1.
3. Write CANCTR1 to select the following options:
    a. enable the MSCAN by setting the CANE bit,
    b. select CAN clock source, crystal or bus clock
    c. select loopback mode, if needed,
    d. select listen mode, if needed,
    e. select bus-off recovery mode, and
    f. select wake-up recovery mode
4. Write CANBTR0 to configure the CAN bus timing and synchronization jump width.
5. Write CANBTR1 to configure bit sample mode and time segments (TSEG1 &TSEG2).
6. Read CANTFL register and write value to CANTBSEL to select the first available TX buffer.
7. Since step 6 select the lowest number available TX buffer, write message ID in the TX buffer IDR0-3 registers.
8. Write the message data into the data segment registers (DSR0-7).
9. Write the data length control (DLC) field based on the number of bytes in this message.
10. Select TX buffer priority by writing the PRIO register.
11. Enable transmitter interrupt by setting TXEIE[2-0] in CANTIER.

Once the MB is ready for transmission, the MSCAN hardware executes the following steps:

1. Arbitrates for the bus and if arbitration is successful, the message is transferred.
2. After the contents of the TX buffer is transmitted, the value of the free running timer is written into the time stamp field in the TSRH/L registers, if stamping is enabled.
3. Based on which TX buffer was transmitted, its associate Exits status flag will set.
4. If enabled, the MSCAN signals an interrupt request to the CPU.

When interrupted, the CPU executes the following steps in the interrupt handler:

1. reads the CANTFLG register to see which TX buffer was transmitted.
2. writes the CANTFLG register value read into the CANBSEL register if there is a need to retransmit another message.
3. writes a logic '1' to the appropriate TXE flag to clear the interrupt request.

## MSCAN Receive Sequence

To begin a message reception sequence, assuming steps 1, 2 and 3 in the transmit sequence above are already performed, the application software should execute the following steps:

1. write IDAM[1:0] into the CANIDAC to select 32, 16 or 8-bit ID acceptance filter mode.
2. write ID acceptance CANIDAR[0-7] registers to filter in desired incoming messages.
3. write ID Mask CANIDMR[7-0] registers to filter multiple incoming messages having similar IDs.
4. Clear all receiver interrupt flags by writing logic '1' to them in the CANRFLG register.
5. If desired, write the CANRIER register to enable the appropriate receiver interrupts.
6. The receive message buffer now is ready to accept incoming messages with matching IDs.

The MSCAN executes the following steps upon reception of an error free message:

1. transfers the received message from the serial message buffer to the receiver FIFO.
2. updates the time stamp field from the free running timer, and the data length field.
3. sets the status flag in the RXF in the CANRFLG register.
4. signals a request to the CPU if the receiver interrupt is enabled.

When interrupted, the CPU executes the following steps in interrupt handler:

1. reads data length control (DLC) field to determine the number of bytes received.
2. reads message in the RxFG buffer and store it to memory.
3. read the time stamp from TSRH/L if needed.
4. write the RXF with a logic '1' to clear the request and to free up the RxFG buffer.

The example below on the next five pages shows detailed C code example that configure the MSCAN module to transmit and receive a message in loopback mode.

```c
/*
 * Copyright (c) 2008, Freescale Semiconductor
 *
 * File name    : main.c
 *
 *
 * Description : a brief description of the module.
 */

#include <hidef.h>              /* for EnableInterrupts macro */
#include <MCF51xx.h>            /* include peripheral declarations */

#pragma LINK_INFO DERIVATIVE "MCF51xx"

/* ID Definition */
#define ST_ID_100 0x20000000    /* Standard Id 0x100 formatted to be loaded  */
                                /* in IDRx Registers in Tx Buffer */

/* Acceptance Code Definitions */
#define ACC_CODE_ID100 0x2000
#define ACC_CODE_ID100_HIGH ((ACC_CODE_ID100&0xFF00)>>8)
#define ACC_CODE_ID100_LOW  (ACC_CODE_ID100&0x00FF)


/* Mask Code Definitions */
#define MASK_CODE_ST_ID 0x0007
#define MASK_CODE_ST_ID_HIGH ((MASK_CODE_ST_ID&0xFF00)>>8)
#define MASK_CODE_ST_ID_LOW  (MASK_CODE_ST_ID&0xFF)

/* Error Flags Definition */
#define NO_ERR 0x00
#define ERR_BUFFER_FULL 0x80

/* Functions Prototypes */
void CANInit(void);
unsigned char CAN0SendFrame(unsigned long id, unsigned char priority, unsigned
char length, unsigned char *txdata );
void Delay(void);

void main () {

    unsigned char errorflag = NO_ERR;
    unsigned char txbuff[] = "ABCDEFGH";

    CANInit();               /* Initialize MSCAN12 Module */

    while(!(CAN0CTL0&0x10)); /* Wait for Synchronization */

    CAN0RFLG = 0xC3;         /* Reset Receiver Flags
                              *
                              *  0b11000011
                              *     |||||||||__ Receive Buffer Full Flag
                              *     |||||||| ___ Overrun Interrupt Flag
                              *     ||||||| ____
                              *     ||||| _____>- Transmitter Status Bits
                              *     |||| _____
                              *     |||_____>- Receiver Status Bits
                              *     ||_____ CAN Status Change Interrupt Flag
                              *     |_____ Wake-Up Interrupt Flag
```

```
CAN0RIER = 0x01;              /* Enable Receive Buffer Full Interrupt
                         *
                         *   0b00000001
                         *   ||||||||__ Receive Buffer Full Int enabled
                         *   |||||||___ Overrun Int disabled
                         *   ||||||____
                         *   |||||_____>- Tx Status Change disabled
                         *   ||||_____
                         *   |||_____>- Rx Status Change disabled
                         *   ||_____ Status Change Int disabled
                         *   |_____ Wake-Up Int disabled
                         */


    EnableInterrupts;

    for (;;) {
        errorflag = CAN0SendFrame((ST_ID_100), 0x00, sizeof(txbuff)-1, txbuff);
        Delay();
                             }

}

/*
 * CANInit: a description of the function functionName.
 *
 * Bit Timing Definitions
 * ----------------------
 *
 * CAN Clock Source (External oscillator) = 16Mhz
 * BitRate = 125Khz
 * Total Time Quanta = 16
 * Sincronization Jump Width = 4 Time Quanta
 * 1 sample
 * Sample point at 75% of Bit Timing
 *
 * CAN_BRP = ((CAN Clock Source)/fTq) - 1
 * fTq = (Bit Rate) * (Total Time Quanta)
 * Total Time Quanta = (SYNCH_SEG+(TSEG1+1)+(TSEG2+1))
 * Total Time Quanta =      1    +    11    +    4    = 16 Time Quanta
 * fTq = 125Khz * 16 Time Quanta = 2Mhz
 * CAN_BRP = (16Mhz/2Mhz) - 1 = 8 - 1 = 7
 * TSEG1 = 10
 * TSEG2 = 3
 * SJW = (Synchronization Jump Width-1) = 3
 *
 * Another line of the description.
 *
 * Parameters: None
 *Return :    None
```

```
void CANInit(void) {

    CAN0CTL0 = 0x01;            /* Enter Initialization Mode
                                *
                                *  0b00000001
                                *      ||||||||__ Enter Initialization Mode
                                *      |||||||___ Sleep Mode Request bit
                                *      ||||||____ Wake-Up disabled
                                *      |||||_____ Time stamping disabled
                                *      ||||_____ Synchronized Status
                                *      |||_____ CAN not affected by Wait
                                *      ||_____ Receiver Active Status bit
                                *      |_____ Received Frame Flag bit
                                */

    while (!(CAN0CTL1&0x01)){}; /* Wait for Initialization Mode acknowledge
                                * INITRQ bit = 1
                                */

    CAN0CTL1 = 0xA0;            /* Enable MSCAN module and LoopBack Mode
                                *
                                *  0b10100000
                                *      ||||||||__ Initialization Mode Acknowledge
                                *      |||||||___ Sleep Mode Acknowledge
                                *      ||||||____ Wake-up low-pass filter disabled
                                *      |||||_____ Unimplemented
                                *      ||||_____ Listen Only Mode disabled
                                *      |||_____ Loop Back Mode enabled
                                *      ||_____ Ext Osc/Xtal as Clock Source
                                *      |_____ MSCAN Module enabled
                                */

    CAN0BTR0 = 0xC7;            /* Synch Jump = 3 Tq clock Cycles
                                *
                                *  0b11000111
                                *      ||||||||__
                                *      |||||||___\
                                *      ||||||____ |
                                *      |||||_____ |_ CAN Clock Prescaler = 7
                                *      ||||_____ |
                                *      |||_____ |
                                *      ||_____/
                                *      |_____>- SJW = 3
                                */

    CAN0BTR1 = 0x3A;            /* Set Number of samples per bit, TSEG1 and TSEG2
                                *
                                *  0b00111010
                                *      ||||||||__
                                *      |||||||__|
                                *      ||||||___|- TSEG1 = 10
                                *      |||||____|
                                *      ||||_____
                                *      |||_____ TSEG2 = 3
                                *      ||_____/
                                *      |_____ One sample per bit
                                */
```

```
CAN0IDAC = 0x10;            /* Set four 16-bit Filters
                             *
                             *   0b00010000
                             *   ||||||||
                             *   ||||||||___
                             *   |||||||___\_ Filter Hit Indicator
                             *   ||||||___/
                             *   |||||_____ Unimplemented
                             *   ||||
                             *   |||_____>- Four 16-bit Acceptance Filters
                             *   ||
                             *   |_____>- Unimplemented
                             */

    /* Acceptance Filters */
    CAN0IDAR0 = ACC_CODE_ID100_HIGH;    //|\   16 bit Filter 0
    CAN0IDMR0 = MASK_CODE_ST_ID_HIGH;   //| \__ Accepts Standard Data Frame Msg
    CAN0IDAR1 = ACC_CODE_ID100_LOW;     //| /   with ID 0x100
    CAN0IDMR1 = MASK_CODE_ST_ID_LOW;    //|/

    CAN0IDAC = 0x10;                    /* Set four 16-bit Filters */

    CAN0IDAR2 = 0x00;                   //|\   16 bit Filter 1
    CAN0IDMR2 = MASK_CODE_ST_ID_HIGH;   //| \__ Accepts Standard Data Frame Msg
    CAN0IDAR3 = 0x00;                   //| /   with ID 0x000
    CAN0IDMR3 = MASK_CODE_ST_ID_LOW;    //|/

    CAN0IDAR4 = 0x00;                   //|\   16 bit Filter 2
    CAN0IDMR4 = MASK_CODE_ST_ID_HIGH;   //| \__ Accepts Standard Data Frame Msg
    CAN0IDAR5 = 0x00;                   //| /   with ID 0x000
    CAN0IDMR5 = MASK_CODE_ST_ID_LOW;    //|/

    CAN0IDAR6 = 0x00;                   //|\   16 bit Filter 3
    CAN0IDMR6 = MASK_CODE_ST_ID_HIGH;   //| \__ Accepts Standard Data Frame Msg
    CAN0IDAR7 = 0x00;                   //| /   with ID 0x000
    CAN0IDMR7 = MASK_CODE_ST_ID_LOW;    //|/

    CAN0CTL0 = 0x00;            /* Exit Initialization Mode Request */
    while ((CAN0CTL1&0x01) != 0){};/* Wait for Normal Mode */

}


/*
 * functionName: a description of the function functionName.
 * Another line of the description.
 *
 * Parameters: param1 - description
 * param2 - description
 *
 * Return : description of the value returned by functionName
 */
unsigned char CAN0SendFrame(unsigned long id, unsigned char priority, unsigned
char length, unsigned char *txdata ){

    unsigned char index;
    unsigned char txbuffer = {0};

    if (!CAN0TFLG)             /* Is Transmit Buffer full?? */
        return ERR_BUFFER_FULL;
```

```
CAN0TBSEL = CAN0TFLG;        /* Select lowest empty buffer */
    txbuffer = CAN0TBSEL;        /* Backup selected buffer */

    /* Load Id to IDR Registers.*/
    *((unsigned long *) ((unsigned long)(&CAN0TXIDR0)))= id;

    for (index=0;index<length;index++) {
        *(&CAN0TXDSR0 + index) = txdata[index];  /* Load data to Tx buffer
                                                  * Data Segment Registers
    }

    CAN0TXDLR = length;                          /* Set Data Length Code */
    CAN0TXTBPR = priority;                       /* Set Priority */
    CAN0TFLG = txbuffer;                           /* Start transmission */

    while ( (CAN0TFLG & txbuffer) != txbuffer);  /* Wait for Transmission
                                                  /* completion */
    return NO_ERR;
}

 * functionName: a description of the function functionName.
 * Another line of the description.
 *
 * Parameters: param1 - description
 * param2 - description
 *
 * Return : description of the value returned by functionName
 */
void Delay (void) {

    unsigned int counter;
    for (counter=0;counter<10000;counter++);
}

 * ModuleISR: a description of the function functionName.
 * Another line of the description.
 *
 * Parameters: None
 * MSCAN0 RECEIVE ISR
 * DESCRIPTION:
 * Interrupt asserted when a message has been received and shifted into
 * the foreground buffer of the receiver FIFO.
 * Return : None
 */
#pragma CODE_SEG NON_BANKED

void interrupt CAN0RxISR(void) {
    unsigned char length, index;
    unsigned char rxdata[8];

    length = (CAN0RXDLR & 0x0F);
for (index=0; index<length; index++)
    rxdata[index] = *(&CAN0RXDSR0 + index); /* Get received data */

    CAN0RFLG = 0x01;   /* Clear RXF */
}

#pragma CODE_SEG DEFAULT
```

# CHAPTER 14

*Universal Serial Bus*

## Introduction to the Universal Serial Bus Modules

The Universal Serial Bus (USB) is an industry standard that extends the architecture of the personal computer to external devices such as keyboards, cameras, printers and disk drives. The USB protocol allows a wide range of peripherals (called devices) to be attached at the same time to a PC (called the host).

Actually, the USB isn't really a "bus" in the strictest sense of the meaning - it is a point-to-point communication path between a single host and a single device. (In contrast, a true electrical bus allows multiple devices to be directly attached to the same physical wires). Communication between the host and multiple devices in a USB topology is achieved by using one or more hubs which are themselves treated as devices, as shown in Figure 14.1.

FIGURE 14.1. **USB Topology**

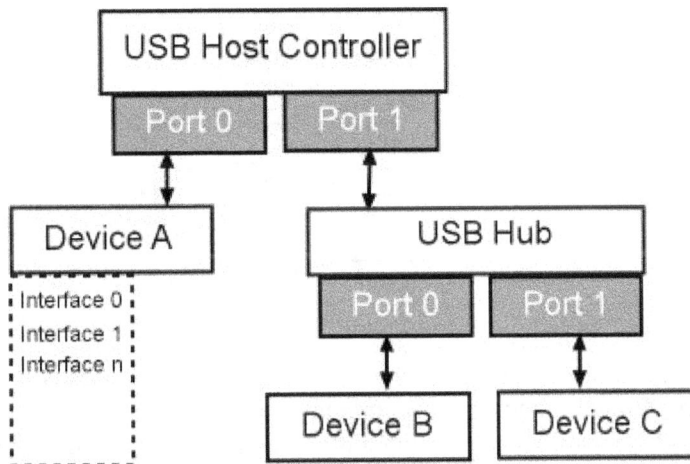

Data transfers on the USB are always initiated by the host, which communicates with one device at a time. The host sends data in packets that it schedules to occur within a time period called a frame (1 ms intervals for low and full speed USB) or microframe (125us intervals for high speed USB). The start of each frame or microframe is indicated by a start of frame (SOF) packet, as shown in Figure 14.2 and is followed by various other packets (labeled P1 to P5) over the duration of the frame or microframe. Note that the

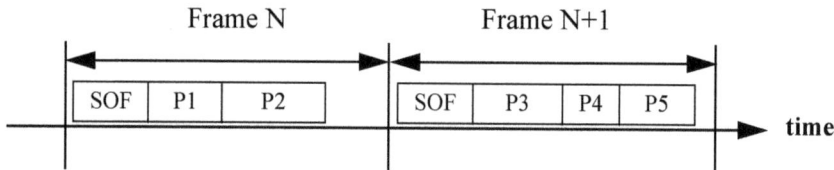

FIGURE 14.2. **USB Frame (or Micro Frame)**

USB is bi-directional and data packets can travel in either direction within any frame (or microframe), depending on the transaction type initially defined by the host.

Each packet is limited to transferring a certain amount of data. If the total amount of data that must be transferred is too large to fit in one packet, the data is distributed in multiple packets, sent sequentially. Moreover, if the total number of packets can't fit in one frame, they are distributed across multiple frames. Packet scheduling is handled in the host controller hardware. The expected behavior of the device, its latency and bus bandwidth requirements are also specified by the transaction type that the host defines.

The logical communication path between the host and a device is called a pipe. The pipe starts at the host software API and ends at a connection called, rather appropriately, the endpoint (EP) in the device, as shown in Figure 14.3. A pipe can be either a stream pipe

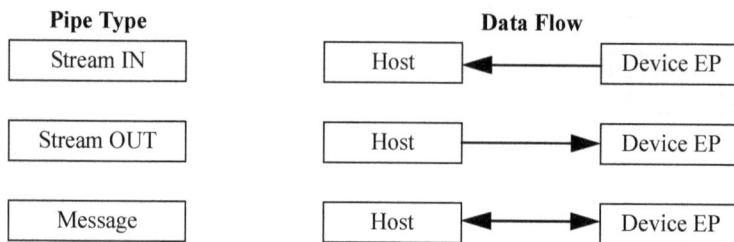

FIGURE 14.3. **USB pipe types**

used to transfer data in any format in one direction, or a message pipe that transfers USB specified format of data in both directions. The data transferred in a stream pipe can be supplied by either the host or the device and is often used to transfer large amounts of data.

A message pipe is generally used to set up the device after power up and to get the status of the device while it is operational.

The USB On The Go (OTG) specification that has been added to USB 2.0 defines a Host Negotiation Protocol (HNP) and a Session Request Protocol (SRP) that allows a single point-to-point connection between only two "devices", which negotiate for host control. More information about OTG is given later in this chapter.

USB devices must support hot insertion and removal, and at the same time notify the host of their operating speed. Moreover, data transfer direction can be dynamically changed after each packet is transmitted. These requirements plus others are met through the physical layer design, which is described next.

In essence the USB physical layer, which consists of a cable with 4 wires (or 5 wires for OTG), could be classified as an asynchronous, differential signalling, bi-directional, half duplex transmission medium.

Data is transferred on two of the wires (D+ and D-) using differential signalling and a Non-Return-to-Zero-Inverted (NRZI) bit protocol shown in Figure 14.4. Data bits are transmitted LSB first. By using NRZI no clock is required to recover the transmitted data.

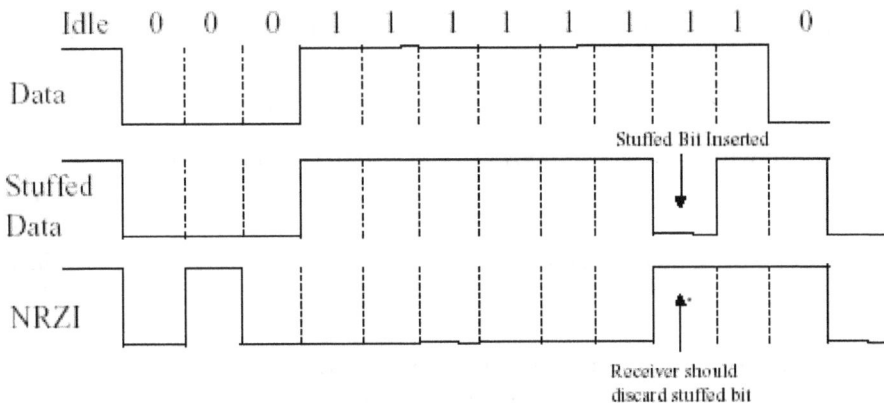

FIGURE 14.4. USB Physical Layer showing bit stuffing

A binary '1' is indicated by no change in the signal level during the expected bit time, while a binary '0' is indicated by a change in the signal level - either from a '0' to a '1' or vice-versa. However, to prevent loss of synchronization between the transmitter and receiver, the USB specification imposes a limit on the number of consecutive '1's that may be indicated. After six consecutive '1's, a transition must be inserted - or stuffed - into the transmitted data stream. This 'stuffed' bit is discarded by the data receiver. The figure shows an example of a data transmission with a single stuffed bit. The trace labelled "Data" shows the raw data pattern being transferred. The "Stuffed Data" trace is the same pattern with an additional "0" bit stuffed after six consecutive "1"s. The "NRZI" trace is the logic level represented on the USB bus.

Data signalling is also bi-directional, with the direction being chosen by the host during the initiation of a bus transaction. As well as transferring data, the D+ and D- wires are used to signal special conditions. These are shown in Table 14.1.

**TABLE 14.1 Physical bus Signalling**

| Signal Type[a] | D+ | D- | Notes |
|---|---|---|---|
| Differential '0' | low | high | Low speed J-state, Full/High speed K-state |
| Differential '1' | high | low | Low speed K-state, Full/High speed J-state |
| SE0 | low | low | Single Ended zero |
| Reset | low | low | For more than 10ms |
| Suspend | high | low | For more than 3ms |
| Resume (Full Speed) | low | high | K-state, for more than 30ms |
| Resume (Low Speed) | high | low | K-state, for more than 30ms |

a. The meanings of Reset, Suspend, Resume, EOP, J and K are described in the USB specification

The remaining two wires of a 4-wire USB bus are Vbus (+5v) and GND to deliver power to the device.

The fifth wire of a 5-wire bus is used to indicate which device is initially the host in an OTG connection.

All transactions on the USB bus are initiated by the host which transmits packets of information to a single device at a time. Prior to transferring data, a host must detect the presence of a device on the bus, and determine its transfer speed. There are three different speed specification for the USB. These are shown in Table 14.2.

TABLE 14.2

| Speed | Typical Application | Attribute | Specification |
|---|---|---|---|
| Low Speed | Keyboard, Mouse, Joystick | Lowest cost. | Low Speed USB 1.1 1.5Mb/s |
| Medium Speed | Printers, Scanners, Audio | Low cost, guaranteed latency and band-width | Full Speed USB 1.1 12Mb/s |
| High Speed | Video, Disk Drive | High bandwidth, guaranteed latency | High Speed USB 2.0 500Mb/s |

The device indicates its presence on the bus and its operating bus speed by pulling either the D+ or D- line to 3.3V as shown in Figure 14.5.

FIGURE 14.5. **Low and Full Speed Detection**

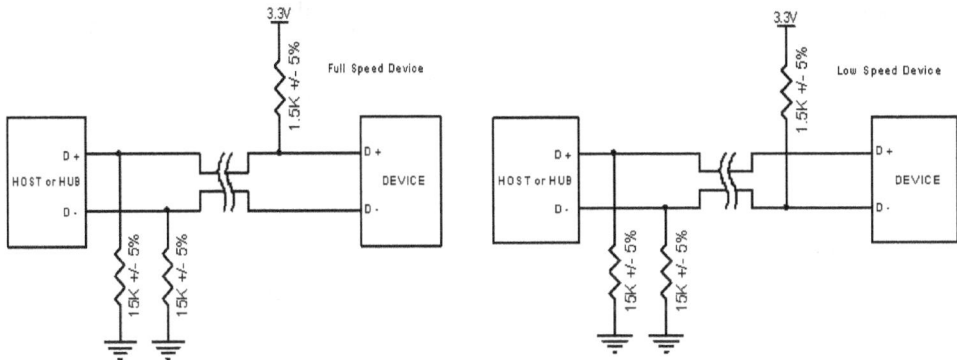

If the pull-up resistor is not present, this indicates to the host that a device is not connected, since at the host end, the D+ and D- lines are pulled down to ground. Because power may be supplied from the host through the connector, once the physical connector is attached, it is important that the host does not try to communicate with the device before it is ready. For example, in the case that the device has an embedded microcontroller that must initialize its USB interface on power up, it must prevent the host from communicating with it until initialization is complete. This can be achieved by making the pull up resistor switchable, so that on initial power up, it is not present on the D+ or D- line and the host therefore is unaware that a physical connection has been made. The device firmware then connects the resistor only after it has initialized its USB interface. Note that, under firmware control, a device that has switchable resistors may automatically disconnect from the bus without unplugging the physical connector.

High speed mode is not set or detected by hardware connections. A high speed device must start up in Full Speed mode and then switch to High Speed mode through a USB protocol.

As has been stated earlier, all transactions on the USB bus are initiated by the host which transmits a packet of information. This means the USB bus is a polled bus, and devices attached to it cannot provide unsolicited responses. Moreover, a device must respond to any request from the host within the turnaround time given in the USB 2.0 specification.

**Low and Full Speed Packet Signalling**

Each packet in a transaction is delimited by a Start of Packet (SOP) and an End of Packet (EOP) bit level. Because the idle state is effectively a J state for Low and Full Speed USB, the SOP is always a K state. The EOP is signaled by an SE0 state for two bit times followed by a J state for one bit time. The bus transceiver then places the D+ and D- physical drivers in a high impedance state. Bus termination resistors hold the bus in the idle state. Figure 14.6 shows the SOP and EOP bus signalling for a Full Speed USB.

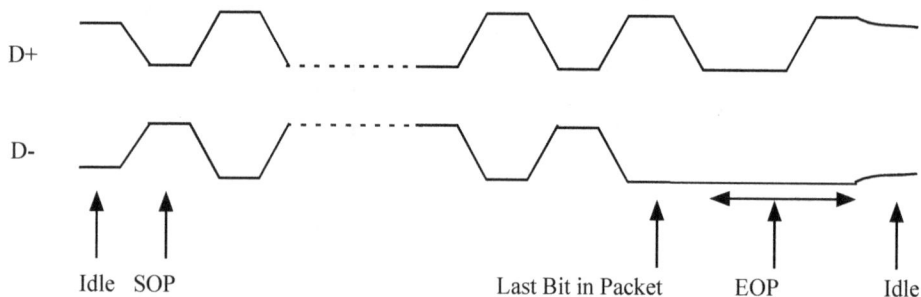

FIGURE 14.6. Full Speed SOP and EOP signalling

## High Speed Packet Signalling

For High Speed USB, the idle state is signalled by driving both D+ and D- to GND. The SOP is a K state. The High Speed EOP is more complex and described in detail in the USB 2.0 specification. To summarize that specification, a High Speed EOP is indicated by generating a bit stuff error in the receiver. For high-speed packets other than SOF's, the transmitted EOP delimiter is a byte value of 01111111 (without bit stuffing). For example, if the last signal state prior to the EOP field is a J, this would lead to an EOP of eight consecutive Ks, where the first K represents the leading 0 and the remaining Ks represent the remaining seven 1s. For high-speed SOF's, the transmitted EOP delimiter is five bytes consisting of a leading 0 followed by 39 consecutive 1s. Thus if the last bit prior to the EOP field is a J, forty consecutive K's would be transmitted, at the end of which the lines return to the high-speed Idle state.

## Packet Identifiers

There are many different forms of packets, and they are scheduled by the host controller to be sent within each frame or microframe. A complete USB transaction between the host and a device consists of three or four packets. Each frame consists first of a Start of Frame (SOF) packet, followed by several transactions, with some transactions being repeated over multiple frames. The host maintains a parameter that defines the latest point in a frame that it can send a transaction without overlapping the start of the next frame. The parameter is called the SOF threshold register. Each transaction targets a particular device endpoint. An endpoint is a uniquely addressable part of a device that transmits (sources) data to the host or receives (sinks) data from the host. The generic format of a packet is shown in Figure 14.7.

| Field Type | Sync | PID | Packet Specific Fields | CRC |
|---|---|---|---|---|
| Number of Bits | 8 | 8 | 0 to 1023 | 0, 5 or 16 |

FIGURE 14.7. Generic Format of a Packet

Each packet consists of the following fields:

- An initial Synchronization (Sync) field where the first bit is the start of packet (SOP) state, as described earlier. The purpose of the Sync field is to provide the maximum density of edges over a short time, to allow the receiver to re-synchronize its timing to the incoming edges. Low and Full Speed hosts transmit a Sync field of 3 K-J state pairs followed by two K states for a total of 8 symbols. Figure 14.8 shows the Full Speed signalling for a Sync field. A High Speed host transmits a Sync field of 15 K-J states fol-

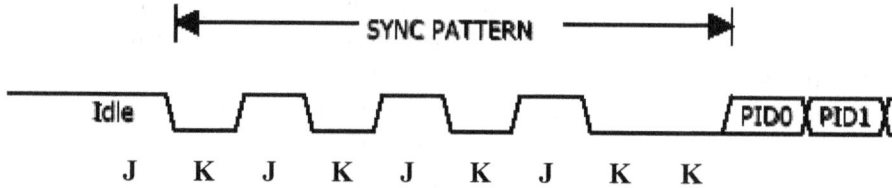

FIGURE 14.8. Sync Field for Full Speed USB

lowed by 2 K states, for a total of 32 symbols.

- A Packet Identifier, or PID indicating the content of the packet. A PID consists of a 4 bit packet type followed by a 4 bit check field of one's complement of the packet type. Figure 14.9 shows the format, while Table 14.3 lists the different packet types.

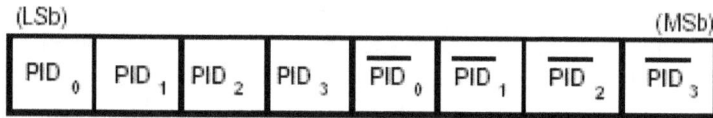

FIGURE 14.9. PID Format

- Packet specific data or control information if appropriate. These fields consist of various combinations of addresses, endpoint number, frame number and byte data.
- A CRC field, whose size depends on the token type. (Token types are described next). Handshake packet types do not have a CRC field. Token packets have a 5 bit CRC and Data packets have a 16 bit CRC.
- An End of Packet delimiter (EOP), as described earlier.

Packet specific data is determined by the Packet ID (PID), which can be one of the following types:

- Token
- Data
- Handshake
- Special

Table 14.3 shows the different types of PIDs and a description of their function.

**TABLE 14.3 PID Types**

| PID Type | PID Name | PID Code[a] | Description |
|---|---|---|---|
| Token | OUT | 0001 | Address + endpoint number in host-to-function transaction |
| | IN | 1001 | Address + endpoint number in function-to-host transaction |
| | SOF | 0101 | Start-of-Frame marker and frame number |
| | SETUP | 1101 | Address + endpoint number in host-to-function transaction for SETUP to a control pipe |
| Data | DATA0 | 0011 | Data packet PID even |
| | DATA1 | 1011 | Data packet PID odd |
| | DATA2 | 0111 | Data packet PID high-speed, high bandwidth isochronous transaction in a microframe |
| | MDATA | 1111 | Data packet PID high-speed for split and high bandwidth isochronous transactions |
| Hand-shake | ACK | 0010 | Receiver accepts error-free data packet |
| | NAK | 1010 | Receiving device cannot accept data or transmitting device cannot send data |
| | STALL | 1110 | Endpoint is halted or a control pipe request is not supported |
| | NYET | 0110 | No response yet from receiver |
| Special | PRE | 1100 | (Token) Host-issued preamble. Enables downstream bus traffic to low-speed devices. |
| | ERR | 1100 | (Handshake) Split Transaction Error Handshake (reuses PRE value) |
| | SPLIT | 1000 | (Token) High-speed Split Transaction Token |
| | PING | 0100 | (Token) High-speed flow control probe for a bulk/control endpoint |
| | Reserved | 0000 | Reserved PID |

a. The four bit PID codes are shown with the least significant bit on the right, which is transmitted first on the bus.

Figure 14.10 shows examples of generic forms of some packet types.

| Token | Sync | PID | Frame | CRC |
|---|---|---|---|---|
| SOF | 8 bits | 01011010 | 11 bits | 5 bits |

| Data | Sync | PID | Data | CRC |
|---|---|---|---|---|
| Data0 | 8 bits | 00111100 | 0 to 1023 | 16 bits |

| Handshake | Sync | PID |
|---|---|---|
| ACK | 8 bits | 00101101 |

**FIGURE 14.10. Examples of Token, Data, and Handshake Packet**

As mentioned earlier, a pipe is a logical abstraction representing the overall connection between host software and the endpoint on the device. The pipe is composed of data structures in the host, the host controller driver and hardware, the data layer consisting of packets of encapsulated data, the physical layer of bit signalling, the device hardware controller and driver, and the data structures stored in the device memory. The host controller schedules data transfers by interrogating a linked list of pointers that locate transfer descriptors and transfers packets of data to an endpoint on the device or requests packets of data from an endpoint on the device.

The first packet sent is a token packet describing:

- the type of the transaction
- the direction of the transaction
- the address of the device
- the endpoint number of the device

Once this packet is decoded by the device, the new data packet is usually transmitted by either the host or the device based on the direction requested by the host in the token packet. In some cases the receiver of the data transaction acknowledges the transfer by sending a handshake packet which is either an ACK, indicating the data was accepted, a NAK, indicating that the data was not accepted, or a STALL, which signals that the endpoint is stalled.

The pipes can operate in one of four modes, specified by the transaction type for the endpoint:

**control transfer.** A control transfer is used to configure a device when it is first attached. Afterwards, software can choose to use control transfers in implementation-specific ways. Data transfer is retried in the event of an error.

**isochronous transfer.** An isochronous transfer occurs continuously and periodically in a unidirectional stream pipe. Typically it contains data that must be processed in real time, such as an audio or video stream. Isochronous transfers are allocated a dedicated percentage of the USB bandwidth to ensure that data can be delivered at the desired rate. The USB is also designed for minimal delay (latency) of isochronous data transfers. To avoid compromising the real time behavior of isochronous pipes, there is no handshaking, so data retries are not performed and there is no guarantee of delivery. Errors are detected using CRC. Isochronous transfers are available in Full and High Speed modes only, with a maximum data payload size of 1024 bytes. Care should be taken when assigning device endpoints to an isochronous pipe. Depending on bus bandwidth allocation, the host may be unable to allocate the optimum bandwidth, and alternative endpoints with varying payload rates should be defined in the design of the device interface.

**interrupt transfer.** Interrupt transfers are typically non-periodic small transfers used to service events that occur in a device and that require a defined response time. A good example is a USB mouse. Note that a USB interrupt request is not like a typical microprocessor interrupt that often causes an immediate response. In a USB system, an Interrupt transaction is buffered in the device until the host polls the device asking for the data. Maximum data payload sizes depend on the USB speed:

- 8 bytes for Low Speed.
- 64 bytes for Full Speed.
- 1024 bytes for High Speed.

**bulk transfer.** A bulk transfer is typically used to transfer large amounts of non-real time data, for example to a printer or to/from a disk drive. Bulk transfers use the bandwidth remaining after all other allocations have been made. While data transfer rate and latency are not guaranteed, delivery is, through CRC, handshaking and retries. Full Speed USB supports packet sizes of 8, 16, 32 or 64 bytes. High Speed USB supports a packet size of up to 512 bytes.

A number of USB tools are available to monitor USB traffic. The diagrams in Figure 14.11 show part of a typical control transfer used to enumerate a device. The diagrams show the setup transaction consisting of the setup, data and ACK packets. Note that the setup and data are output from the host and the ACK is an input to the host. Also, the

| Packet | Dir | Sync | Setup | ADDR | EP[a] | CRC5 | EOP |
|--------|-----|------|-------|------|-----|------|-----|
| 1 | O | 00000001 | 0xB4 | 0 | 0 | 0x08 | 3.00 |

| Packet | Dir | Sync | DATA0 | Data | CRC16 | EOP |
|--------|-----|------|-------|------|-------|-----|
| 2 | O | 00000001 | 0xC3 | 80 06 00 01 00 00 40 00 | 0xBB29 | 3.00 |

| Packet | Dir | Sync | ACK | EOP |
|--------|-----|------|-----|-----|
| 3 | I | 00000001 | 0x4B | 2.80 |

a. EP is the Endpoint address

FIGURE 14.11. Typical Setup Transfer

token values are shown least significant bit first (the bit order is reversed compared with the values given in Table 14.3).

At least 10 per cent of every frame is reserved for use by control transfers. This proportion can be increased by the system software if performance is found to be suffering through control packets being unduly delayed. The maximum continuous throughput over USB must therefore be less than 90 per cent of the signalling rate.

Part or all of the remaining time in each frame can be reserved by pipes serving isochronous devices. The actual portion allocated to each pipe is pre-negotiated when the pipe is set up. This ensures that a specific amount of data can be transferred every millisecond. Any bandwidth remaining is available for other types of transfer.

Isochronous devices must buffer data one frame's worth at a time, and send each block over the bus as a single transaction. At the receiving end the data is unbuffered and restored to real time. For example, an audio device operating at a CD-quality 44.1KHz sampling rate would send nine frames with 44 samples per frame, followed by one frame with 45 samples. After buffering at the source and unbuffering at the destination there will be a delay of a couple of milliseconds in delivering the data, but the rate of delivery will be preserved.

Interrupt transfers are also to an extent time critical. When a pipe is created for an interrupt endpoint, a desired bus access period of between 1 and 255ms (10 and 255ms in the case of low speed devices) is specified. The system software polls the interrupt endpoint at an interval which ensures that if an interrupt transaction is pending it is dealt with within the desired time-frame.

The behaviors of the device controller for interrupt and bulk endpoints are identical. All valid IN and OUT transactions to bulk pipes will handshake with a NAK unless the endpoint had been configured. Once the endpoint has been configured, data delivery will commence.

### Reset

The host controller issues a bus reset to initialize devices. When a bus reset is detected, the USB device will renegotiate its attachment speed, reset the device address to 0, and if enabled, issue an interrupt to the device driver software. After a reset is received, all endpoints except endpoint 0 are disabled and any existing transactions will be cancelled by the device controller.

### Suspend

All devices support the suspend state, which is entered when a constant idle signal is present for more than 3.0 ms. In the suspended state, the device must draw only the suspend current. A full or high speed host can prevent a device entering the suspended state by simply issuing a SOF token with no further bus traffic, every frame or microframe. Note that while suspended, a device must continue to provide power to its D+ (Full/High Speed) or D- (Low Speed) pull-up resistor to provide correct connectivity status.

### Resume

A device exits from its suspended state when it detects any non-idle signalling on its D- an D+ pins. Moreover, devices that support the optional remote wakeup capability can signal the host to resume normal bus operation. Resume signalling for a low/full speed bus is different to high speed, but in essence they both involve applying a non-idle signal (K-state) for at least 20ms, followed by a terminating signal. Devices that support remote wakeup capability must also allow the host to enable and disable the capability.

### Error Handling

Considerable error checking and error handling features have been built in to the USB to ensure that it is a reliable method of connecting peripherals to a PC.

Physical layer noise immunity is provided by differential signalling and shielded cabling. When errors do occur, cyclic redundancy checks (CRCs) performed separately on both the control and data fields of packets provide recovery from bit errors. Unrecoverable errors are also detected with a high degree of confidence.

A self-recovery mechanism is built into the messaging protocol, with timeouts for lost and invalid packets. Some error recovery is built into the hardware. The host controller will retry a failed transaction three times before reporting an error to the client software. How a reported error is dealt with is the responsibility of the software layer.

The receiver of interrupt and bulk data transfers responds with a handshake packet to confirm that data was received, or request a retransmission. Delivery of this type of data is therefore guaranteed, even if the time taken to deliver it is not.

Both the host and peripheral contain data toggle bits, one per endpoint. The toggle bits determine which of two data PIDs (DATA0 or DATA1) should be used for data transfer. When the peripheral comes out of reset, both the host and device reset their internal data toggle bits to zero. The first data packet is, therefore, sent using the DATA0 PID.

When an error-free data packet is transferred (signaled when the transmitter receives an ACK), both sides complement their data toggle values. The second data packet for the endpoint is then sent using the DATA1 PID. As successful transfers progress, the data packets use alternating (or toggling) DATA0 and DATA1 PIDs.

For test purposes only, the USB HS DR module described later in this chapter has a data toggle inhibit bit which causes it to ignore the data toggle pattern and accept all incoming data packets regardless of the data toggle state.

With isochronous data it is not possible to retry a failed transaction. Since only one slot is allocated to the pipe during each frame, resending the data would delay transmission of the succeeding data samples, upsetting the timeliness of the data delivery. As a result no handshake packet is sent and the data must be accepted as-is.

## USB On The Go (OTG)

The USB On-The-Go specification defines new mini connectors, negotiation rules and power requirements used in both the host and the device. The specification also defines a 'dual-role device' which can act as either a USB host or USB device depending on how the cable is inserted into the new mini-AB connector. When the OTG device is connected to a mini-A plug, it becomes a host. If instead it is connected to a mini-B plug it becomes a device. The dual role device determines the difference between the insertion of an A or B connector by monitoring an additional fifth connector pin which supplies either GND or +5V.

If both devices are USB On-The-Go devices, they will automatically negotiate so that one will act as a host and the other will act as a device, regardless of the connection type. This behavior is defined as the Host Negotiation Protocol (HNP).

It is worth summarizing here the functions that a host provides but are excluded from a device:

- 15k pull-down resistors on D+ and D-
- A means to supply, rather than draw, power on VBUS
- Send SOF (Start of Frame) packets
- Send SETUP, IN, and OUT packets
- Schedule transfers within USB 1 msec frames
- Signal USB reset
- USB power management

As well as requiring a dual role host/device USB controller, OTG requires additional hardware circuits to support HNP and Session Request Protocol (SRP). Once connected, OTG dual-role devices can exchange roles using HNP.

### Host Negotiation Protocol

Two OTG enabled devices use the Host Negotiation Protocol (HNP) to transfer host control from the default host A-device to the default B-device. The default operation is set by the level of the ID signal that is input to the OTG device ID pin from the fifth connection on the mini-AB USB socket (remember that a standard USB connector has only 4 signals). On a mini-A plug the ID signal is tied to GND, so when the plug is inserted it tells the OTG device to become a host. On a mini-B plug, the ID signal is either not connected or connected to GND through a resistor. In this case, the signal should be treated as an ID input of logic "1" telling the device to be a USB device.

After the default connection is established when the USB cable is attached, HNP enabled devices can participate in the protocol to swap roles. The A-device initiates the protocol. The following steps execute HNP:

- A-device sends a SetFeature command to B-device to indicate the start of the exchange. If the B-device doesn't support HNP, it will STALL this command.
- A-device idles the bus (by not sending any more transactions).
- B-device detects that the bus is idle for more than the prescribed time. If the B-device is in High Speed mode, it first switches to Full Speed. The B-device then turns off its D+ pull-up so that the bus can discharge to the SE0 state.

- The A-device recognizes the SE0 as a request from the B-device to become the host. The A-device turns on its D+ pull-up within the prescribed time.

- After detecting the D+ pull-up for a minimum of the prescribed time (to allow for residual leakage effects) the B-device knows that the A-device has relinquished host control. The B-device then issues a bus reset within the prescribed time, and assumes the role of bus host.

- To release host control, the B-device simply idles the bus and turns on its D+ pull-up.

- When the A-device detects the idle state for more than the specified time, it turns off its D+ pull-up. (The A-device can alternatively turn off Vbus to end the session.)

- After detecting the D+ pull-up for a minimum of the prescribed time (to allow for residual leakage effects) the A-device knows that the B-device has relinquished host control. The A-device then issues a bus reset within the prescribed time, and restores its role as bus host.

Low speed devices do not support HNP.

**Session Request Protocol**

An OTG session is defined as the duration that the A-device supplies VBUS power. The A-device always supplies VBUS power, even if HNP caused it to switch roles. The A-device can end a session by turning off VBUS to conserve power, a very important requirement in a battery powered device such as a cell phone. SRP allows a B-device to request an A-device to turn on VBUS power and start a session. A B-device must not attempt SRP until it is sure the A-device has detected the end of the previous session. The OTG specification allows the B-device to discharge the Vbus through a current limiting resistor to speed up the detection process. Additionally, the B-device must detect that an SE0 is present on the bus for at least the minimum prescribed time.

Any OTG host can request a descriptor from a B-device describing whether or not it can handle SRP and/or HNP. Any B-device that supports either HNP or SRP must respond to this request.

SRP is performed by the B-device using both the "data-line pulsing" and "VBUS pulsing" methods. The OTG host needs only support one of these methods. In data-line pulsing the B-device must turn on its D+ pull up resistor for a period of 5 to 10mS.

If an OTG complaint B-device is connected to a standard host such as on a personal computer, it is important that Vbus pulsing will not cause damage to a powered-down computer. This can be ensured by limiting the duration of the pulse used to turn on Vbus at the B-device side. The non-OTG host controller Vbus capacitor is specified to be at least 120uF compared with a maximum of 6.5uF for an OTG host/device. As a result, the rise time of Vbus during SRP is sufficiently slow enough to prevent damage to the powered-down non-OTG host device. Conversely, when two OTG host/devices are connected, the rise time is fast enough for the A-device to become powered and respond to the B-device's request.

### A-Device Session

An A-device session is the duration that the Vbus is above the specified A-device Vbus threshold. The A-device uses this threshold, combined with a power-up time duration to detect whether the B-device is drawing too much current at which point the session is terminated. The A-device has a second, lower threshold level that it uses to detect SRP Vbus pulsing. Additionally, when Vbus falls below this lower threshold, it indicates an A-device session end.

### B-Device Session

A B-device session is the duration that Vbus is above the specified B-device session valid threshold. A second lower voltage threshold defines the end of the B-device session, at which point the B-device is able to initiate an SRP.

## *USB OTG Module*

The OTG module supports USB 2.0 Full and Low Speed host and device operation, and with additional hardware, the OTG supplement to the USB 2.0 specification. A block diagram of this module is shown in Figure 14.12. The module contains a DMA controller, receive and transmit FIFOs, a Serial Interface Engine (SIE) and integrated transceiver. The SIE logic performs the low level operations such as bit stuffing, CRC generation and checking, and error reporting. The main role of the SIE is to perform the bi-directional processing of the serial signals on the physical bus and the bytes accessed by the Cold-Fire® core. The DMA controller moves USB data between the FIFOs in the module and buffers in memory. Note that there are two variants of the USB OTG module. One makes use of general purpose system memory while the other has memory integrated into the USB module itself. The FIFOs decouple the system memory transfers from the extremely tight timing required by USB and compensate for the latencies in processing data transfer requests.

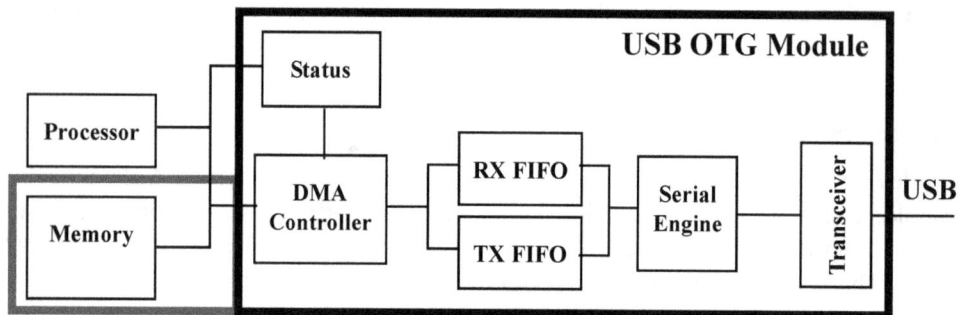

FIGURE 14.12.  USB module

When operating as a host the OTG module uses a single 16 byte transmit buffer and 16 byte receive buffer. When operating as a device it uses a 16 byte receive buffer and 16 byte transmit buffer for each enabled endpoint.

The OTG module can support host and device modes with little external hardware since the module contains an integrated transceiver. OTG operation requires the addition of an external interface that can control and supply current on Vbus, as well as sense the levels of Vbus to signal the state of both A and B sessions. The external interface may be connected through the 2 wire I²C serial bus interface on the ColdFire® device. A block diagram of a minimal interface is shown in Figure 14.13.

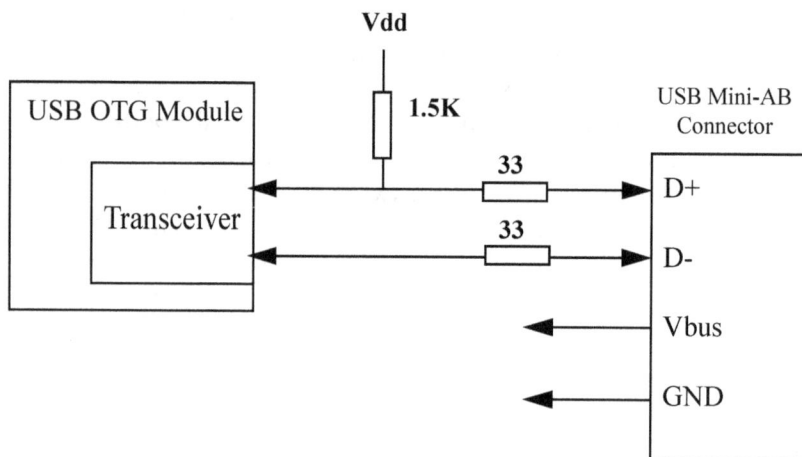

FIGURE 14.13.  Minimum Component count for a Full Speed USB device

To ensure driver output impedance, eye diagram, and Vbus cable fault tolerance requirements are met, each of the D- and D+ signals should have a 33 ohm resistor placed in series between its device pin and the USB connector. To maintain signal integrity, the resistors should be placed as close as possible to the device pins.

Internal, weak, pull-down resistors are included which keep the D+ and D- ports in a known quiescent state when the USB port is disabled, or when a USB cable is not connected.

To support host operation, 15K ohm external resistors should be connected from the D+ and D- pins to GND. For full speed device operation a 1.5K ohm pull-up resistor should be connected between D+ and the USB_PULLUP pin on the device. This allows the device driver software to disable the pull-up to cause a disconnection from the bus.

All registers in the OTG module are implemented as 8 bit registers, but are spaced in the memory map by 32 bits. Certain registers and some bits in certain registers have dual purposes, and provide different functionality depending on whether the module is operating as a host or a device.

**Host Operation**

Host operation is supported through a buffer descriptor table (BDT) pointer, control, address and status registers, and optional interrupts from level changes on certain input pins, and from a 1ms timer. The two variants of USB OTG modules have slightly different BDT implementations.

In the first variant, the BDT addressing is hardwired into the module. The BDT occupies the first portion of the USB RAM as shown in Figure 14.16. There are a total of ten, 3 byte buffer descriptors (BDs) in the table. Endpoints 0, 5 and 6 each have 2 descriptors, while endpoints 1, 2, 3 and 4 each have one descriptor.

**FIGURE 14.14. USB RAM Organization**

| USB RAM Address | USB RAM Contents |
|:---:|:---:|
| 0x00 | Endpoint 0, IN |
| 0x03 | Endpoint 0, OUT |
| 0x06 | Endpoint 1 |
| 0x09 | Endpoint 2 |
| 0x0C | Endpoint 3 |

**FIGURE 14.14. USB RAM Organization**

| USB RAM Address | USB RAM Contents |
|---|---|
| 0x0F | Endpoint 4 |
| 0x12 | Endpoint 5, Buffer EVEN |
| 0x15 | Endpoint 5, Buffer ODD |
| 0x18 | Endpoint 6, Buffer EVEN |
| 0x1B | Endpoint 6, Buffer EVEN |
| 0x1E-0x1F | Reserved |
| 0x20 - 0xFF | Endpoint Buffer Space |

The BD organization is shown in Figure 14.15.

| Offset | 7 | 6 | 5 | 4 | 3 | 2 | 1 | 0 |
|---|---|---|---|---|---|---|---|---|
| 0x00 | OWN | DATA0/1 | BDTKPID[3] | BDTKPID[2] | DTS/BDTKPID[1] | BDT_STALL/ BDTKPID[0] | 0 | 0 |
| 0x01 | BC[7:0] | | | | | | | |
| 0x02 | EPADR[9:2] | | | | | | | |

**FIGURE 14.15. Buffer Descriptor Attributes (variant 1)**

The bit descriptions are provide below

- OWN

  This bit determines who currently owns the buffer. The USB SIE generally writes a 0 to this bit when it has completed a token. The USB module ignores all other fields in the BD when OWN=0. Once the BD has been assigned to the USB module (OWN=1), core software should not change it in any way. This byte of the BD should always be the last byte that the core updates when it initializes a BD. Although the hardware will not block the core from accessing the BD while owned by the USB SIE, doing so may cause undefined behavior and is generally not recommended.

- DATA0/1

  This bit defines if a DATA0 field (DATA0/1=0) or a DATA1 (DATA0/1=1) field was transmitted or received. It is unchanged by the USB module.

- DTS

  When set, this bit enables data toggle synchronization.

- BDTSTALL

  Setting this bit will cause the USB module to issue a STALL handshake if a token is received by the SIE that would use the BDT in this location. The BDT is not consumed by the SIE (the OWN bit remains and the rest of the BD is unchanged) when the BDT-STALL bit is set.

- BC[7:0]

  The Byte Count bits represent the 8-bit byte count. The USB module serial interface engine (SIE) will change this field upon the completion of a RX transfer with the byte count of the data received. Note that while USB supports packets as large as 1023 bytes for isochronous endpoints, this module limits packet size to 64 bytes.

- EPADR[9:2]

  The endpoint address bits represent the upper 8 bits of the 10-bit buffer address within the module's local USB RAM. Bits [1:0] of EPADR are always zero, therefore the address of the buffer must always start on a four-byte aligned address within the local RAM. These bits are unchanged by the USB module. This is NOT the address of the memory on the system bus. EPADR is relative to the start of the local USB RAM.

- BDTKPID[3:0]

  The current token PID is written back to the BD by the USB module when a transfer completes. The values written back are the token PID values from the USB specification: 0x1 for an OUT token, 0x9 for and IN token or 0xd for a SETUP token.

In the second variant, the buffer pointer is segmented across three 8 bit registers, BDT_PAGE_01 to BDT_PAGE_03. In host mode, the values stored in these three registers are concatenated to form the address in system memory of the Buffer Descriptor Table (BDT) that may contain one or more Buffer Descriptors (BD) that define the location of the payload and its attributes. A graphical representation of the pointer and descriptor layout is shown in Figure 14.16. Note that the Buffer Descriptor format is 32 bits wide.

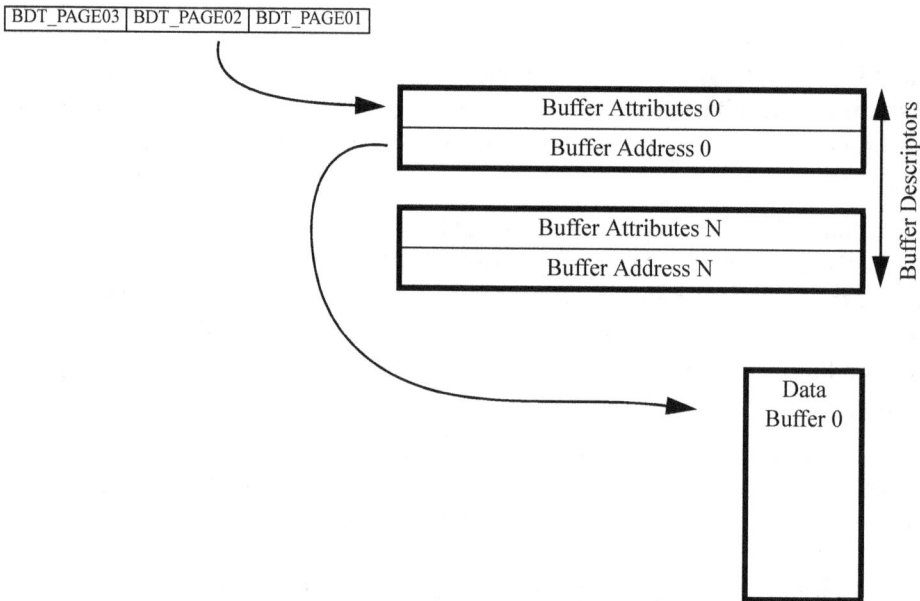

FIGURE 14.16. **Buffer Descriptor Table Organization**

The attributes associated with each BD are shown in Figure 14.17.

| 31:26 | 25:16 | 15:8 | 7 | 6 | 5 | 4 | 3 | 2 | 1 | 0 |
|-------|-------|------|-----|--------|------------------|------------------|------------------|-----------------------|---|---|
| - | BC | - | OWN | DATA0/1 | KEEP/ TOK_PID[3] | NINC/ TOK_PID[2] | DTS/ TOK_PID[1] | BDT_STALL/ TOK_PID[0] | 0 | 0 |

FIGURE 14.17. **Buffer Descriptor Attributes (variant 2)**

The purpose of each bit is as follows:

**BC**

A 10 bit field containing the count of bytes stored in the Data Buffer. The SIE updates this field.

**OWN**

This bit is a semaphore indicating whether the SIE or the ColdFire® core has exclusive access to the BD. If OWN is set, the OTG module has exclusive access, otherwise the core has exclusive access. When the SIE finishes processing a token, it clears this bit, unless KEEP is set. When the core initializes the BD, it should clear this value last to allow the SIE to process the BD and perform the Data Buffer reads and writes necessary for a USB transaction.

### DATA0/1
This bit indicates whether a DATA0 or DATA1 packet was transferred.

### KEEP
If KEEP is set and the OWN bit is set, the OTG module will retain permanent, exclusive access to the BD. KEEP should be cleared so that the OTG module can release the BD when a token has been processed. Typically this bit is set for isochronous endpoints that are feeding a FIFO. In this case the NINC is normally also set to inhibit address increment. If KEEP is set, the OTG module will not change it.

### NINC
When set, the DMA engine will not increment its address when accessing the Data Buffer. This is useful for endpoints that access data at a single memory location such as a FIFO. If KEEP is set, the OTG module will not change NINC

### DTS
If DTS is set, the OTG module will perform data toggle synchronization.
If KEEP is set, the OTG module will not change DTS.

### BDT_STALL
Device software may set this bit, which will then cause the OTG module to issue a STALL handshake if it receives a token that should be processed by this BD. The BD contents and its OWN bit are unchanged in this situation. If KEEP is set to 1, BDT_STALL is not changed by the OTG module, otherwise bit 0 of the current token is written back.

### TOK_PID[3:0]
TOK_PID[3:0] are shared with the previously described bits, KEEP, NINC, DTS and BDT_STALL. The OTG module updates these 4 bits with the value of the current PID once the USB transfer completes. In host mode this bit field is updated with the last returned PID or a transfer status indicator with the following meanings:
0x3 - DATA0
0xB - DATA1
0x2 - ACK
0xE - STALL
0xA - NAK

0x0 - Bus Timeout
0xF - Data Error

In addition to configuring the BD, host driver software may update the following registers:

- OTG_CTRL to enable data line pull down resistors. This is the default after reset.
- CTL, to enable the entire OTG module, enable host mode, issue RESET and RESUME signalling, clear all ping-pong buffer control bits (described later), report whether SIE has processed last written token, detect speed of attached device.
- ADDR, to specify the device address transmitted with the token packet, and whether Low Speed or Full Speed mode should be used.
- TOKEN, to specify the token PID and endpoint address. The SIE transmits a packet when host software writes to this register.
- SOF_THLD, to specify the deadline for starting a new transaction on the USB. The worst case condition (i.e. earliest time for the deadline) is normally for an IN token followed by a data packet from the target followed by the host response. Typical values for this register are:
  74 for 64 byte data packets
  42 for 32 byte data packets
  26 for 16 byte data packets
  8 for 8 byte data packets
- ENDPT0, to configure the handshake, retry and low speed characteristics of the transaction. Typical values for this register depend on the pipe transaction type:
  0x4D for Control, Bulk and Interrupt
  0x4C for Isochronous
- INT_ENB, to enable interrupts on Stall, Attach events, Remote Wakeup (Resume) request from a B-device, and when SOF threshold is reached.
- ERR_ENB, to enable interrupt on the occurrence of OTG module errors.

Status information that may be accessed in host mode is available from the following registers:

- INT_STAT, to detect Stall, Attach events, Remote Wakeup (Resume) request from a B-device, and when the SOF threshold is reached.
- ERR_STAT, to detect Data Buffer overflow and end of frame errors as a result of a transaction still being in progress across a frame boundary.
- CTL, to check when a new token may be written, and detect the live state of the bus signals.

## Device Operation

Device operation is supported through a similar set of resources as the host. The notable difference is that all 16 endpoint control registers are available. This means the buffer descriptor table is generally much larger. The values stored in the three BDT_PAGE0X registers are now concatenated with the endpoint number, endpoint direction bit and additional 1 bit flag, to form the address in system memory of the Buffer Descriptor Table (BDT) that corresponds to the endpoint.

The additional concatenation of the endpoint attributes to the buffer descriptor table pointer reduces the OTG module's latency in processing transactions. The details of the relationship between the endpoint attributes and the pointer address is shown in Figure 14.18. The layout of the descriptors is the same as shown in Figure 14.16 and Figure 14.17.

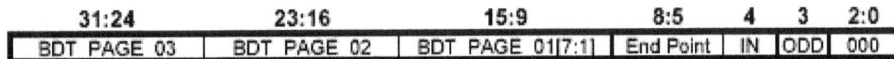

| 31:24 | 23:16 | 15:9 | 8:5 | 4 | 3 | 2:0 |
|---|---|---|---|---|---|---|
| BDT_PAGE_03 | BDT_PAGE_02 | BDT_PAGE_01[7:1] | End Point | IN | ODD | 000 |

FIGURE 14.18. Device mode Buffer Descriptor Table Pointer Format

In device mode, the buffers and descriptors must be managed more carefully since the OTG module will use and update their contents asynchronously to the driver software executing on the core. Two mechanisms are implemented that allow optimal coherent accesses of resources shared between the core and the OTG module. Firstly, the SIE toggles bit 3 in the buffer descriptor page pointer address, so that the buffer descriptor it selected is "ping-ponged" between two entries in the table. So, for any given endpoint, this allows the SIE to work on one buffer while the core processes the other. Additionally, each BD has an OWN bit that is used to reserve the buffer for either the core or the SIE.

In addition to configuring the BDs for each endpoint necessary to implement the application embedded in the ColdFire® microcontroller, device software must also access a number of control and status registers. The control register set is similar to host mode, but in general, different bits, and in some cases additional registers are accessed:

- CTL, to enable the OTG module, set device mode, clear suspended transfer which occurred when a Setup Token was received.
- INT_ENB, to enable interrupt on a Reset, Error, SOF token, Token ready, Idle detect (Sleep), Remote Wakeup (Resume) request from A-device (host) and Stall event.
- ERR_ENB, to enable interrupts on various error conditions in the received Packet PID, bus turnaround time, OTG module DMA transfer error and bit stuff error.

- ADDR, to set the device's USB bus address. The host supplies the value during enumeration.

- ENDPT0-15, to set up the pipes established between the host and the device, and define whether handshaking should be enabled, (it should only be disabled for isochronous endpoints) various error conditions in the received Packet PID, bus turnaround time, OTG module DMA transfer error and bit stuff error. Note that while all these conditions have their own enable bit, they are wired-or to the same interrupt signal INT_STAT[ERROR]. Each endpoint that is enabled selects a separate buffer descriptor location in the buffer descriptor table shown in Figure 14.16. After device software detects a bus reset it should enable only endpoint 0, to allow host enumeration to take place.

The status registers that the embedded device software accesses are:

- INT_STAT, to detect the presence of a Reset, Error, SOF token, Token ready, Idle detect (Sleep), Remote Wakeup (Resume) request from A-device (host), a Stall event, and all errors generated by events enabled in the ERR_STAT register.

- ERR_STAT, to detect various error conditions in the received Packet PID, bus turnaround time, OTG module DMA transfer error and bit stuff error. Note that while all these conditions have their own enable bit, they are wired-or to the same interrupt signal INT_STAT[ERROR].

- STAT, to detect the transaction status. This register should be read when a Token is ready, indicated by a set INT_STAT[TOK_DNE]. This register contains the endpoint address of the previously transferred token, combined with its direction and ping-pong bit value. This is sufficient for the device driver software to identify which buffer descriptor and data buffer were updated by the last USB transaction.

### OTG Operation

The OTG behavior of this module is specifically related to the physical bus signalling and protocol of SRP and HNP. Once SRP, and optionally HNP are completed, the device will operate in either host mode or device mode. The registers that need to be accessed to perform SRP are:

- OTG_STAT, to detect changes in the A- and B-device session voltages, changes in the ID signal indicating A- or B-device connector attachment and the state of the physical bus. Note that the bits in this register are updated simply when transitions occur on the external pins USB_SESSVLD, USB_SESSEND, USB_VBUSVLD and USB_ID.

- OTG_INT_EN, to enable interrupts on the occurrence of events reported in OTG_STAT.

- OTG_CTRL, to control Vbus power, charge and discharge control pins, and the D+, and D- pull-up and pull-down control pins.

# CHAPTER 15

*Fast Ethernet Controller*

## *Introduction to the Fast Ethernet Controller*

The Fast Ethernet Controller (FEC) is a module on a number of ColdFire® devices that implements 10M and 100M bits per second ethernet protocols. The FEC is an Ethernet Media Access Controller (MAC) implemented with a combination of hardware and micro-code.

The FEC communicates with an external PHY through a single industry standard Media Independent Interface (MII), and communicates with the core through a set of buffers that would typically reside in system SRAM, plus control and status registers and counters in the FEC memory map. The FEC transfers data to and from the buffers using its integrated and dedicated DMA controller. The FEC is clocked by the on-chip system clock.

The FEC handles the CSMA/CD (Carrier Sense Multiple Access with Collision Detection) protocol for transmission and reception of frames in accordance with IEEE 802.3 standards. It performs frame data encapsulation and de-encapsulation, frame transmission, and frame reception.

### FEC interfaces

The FEC provides 2 different physical interconnections to an external ethernet PHY transceiver:

- A 7 wire connection consisting of the subset of signals shown in Table 15.1 (TxData, TxCLK, TxEN, RxData, RxCLK, COL, & CRS). It uses single serial receive and transmit data lines clocked at 10MHz to provide a 10Mbps transfer rate. This connection is provided to support older PHY devices and is sometimes referred to as the "AMD mode", GPSI (General Purpose Serial Interface) mode or SNI (Serial Network Interface) mode.

- An 18 wire connection consisting of the signals shown in Table 15.1. This connection has 4 serial transmit and 4 serial receive data lines. The data is clocked at 2.5MHz for 10Mbs operation, and at 25MHz for 100Mbs operation. This connection is called the MII, and supports the IEEE 802.3 standard.

The 7 wire connection uses a subset of the signals defined for the MII connection. The 7 wire interface uses a single transmit and receive line, and does not use the Carrier Sense, Transmit Error, Receive Error, Management Data Clock nor Management Data I/O pins. The latter two pins provide a control and data path to the external PHY transceiver, to allow the FEC to configure PHY parameters.

## Summary of the FEC Features

- 10 Mbps 7 wire interface and 10/100 Mbps compliant with IEEE 802.3, 1998 edition.
- Built-in FIFO and DMA controller
- Programmable maximum frame length supporting IEEE 802.1 VLAN tags and priority
- IEEE 802.3 full duplex flow control
- Full-duplex operation (200 Mbps throughput) with a minimum system clock rate of 50MHz
- Half-duplex operation (100 Mbps throughput) with a minimum system clock rate of 25MHz
- Retransmission from transmit FIFO following a collision without processor bus utilization
- Automatic internal flushing of the receive FIFO for runts (collision fragments) and address recognition rejects without processor bus utilization
- Address recognition
  Frames with broadcast address may be always accepted or always rejected
  Exact match for single 48-bit individual (unicast) address
  64 bit hash table for individual (unicast) addresses
  64 bit hash table for group (multicast) addresses
  Promiscuous mode
- RMON (RFC 1757) and IEEE 802.3 statistics
- Interrupts for network activity and error conditions

**TABLE 15.1 MII Pins**

| Name | Dir | Description | 7-wire usage |
|------|-----|-------------|--------------|
| MDIO | I/O | MII Data Input/Output | - |
| MDC | O | MII Data Clock | - |
| RxD | I | Rx Data | Y |
| RxD | I | Rx Data | - |
| RxD | I | Rx Data | - |
| RxD | I | Rx Data | - |
| Rx_DV | I | Rx Data Valid | Y |
| Rx_CLK | I | Rx Clock | Y |
| Rx_ER | I | Rx Error | - |
| Tx_ER | O | Tx Error | - |
| Tx_CLK | I | Tx Clock | Y |
| Tx_EN | O | Tx Enable | Y |
| TxD | O | Tx Data | Y |
| TxD | O | Tx Data | - |
| TxD | O | Tx Data | - |
| TxD | O | Tx Data | - |
| COL | I | Collision | Y |
| CRS | I | Carrier Sense | - |

## *Fast Ethernet Controller Architecture*

The block diagram of the FEC is shown in Figure 15.1. The FEC is implemented with a combination of hardware and microcode. The hardware consist of transmit and receive registers, transmit and receive FIFOs, MII registers, Control, Status and MIB registers, a 32 bit host interface, a DMA controller and a RISC controller. Microcode is executed by the RISC controller embedded in the FEC module. Its job is to interpret the contents of the transmit and receive descriptor blocks stored in system memory external to the FEC, and in combination with the DMA controller transfer data packets from transmit and to receive buffers placed in system memory. The descriptor block functionality is detailed in a later section of this chapter.

In order to use the FEC, software executing on the ColdFire® host must first initialize several control registers in the FEC. Host software must also initialize the transmit and receive buffer descriptors and load the transmit buffers with appropriate data before it enables the FEC. To enable the FEC, the host controller should set the ECR[ETHER_EN] bit. This causes the FEC RISC controller to start executing microcode that performs the following operations internal to the FEC module:

- Initialize the BackOff Random Number Seed
- Activate the Receiver
- Activate the Transmitter
- Clear Transmitter FIFO
- Clear Receiver FIFO
- Initialize Transmitter Ring Pointer
- Initialize Receiver Ring Pointer
- Initialize FIFO Count Registers

Once enabled, the FEC is independent of the host and can autonomously transfer data from the transmit buffer in system memory to the transmit hardware via the Tx FIFO, and transfer data from the receive hardware, via the Rx FIFO, to the receive buffers in system memory. Host software can also write to FEC control registers to enable FEC interrupts that signal completion of transmitted and received data, and that signal transmit and receive error conditions in the FEC module.

To allow for different system latencies, the FEC's transmit and receive FIFO sizes may be somewhat adjusted. The total memory space allocated for both FIFOs is 512 bytes. This memory is contiguous, and the control register R_FSTART, defines the point that splits the memory between transmit and receive FIFO partitions. Additionally, the transmit FIFO has a programmable watermark, specified in register X_WMRK, that defines how much data must be loaded into it before the hardware starts transmitting. There is no equivalent watermark for the receive FIFO.

The Message Information Block (MIB) contains a set of counters that provide information about network events and statistics. These counters are not needed for correct FEC operation, but are provided to comply with the RMON (RFC 1757) Ethernet Statistics group and some of the IEEE 802.3 counter definitions.

The FEC receiver, in combination with RISC microcode, supports individual (unicast), group (multicast) and broadcast (all-ones group address) destination address types. To accelerate the acceptance of received multicast addresses when flow control is disabled, the FEC implements a hashing algorithm to map a multicast address into a hash table contained in a control register.

Because the FEC DMA controller is a bus master, it must arbitrate with other bus masters for memory resources. Host software accesses the internal control registers through the SIF interface.

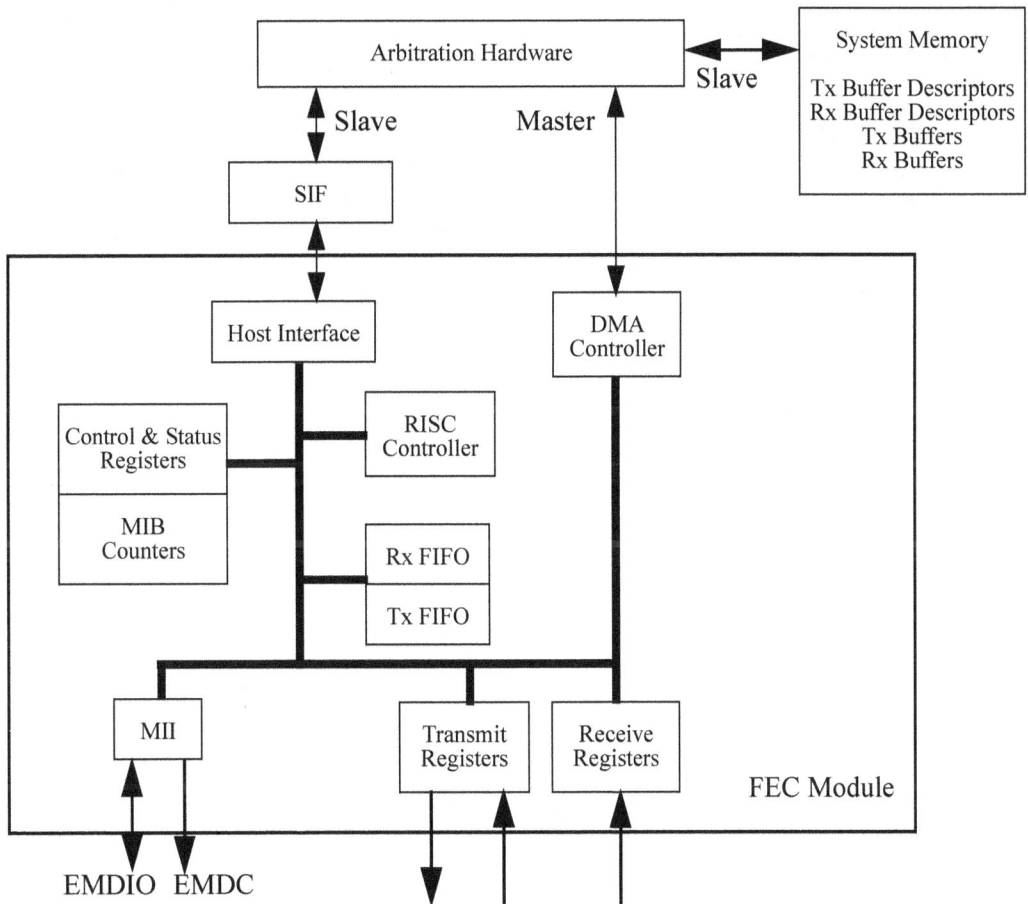

**FIGURE 15.1. FEC Block Diagram and connection to system memory**

## Fast Ethernet Controller Functional Details

### Transmit FIFO and Buffers

To transmit a data frame out of the FEC, the FEC DMA engine fetches the required data from one or more transmit buffers located in system memory external to the FEC. Normally this memory would be on-chip SRAM, but it could be any memory location that the FEC DMA has access to. The location and size of each buffer is defined by a buffer descriptor. To maximize flexibility for the user, each buffer descriptor is also located in memory external to the FEC. A buffer descriptor has a fixed size, and contains status and control information relating to a single buffer, a pointer to the start address of the buffer and the buffer length. The FEC uses the contents of the buffer descriptor to manage its corresponding data buffer. Since all buffer descriptors are the same size, host software places them in contiguous memory locations, with the location of the first buffer defined by the register X_DES_START. Figure 15.2 shows an example of the relationship between the register in the FEC, plus the buffer descriptors and their respective data buffers in system RAM.

Because the least significant 2 bits in X_DES_START are always zero, the address of the

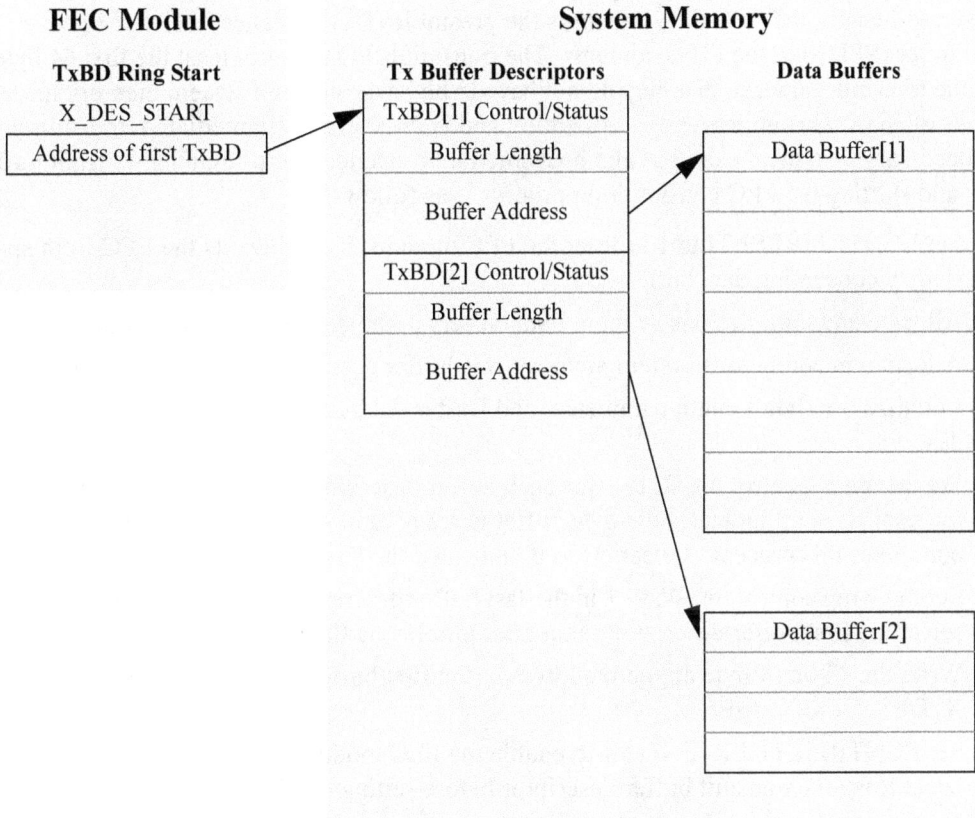

**FIGURE 15.2. Buffer Descriptor Ring and Data Buffers**

first buffer descriptor must be aligned to 4 bytes. Each buffer descriptor has a size of 8 bytes. Host software may place the start of data buffers in any valid memory location by initializing the 32 bit 'Buffer Address' parameter shown in Figure 15.2. The size of the data buffer is defined, in number of bytes, by the 16 bit 'Buffer Length' parameter.

Host software creates (produces) a transmit buffer by allocating and initializing memory, and initializing a transmit buffer descriptor. A sequential group of buffer descriptors is refer to as a "ring".

The FEC module processes (consumes) buffers by sequentially accessing each buffer descriptor in turn until all are consumed. The FEC DMA consumes a buffer by transferred its data into the transmit FIFO. Once the FIFO fills to the watermark, the MAC transmit logic asserts the output signal ETXEN. Provided the network is not busy (indicated by an asserted ECRS), the FEC then transmits the preamble (PA) sequence, the start frame delimiter (SFD) and the FIFO contents. The transmit FIFO stores at least the first 64 bytes of the transmit frame, so that they do not have to be retrieved from system memory in case of a collision. This improves bus utilization and latency in case immediate retransmission is necessary. A typical example of a host software procedure for producing transmit buffers and starting the FEC consumption process is as follows:

- Set ECNTRL[RESET] to 1 to reset the FEC module. This prevents the FEC from spuriously consuming data buffers.
- Allocate and initialize one or more data buffers in the desired system memory.
- Allocate, in contiguous system memory, one buffer descriptor for each data buffer.
- Initialize the Data Length parameters and Buffer Address pointers in each buffer descriptor.
- Set the ready control bit, R, to 1 for each buffer descriptor. Once the FEC is started by the host, a set bit indicates the data buffer is ready to be transmitted. When the FEC completes this process, it clears R to 0, indicated the buffer has been consumed.
- Set the wrap control bit, W, to 1 in the last buffer descriptor. This causes the FEC to select the first buffer descriptor again after processing the one with the wrap bit set.
- Write the 32 bit (4 byte aligned) address of the first buffer descriptor to the X_DES_START register.
- Set ECNTRL[ETHER_EN] to 1 to enable the FEC module. Host software must initialize at least one transmit buffer descriptor before setting this bit.
- Write any value to the TDAR[X_DES_ACTIVE] to start the FEC transmit process.

To detect when the buffers have been consumed, host software may poll the R bit in each buffer descriptor or may rely on buffer or frame interrupts.

The FEC architecture allows a transmit frame to be divided between multiple buffers. For example, the application payload can reside in one buffer, the TCP header in a second buffer, the IP header in a third buffer and the Ethernet/IEEE 802.3 header in a fourth buffer. The Ethernet MAC does not prepend the Ethernet header (destination address, source address, length/type field(s)), so host software must provide this in one of the transmit buffers. The Ethernet MAC can append the Ethernet CRC to the frame. Whether the CRC is appended by the MAC or by host software is determined by the TC bit in the transmit buffer descriptor which must be set by the host. When a frame is divided in this way, between multiple buffers, the L (Last) bit in the buffer descriptor control word must be set to 1 only in the last buffer associated with the frame. This ensures that the frame preamble is prepended only to the contents of the first buffer, and the CRC is appended (if enabled) only to the contents of the last buffer. If the entire frame is defined by a single buffer, then the L bit in its buffer descriptor should be set to 1.

The buffer descriptor control/status word contains two bits, TO1 and TO2 for exclusive use of host software. The bits may be read at any time, but should only be updated when the R bit is 0, indicating that the FEC is no longer accessing the buffer descriptor. If this is not done, the FEC could rewrite prior values of these bits to the control/status word after host software had modified them.

In most cases, host software will dynamically reuse already consumed buffer descriptors in the ring while the FEC is actively processing the ring. The software should ensure that the first buffer descriptor in the set of dynamically allocated descriptors is the last one to be made ready. One way to do this is to set the R bit for each buffer descriptor in reverse order of consumption. This ensures the complete frame is ready in memory before the FEC starts consuming the new data. If the buffer descriptors are made ready in forward order, the FEC could consume the first buffer descriptor before the second was made available, potentially causing a transmit FIFO underrun.

Once the FEC reads a transmit buffer descriptor with its R bit cleared to 0, it stops processing any further buffer descriptors until host software writes to X_DES_START.

Transmit frame status is indicated via individual interrupt bits and in statistic counters in the MIB block.

### Receive FIFO and Buffers

Provided they are not rejected by the FEC's address recognition algorithms, the FEC DMA controller transfers received data frames from the FEC RxFIFO into one or more receive buffers that typically reside in on-chip SRAM. However, the receive buffers may be placed in any volatile memory location that the FEC DMA controller can access. Because the received frame length is usually unknown during run time initialization of the system, the FEC contains a field in two configuration registers that define the maximum size of a received frame and the maximum amount of data that may be transferred into any receive buffer. The maximum frame size is defined by the R_CNTRL[MAX_FL] register field, while the maximum buffer size is defined by the R_BUFF_SIZE register. The values are global for all received frames and buffers. The buffer size is variable between 16 and 2047 bytes with a granularity of 16 bytes. To minimize the number of buffers need to store a complete received frame, the buffer size should be made as large as practicable. When the buffer size is equal to or greater than the maximum frame size, the entire received frame will fit in a single buffer. The maximum allowed frame size is 2047 bytes, while the default value after reset is 1518 bytes. Frame size granularity is 1 byte. A frame length is measured from the received destination address and includes the CRC appended to the end of the frame. Note that frames that exceed the value defined by R_CNTRL[MAX_FL], but are less than the maximum of 2047 bytes are still received without truncation. In this case the BABR (Babbling receiver) interrupt will occur if enabled, and the LG bit in the end of frame buffer descriptor status word is set to 1. Frames that exceed 2047 bytes are truncated and the TR bit is set to 1 in the corresponding buffer descriptor's status word.

Host software creates (produces) an empty receive buffer by initializing a receive buffer descriptor. A sequential group of buffer descriptors is refer to as a "ring".

The FEC module processes (consumes) buffers by sequentially accessing each buffer descriptor in turn until all are consumed. A typical example of a host software procedure for producing receive buffers and starting the FEC consumption process is as follows:

- Set ECNTRL[RESET] to 1 to reset the FEC module. This prevents the FEC from spuriously consuming data buffers.
- Allocate, in contiguous system memory, one buffer descriptor for each receive buffer.
- Initialize the Buffer Address pointers in each buffer descriptor.
- Set the empty control bit, E, to 1 for each receive buffer descriptor. Once the FEC is started by the host, a set bit indicates the buffer is ready to receive data. When the FEC completes this process, it clears E to 0, indicating the buffer has been consumed.
- Set the wrap control bit, W, to 1 in the last buffer descriptor. This causes the FEC to select the first buffer descriptor again after processing the one with the wrap bit set.
- Write the 32 bit (4 byte aligned) address of the first buffer descriptor to the R_DES_START register.
- Write the maximum expected frame size to R_CNTRL[MAX_FL].
- Write the maximum size allocated for all receive buffers to R_BUFF_SIZE.
- Set ECNTRL[ETHER_EN] to 1 to enable the FEC module.
- Write any value to the R_DES_ACTIVE to start the FEC receive process.

As frames are received, the FEC will fill empty receive buffers and update their corresponding buffer descriptors to indicate that the buffers are full, how much data is contained in each buffer (or the total frame size), which buffer contains the end of frame data, and any error conditions associated with frame size, data integrity and address recognition.

When a received data frame spans multiple buffers, all buffers but the last will have a data length equal to the value contained in R_BUFF_SIZE, and their buffer descriptor L bits set to 0. The last buffer will contain the end of frame data, its buffer descriptor data length will be set to the total frame size, the buffer descriptor L bit set to 1, and all status bits are updated.

A summary of the receive buffer descriptor error status bits that may be set by the FEC on completion of a received frame is given below.

- **M**

  The frame was accepted in promiscuous mode, but failed internal address recognition.

- **LG**

  The frame length exceeds the value defined in R_CNTRL[MAX_FL].

- **NO**

  The frame size does not have a granularity of 8 bits. If this bit is set, the CR bit will not be set.

- **CR**

  The frame contains a CRC error.

- **OV**

  An RxFIFO overrun occurred during frame reception.

Additional buffer descriptor status bits set by the FEC are:

- **BC**

  The received destination address is a broadcast address.

- **MC**

  The received destination address is a multicast address.

- **TR**

  The frame is truncated.

As in the transmit buffer descriptor, the receive buffer descriptor control/status word contains two bits, RO1 and RO2 for exclusive use of host software. The bits may be read at any time, but should only be updated when E is 0, indicating that the FEC is no longer accessing the buffer descriptor. If this is not done, the FEC could rewrite prior values of these bits to the control/status word after host software had modified them.

Once the FEC reads a receive buffer descriptor with its E bit cleared to 0, it stops processing any further buffer descriptors until host software writes to R_DES_START.

To detect when the buffers have been consumed, host software may poll the E bit in each buffer descriptor or may rely on buffer or frame interrupts.

## Ethernet Address Recognition

The FEC filters the received frames based on destination address (DA) type ó individual (unicast), group (multicast), or broadcast (all-ones group address). The difference between an individual address and a group address is determined by the I/G bit in the destination address field. A flowchart for address recognition on received frames is illustrated in the figures below.

Address recognition is accomplished through the combination of the receive block and microcode running on the microcontroller. The flow chart shown in Figure 15.3 illustrates the address recognition decisions made by the receive block, while Figure 15.4 illustrates the decisions made by the microcode executing on the FEC RISC engine.

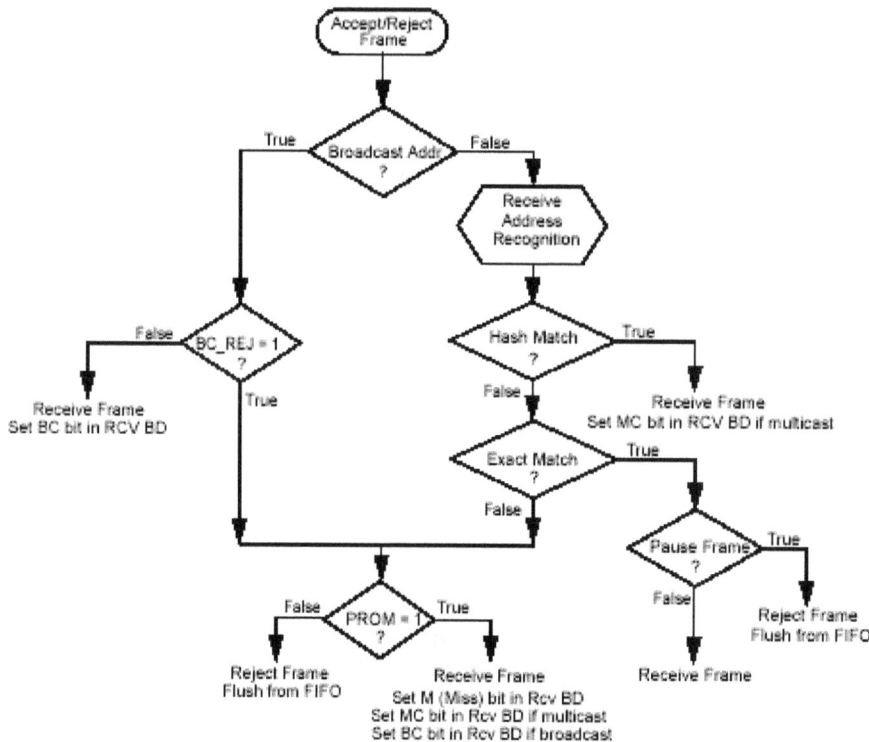

FIGURE 15.3. Ethernet Address Recognition - Receive block Flow Chart

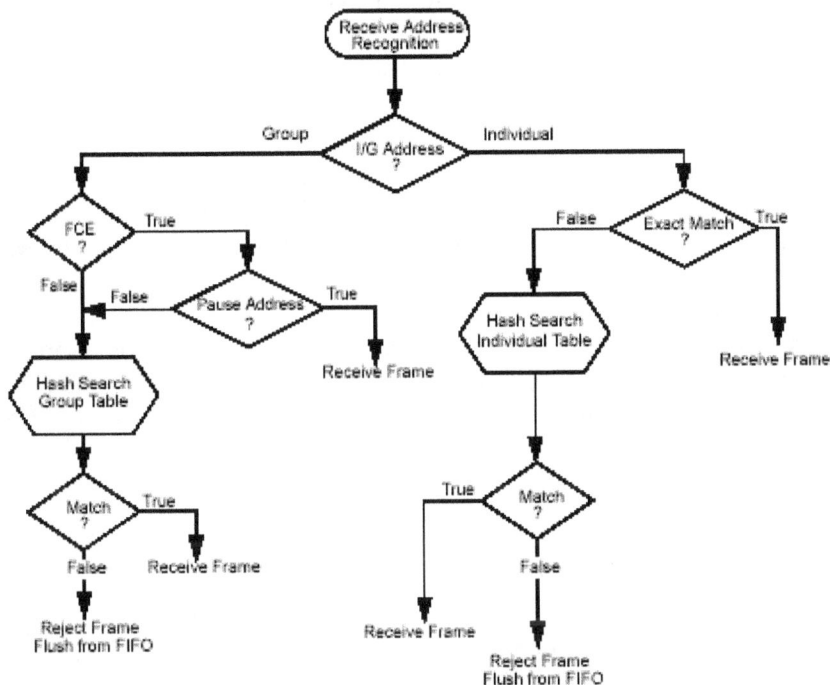

**FIGURE 15.4. Ethernet Address Recognition - Microcode Flow Chart**

If the DA is a broadcast address and broadcast reject (RCR[BC_REJ]) is de-asserted, then the frame will be accepted unconditionally, as shown in Figure 15.3. Otherwise, if the DA is not a broadcast address, then the microcontroller runs the address recognition subroutine, as shown in Figure 15.4.

If the DA is a group (multicast) address and flow control is disabled, then the microcontroller will perform a group hash table lookup using the 64-entry hash table programmed in GAUR and GALR. If a hash match occurs, the receiver accepts the frame.

If flow control is enabled, the microcontroller will do an exact address match check between the DA and the designated PAUSE DA (01:80:C2:00:00:01). If the receive block determines that the received frame is a valid PAUSE frame, then the frame will be rejected. Note the receiver will detect a PAUSE frame with the DA field set to either the designated PAUSE DA or the unicast physical address.

If the DA is the individual (unicast) address, the microcontroller performs an individual exact match comparison between the DA and 48-bit physical address that the user programs in the PALR and PAUR registers. If an exact match occurs, the frame is accepted; otherwise, the microcontroller does an individual hash table lookup using the 64-entry hash table programmed in registers, IAUR and IALR. In the case of an individual hash match, the frame is accepted. Again, the receiver will accept or reject the frame based on PAUSE frame detection, shown in Figure 15.3.

If neither a hash match (group or individual), nor an exact match (group or individual) occur, then if promiscuous mode is enabled (RCR[PROM] = 1), then the frame will be accepted and the MISS bit in the receive buffer descriptor is set; otherwise, the frame will be rejected.

Similarly, if the DA is a broadcast address, broadcast reject (RCR[BC_REJ]) is asserted, and promiscuous mode is enabled, then the frame will be accepted and the MISS bit in the receive buffer descriptor is set, otherwise, the frame will be rejected.

In general, when a frame is rejected, it is flushed from the FIFO.

**MII Control and Data Registers and Interface to the EPHY**

The MII is an interface that allows the FEC to configure and obtain status on MII compatible PHY devices attached to the FEC I/O pins.

When connected to an external PHY, the MII configuration and status data is transferred using a clock signal (MDC) and a single bidirectional data signal (MDIO). The MII clock rate (FEC_MDC pin) is set in the MII Speed Control Register, MSCR[MII_SPEED] 6 bit field. The register also contains a control bit MSCR[DIS_PREAMBLE] to disable a 32 bit preamble that occurs after a hardware reset, if not required by the attached PHY.

MSCR[MII_SPEED] must be programmed with a value that results in an FEC_MDC frequency of less than or equal to 2.5 MHz to be compliant with the IEEE 802.3 MII specification. MII_SPEED can be calculated using the following equation:

MII_SPEED = System Clock Frequency/(FEC_MDC * 2)

The result of the calculation must be rounded to a value that ensures FEC_MDC does not exceed 2.5MHz. Table 15.2 shows actual FEC_MDC frequencies obtained from certain integer values of MII_SPEED.

**TABLE 15.2 Actual MII clock rates based on MII_SPEED value**

| System Clock Frequency | MII_SPEED field | FEC_MDC frequency |
|---|---|---|
| 25MHz | 0x5 | 2.5MHz |
| 33MHz | 0x7 | 2.36MHz |
| 40MHz | 0x8 | 2.5MHz |
| 50MHz | 0xA | 2.5MHz |
| 66MHz | 0xD | 2.5MHz |

Host software may read and write data to the external PHY through the 32 bit MMFR register. The 16 most significant bits in this register contain fields that define frame delimiters, types of frames, register addresses and data direction, as shown in Figure 15.5. The bit

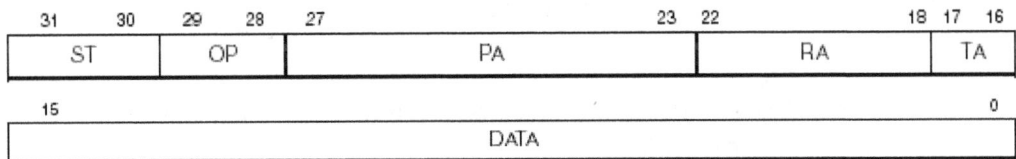

**FIGURE 15.5. MII Management Frame Register**

fields are defined as follows:

**ST**
These 2 bits define the Start of frame delimiter and must be set to 0b01 for a valid MII frame.

**OP**
These 2 bits actually define only 2 valid MII frame types: 0b10 for a register read and 0b01 for a register write. When a read or write operation has completed, the EIR[MII] bit is set, and an optional interrupt is generated.

**PA**
This 5 bit field defines the address of the attached physical transceiver.

**RA**
This 5 bit field defines the address of a register internal to the physical transceiver.

**TA**

These 2 bits define the "Turn Around" and must be programmed to 0b10 to generate a valid MII frame.

The least significant 16 bits contain the actual data to be written or read. Usage of this register for external PHYs depends on the particular PHY connected to the FEC, and is outside the scope of this book.

## *Fast Ethernet Controller Initialization Example*

The code shown in the following example initializes two transmit buffers, eight receive buffers, enables MII mode, and sets the MII clock rate to 2.5MHz, with an assumed system clock of 60MHz. After configuring the buffer descriptors, the first transmit buffer is loaded with a 64 byte packet and all bytes in the first receive buffer are initialized for test purposes to a value of 0x55. All eight receive buffers are marked as empty, and the first transmit buffer, with the 64 byte packet is marked ready. The data transfer may be tested without an external ethernet received by either enabling loopback mode internally, or connecting the external twisted pair transmit and receive signals together at the RJ45 conector.

After the first frame has been transmitted, the first receive buffer should contain the same data, defined in the array packet[].

The 64 byte packet contains the unicast address 00:CF:52:82:C3:01, which is the value written to the physical address registers PALR/PAUR to allow the FEC receiver to accept the frame when it loops back.

Because the example code employs exact address matching, the hash table registers IALR/IAUR and GALR/GAUR are unused and are cleared to zero.

Note that at the time of writing, most of the FEC register names were in the process of being changed. Please refer to the latest Freescale documentation for updates to the register names.

The example code simply waits for the first receive buffer to fill, then verifies that received data matches the transmitted packet..

```c
#define RX_BUFFER_SIZE 576        /* must be divisible by 16 */
#define TX_BUFFER_SIZE 576

/* Number of Receive and Transmit Buffers and Buffer Descriptors */
#define NUM_RXBDS 8
#define NUM_TXBDS 2

/* Buffer Descriptor Format */
typedef struct
{
 uint16 status;                   /* control and status */
 uint16 length;                   /* transfer length */
 uint8  *data;                    /* buffer address */
} NBUF;

/* Bit level Buffer Descriptor definitions */
#define TX_BD_R           0x8000
#define TX_BD_INUSE       0x4000
#define TX_BD_TO1         0x4000
#define TX_BD_W           0x2000
#define TX_BD_TO2         0x1000
#define TX_BD_L           0x0800
#define TX_BD_TC          0x0400
#define TX_BD_DEF         0x0200
#define TX_BD_HB          0x0100
#define TX_BD_LC          0x0080
#define TX_BD_RL          0x0040
#define TX_BD_UN          0x0002
#define TX_BD_CSL         0x0001

#define RX_BD_E           0x8000
#define RX_BD_INUSE       0x4000
#define RX_BD_R01         0x4000
#define RX_BD_W           0x2000
#define RX_BD_R02         0x1000
#define RX_BD_L           0x0800
#define RX_BD_M           0x0100
#define RX_BD_BC          0x0080
#define RX_BD_MC          0x0040
#define RX_BD_LG          0x0020
#define RX_BD_NO          0x0010
#define RX_BD_SH          0x0008
#define RX_BD_CR          0x0004
#define RX_BD_OV          0x0002
#define RX_BD_TR          0x0001
```

**Example 15.1.  FEC Initialization (part 1)**

```
INT16 fec_test(void)
{
 uint32 i;
 NBUF *pNbuf;
 UINT32 fail = 0;
}

/* Ethernet data to be transmitted */
const uint8 packet[] =
{
 0x00, 0xCF, 0x52, 0x82, 0xC3, 0x01, 0x00, 0xCF,
 0x52, 0x82, 0xC3, 0x01, 0x08, 0x00, 0x45, 0x00,
 0x00, 0x3C, 0x2B, 0xE8, 0x00, 0x00, 0x20, 0x01,
 0xA6, 0x1B, 0xA3, 0x0A, 0x41, 0x55, 0xA3, 0x0A,
 0x41, 0x54, 0x08, 0x00, 0x0C, 0x5C, 0x01, 0x00,
 0x40, 0x00, 0x61, 0x62, 0x63, 0x64, 0x65, 0x66,
 0x67, 0x68, 0x69, 0x6A, 0x6B, 0x6C, 0x6D, 0x6E,
 0x6F, 0x70, 0x71, 0x72, 0x73, 0x74, 0x75, 0x76
};

/* Buffer Descriptors are aligned on a 16 byte boundary */
uint8 unaligned_txbd[(sizeof(NBUF) * NUM_TXBDS) + 16];
uint8 unaligned_rxbd[(sizeof(NBUF) * NUM_RXBDS) + 16];

NBUF *TxNBUF;
NBUF *RxNBUF;

/* Data Buffers are aligned on a 16-byte boundary. */
uint8 unaligned_txbuffer[(TX_BUFFER_SIZE * NUM_TXBDS) + 16];
uint8 unaligned_rxbuffer[(RX_BUFFER_SIZE * NUM_RXBDS) + 16];

uint8 *TxBuffer;
uint8 *RxBuffer;

  TxNBUF = (NBUF *)((uint32)(unaligned_txbd + 16) & 0xFFFFFFF0);
  RxNBUF = (NBUF *)((uint32)(unaligned_rxbd + 16) & 0xFFFFFFF0);

  TxBuffer = (uint8 *)((uint32)(unaligned_txbuffer + 16) & 0xFFFFFFF0);
  RxBuffer = (uint8 *)((uint32)(unaligned_rxbuffer + 16) & 0xFFFFFFF0);
```

Example 15.1 **FEC Initialization (part 2)**

```
/* Initialize receive descriptor ring */
  for (i = 0; i < NUM_RXBDS; i++)
  {
  RxNBUF[i].status = RX_BD_E;
  RxNBUF[i].length = 0;
  RxNBUF[i].data = &RxBuffer[i * RX_BUFFER_SIZE];
  }
  /* Set the Wrap bit on the last one in the ring */
  RxNBUF[NUM_RXBDS - 1].status |= RX_BD_W;

  /* Initialize transmit descriptor ring */
  for (i = 0; i < NUM_TXBDS; i++)
  {
  TxNBUF[i].status = TX_BD_L | TX_BD_TC;
  TxNBUF[i].length = 0;
  TxNBUF[i].data = &TxBuffer[i * TX_BUFFER_SIZE];
  }
  /* Set the Wrap bit on the last one in the ring */
  TxNBUF[NUM_TXBDS - 1].status |= TX_BD_W;

/* Set the source address for the controller */
  FEC.PALR.R = 0x00CF5282;
  FEC.PAUR.R = 0xC3010000;

  FEC.IALR.R = 0x00000000;
  FEC.IAUR.R = 0x00000000;

  FEC.GALR.R = 0x00000000;
  FEC.GAUR.R = 0x00000000;

/* Set Receive Buffer Size */
  FEC.EMRBR.R = (uint16)RX_BUFFER_SIZE;

/* Point to the start of the circular Rx buffer descriptor queue */
  FEC.ERDSR.R = (uint32)RxNBUF;

/* Point to the start of the circular Tx buffer descriptor queue */
  FEC.ETDSR.R = (uint32)TxNBUF;

/* Set the transceiver interface to MII mode */
  FEC.RCR.R = 4;                   /* MII mode - 18 signals */


  FEC.RCR.B.LOOP = 1;              /* and enable internal loop mode */

/* Operate in full-duplex mode, no heart beat control */
  FEC.TCR.R = 0x0004;

/* Set MII speed to be 2.5Mhz with a 60MHz system clock */
  FEC.MSCR.R = 12;

/* Grab buffer in ring */
  pNbuf = TxNBUF;

/* Copy constant data into the data buffer */
  memcpy(pNbuf->data, packet, 64);

/* Set the length of the packet */
  pNbuf->length = 64;

/* Enable FEC */
  FEC.ECR.B.ETHER_EN = 1;
```

Example 15.1 **FEC Initialization (part 3)**

```
/* Initialize the EPHY */
  ephy_init();

  for (i = 0; i < 64; i++)
  {
  RxNBUF[0].data[i] = 0x55;         /* wipe out receive data before transfer */
  }


/* Set the Wrap bit on the last one in the ring */
  RxNBUF[NUM_RXBDS - 1].status |= RX_BD_W;
// receive control reg - loopback
/* Indicate Empty buffers have been produced */
  FEC.RDAR.B.R_DES_ACTIVE = 1;

/* Mark packet as ready to send */
  pNbuf->status |= TX_BD_R;

/* Indicate to FEC that transmit buffer is ready to send */
  FEC.TDAR.B.X_DES_ACTIVE = 1;

  for (i=0; i < 100000; i++)
  {
     if (FEC.EIR.B.RXF)
     {
        break;
     }
  }

  if (i == 100000)
  {
     /* Timed-out */
     fail++;
  }

  for (i = 0; i < 64; i++)
  {
     if (TxNBUF[0].data[i] != RxNBUF[0].data[i])
     {
        fail++;
     }
  }

  if (fail)
     return (FAIL);
  else
     return (PASS);
}    /* end of fec_test */
```

Example 15.1 **FEC Initialization (part 4)**

```
void ephy_init(void)
{
// External PHY specific initializtion can be placed here.
/* } // end of void ephy_init()
```

**Example 15.2. EPHY Initialization Routine**

# CHAPTER 16

# *Timer Systems*

## *Introduction to Timer Systems*

Timer systems are essential components of real time embedded control. They are used to measure external signals and values, generate output signals to drive actuators, motors, or to communicate parameter values. Timer systems may also be used for internal software timing to help schedule tasks in a real time scheduler or operating system, or to provide watchdog capability. Timers are also used to monitor and control virtually everything that moves, from the internal combustion engine to the arms of an industrial robot.

S08 and ColdFire V1 products offer several timer systems:

- Timer/PWM Module (TPM)
- Flex Timer Module (FTM)
- Real Time Counter (RTC)
- Independent Real Time Clock (IRTC)

## *Timers Architecture*

The basic building block of any timer is a counter. A counter may be used directly to count incoming pulses, but typically a counter is clocked at a fixed frequency and used as a timing reference. Incoming signals are measured by capturing the value of the reference counter when an external event occurs, and outgoing events are generated when the reference counter equals a predetermined value. This allows the monitoring and control of the individual events, and the relationship between many events. Two output compare registers and a reference counter with the appropriate output logic may be used to generate pulse width modulated (PWM) outputs, that with the use of a low pass filter provide an analog output.

Flexis and ColdFire V1 products offer a range of timer modules to provide functions from general real time control, to time and calendar keeping, and specialized motor control.

## Timer/Pulse-Width Modulator (TPM)

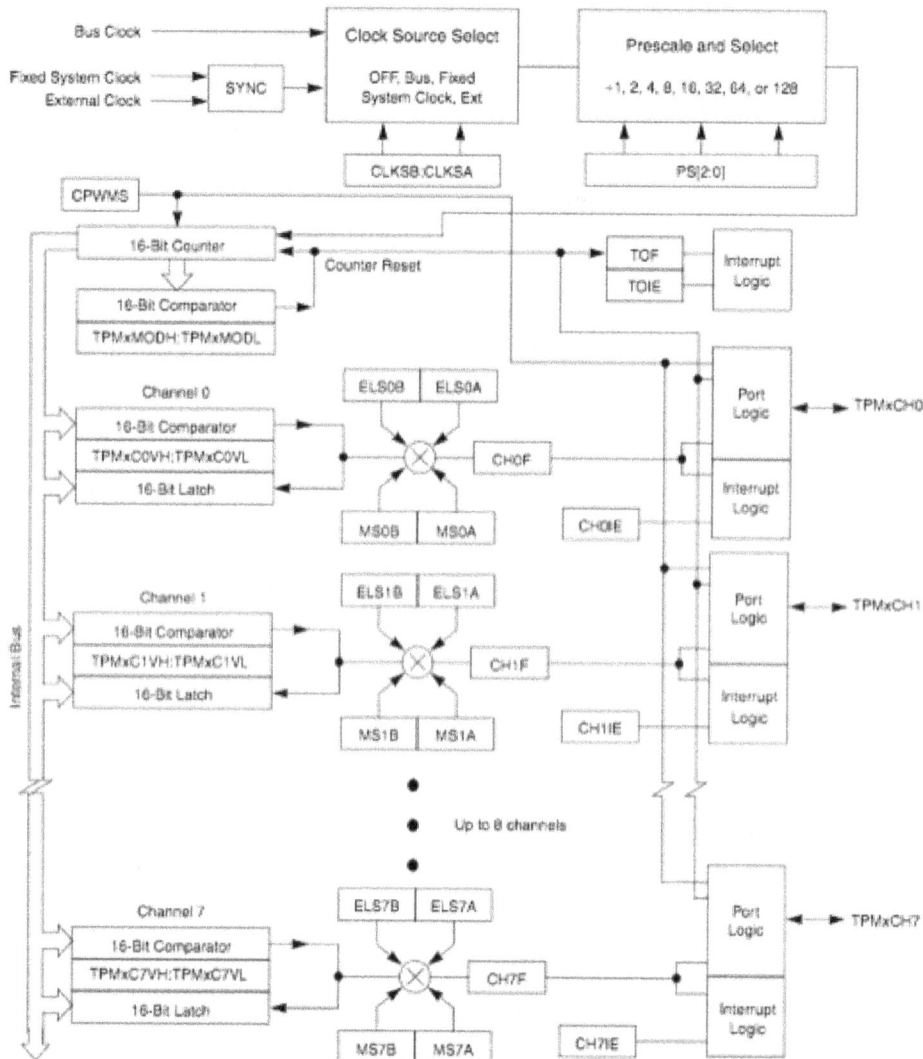

FIGURE 16.1. TPM Block Diagram

The TPM is a one-to-eight-channel 16-bit timer system that supports traditional input capture, output compare, and PWM on each channel. The TPM uses one 16-bit modulo counter as the timing reference for all the timing channels within one TPM module. The reference counter counts up from 0 until it's value matches that in it modulo register, and then it resets to 0, and restarts the cycle. The reference counter clock source includes a number of options, the direct system clock, the bus clock, or an external clock. The selected clock source may then be further divided by a binary prescaler with divide ratios from 1 to 128.

Each TPM channel consists of a 16-bit comparator for the output compare function, and a 16-bit latch for the input capture function. For output compare, when the reference counter value matches that in the respective output compare register an output event is generated on the associated channel output pin. The output event generated may be selected to be a set, clear, or toggle. For the input capture function, the selected external event will result in the current value of the reference counter to be latched in the input capture register. The available trigger events are rising edge, falling edge, or any edge.

Each TPM channel may also be configured for edge aligned PWM, with selectable polarity outputs, or the entire module may be configured for center aligned PWM on all channels. Therefore, when the center aligned PWM mode is selected, output compare, input capture and edge aligned PWM functions are not available on that TPM module.

In the edge aligned PWM mode, the value in the reference counter's modulo register (plus 1) sets the period. This same period applies to all channels configured for the edge aligned PWM function. The individual channel value register (the output compare register) sets the duty cycle for that channel.

Edge Aligned PWM

Center Aligned PWM

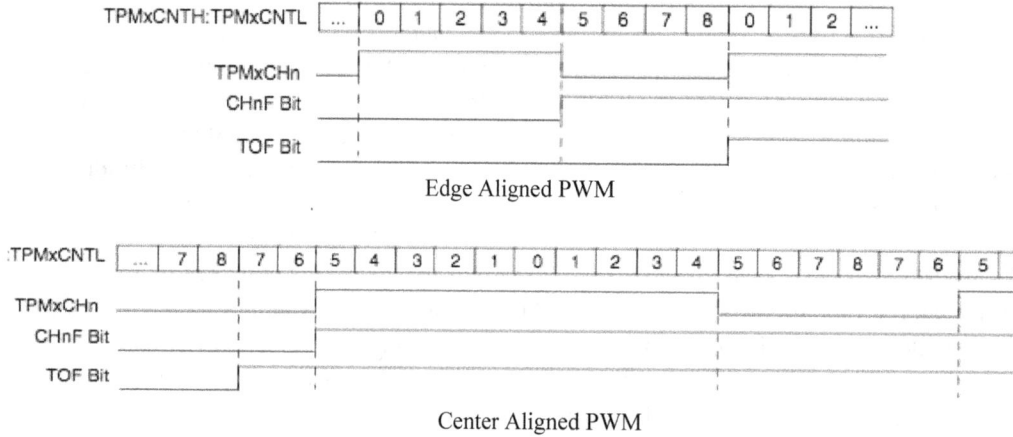

**FIGURE 16.2. Edge and Center Aligned PWMs**

In the center aligned PWM mode, the reference counter is again used to set the PWM period for all channels, only this time the reference counter counts up to the modulus value, but then instead of resetting, the reference counter counts down to zero. At zero it starts to count up again, towards the modulus value. In this mode the PWM period will be twice the value written into the modulus register. The individual channels output compare register is used to set the duty cycle, only this time when the value in the compare register matches the reference counter while counting down, the channels output will be set to active, and on matches while counting up the output will be set to inactive. The active level is selectable to be either high or low. Center aligned PWM is particularly suitable for driving electric motors.

# Flex Timer Module (FTM)

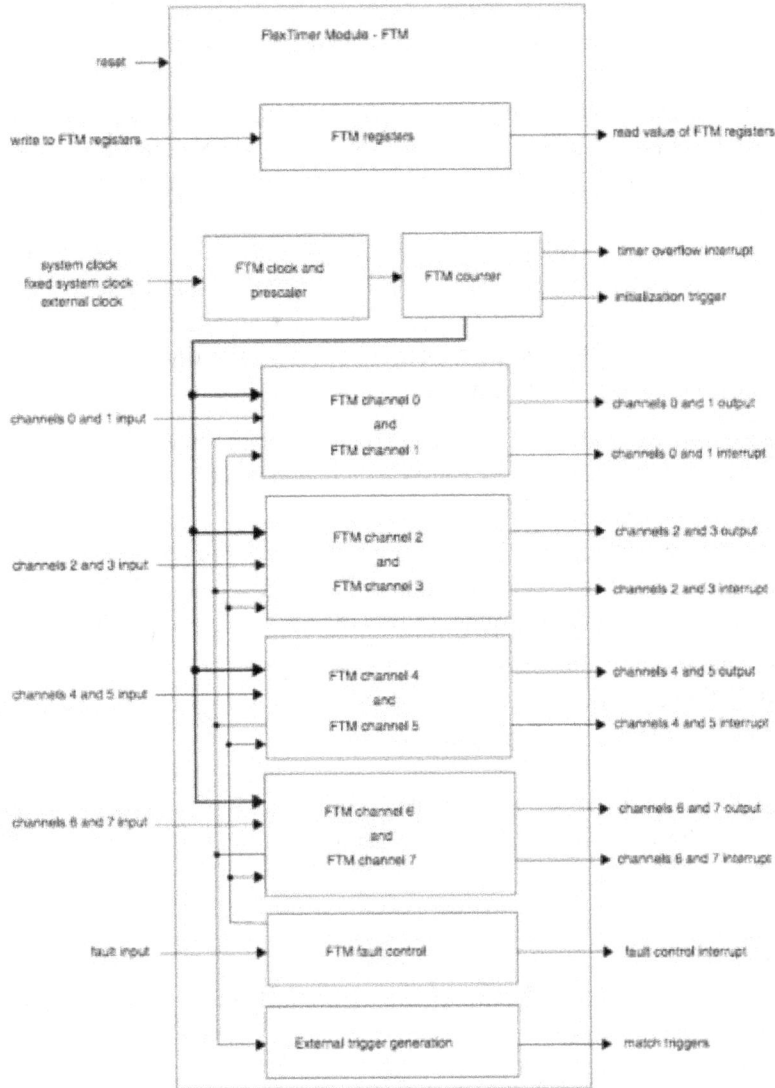

FIGURE 16.3. **FTM Block Diagram**

The Flex Timer Module (FTM) was specifically designed for electric motor control, in particular brushless DC and AC motors. This timer functionality is also particularly suitable for energy conversion applications, such as DC to AC inverters. Since the control of different motor types requires a wide range of timer resources, the FTM was made flexible to make any resources not required for motor control suitable for other timing functions. The FTM therefore supports input capture and output compare function in addition to PWM for motor control and power management. To facilitate the control of sensorless brushless DC and AC motors, the FTM supports synchronization with the Analog to Digital Converter. The FTM maintains backwards compatibility with the Timer/ Pulse-Width Modulator (TPM) timer system to ease use of, and migration between, both timer systems. This includes a TPM emulation mode, enabled by the FTMEN control bit in the FTMx-MODE register. When the FTMEN bit is 0, only TPM compatible registers may be used without any restrictions.

The FTM uses a 16-bit reference counter for all timing functions. This counter is clocked by either the direct system clock, the bus clock, or an external clock. The selected clock source may then be further divided by a binary prescaler with divide ratios from 1 to 128. The reference counter supports three modes of operation:

- Unsigned Modulo up counter, in this mode the counter counts up from 0 to the modulus value stored in the Modulo register FTMxMOD, then resets back to zero and counts up again.(mode compatible with TPM)

- Signed Modulo up counter, in this mode the counter counts up from the negative modulo value through 0 to the positive modulo value, then reloads with the negative modulo value and starts to count up again. (mode incompatible with TPM)

- Up/Down counter, in this mode the counter counts up from 0 to the modulus value, reverses direction and counts down to 0, reverses direction and so on. (compatible with TPM)

The signed modulo counter mode is particularly suitable for providing the reference for controlling multi-phase brushless DC and AC motors.

## Real Time Counter (RTC)

The Real Time Counter provides a programmable periodic interrupt that can be used for any of the following functions:

- Time of day and calendar keeping
- Task scheduling
- Cyclic wake up from low power modes
- Hardware trigger for the Analog to Digital Converter (ADC)

The RTC consists of one 8-bit counter, one 8-bit comparator, several binary-based and decimal-based prescaler dividers, three clock sources, and one programmable periodic interrupt.

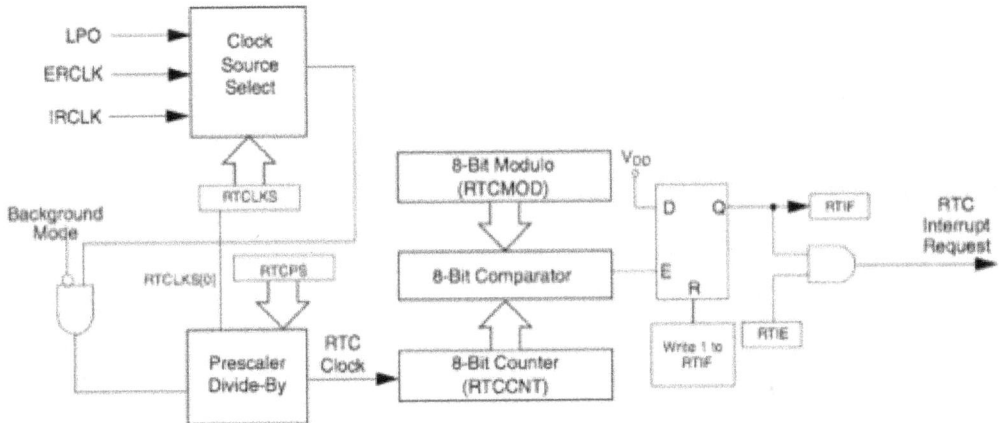

**FIGURE 16.4. Real Time Counter (RTC) Block Diagram**

Typical use of the Real Time Counter (RTC) would be to initialize the clock source, prescaler divider, and the modulo register so as to generate one interrupt per second. This can then be used by a time of day and calendering routine to track both time and date. Once enabled, the RTC will continue to function during wait mode and stop3 mode with any clock source, and during stop2 with the Low Power Oscillator (LPO).

The following software example is for initialization and time of day keeping using the RTC with internal 1 kHz Low Power Oscillator (LPO).

```
/* Initialize the elapsed time counters */
Seconds = 0;
Minutes = 0;
Hours = 0;
Days=0;

/* Configure RTC to interrupt every 1 second from 1-kHz clock source */
RTCMOD.byte = 0x00;
RTCSC.byte = 0x1F;

/********************************************************************
Function Name: RTC_ISR
Notes: Interrupt service routine for RTC module.
********************************************************************/
#pragma TRAP_PROC
void RTC_ISR(void)
{
          /* Clear the interrupt flag */
          RTCSC.byte = RTCSC.byte | 0x80;
          /* RTC interrupts every 1 Second */
```

```
Seconds++;
/* 60 seconds in a minute */
if (Seconds > 59){ Minutes++; Seconds = 0; }
/* 60 minutes in an hour */
if (Minutes > 59){ Hours++; Minutes = 0;
}
/* 24 hours in a day */
if (Hours > 23){ Days ++; Hours = 0;
}
```

## Independent Real Time Clock (IRTC)

The Independent Real Time Clock is a subsystem designed to operate from a backup battery when the main microcontroller power is not available, and includes the following functionality:

- Fully hardware implemented time of day clock with configureable auto daylight saving adjustment
- Fully hardware implemented calendar system with automatic month and leap year adjustment
- Programmable alarm
- Periodic interrupts
- Minute countdown timer
- Tamper protection
- Standby RAM (battery backed up)

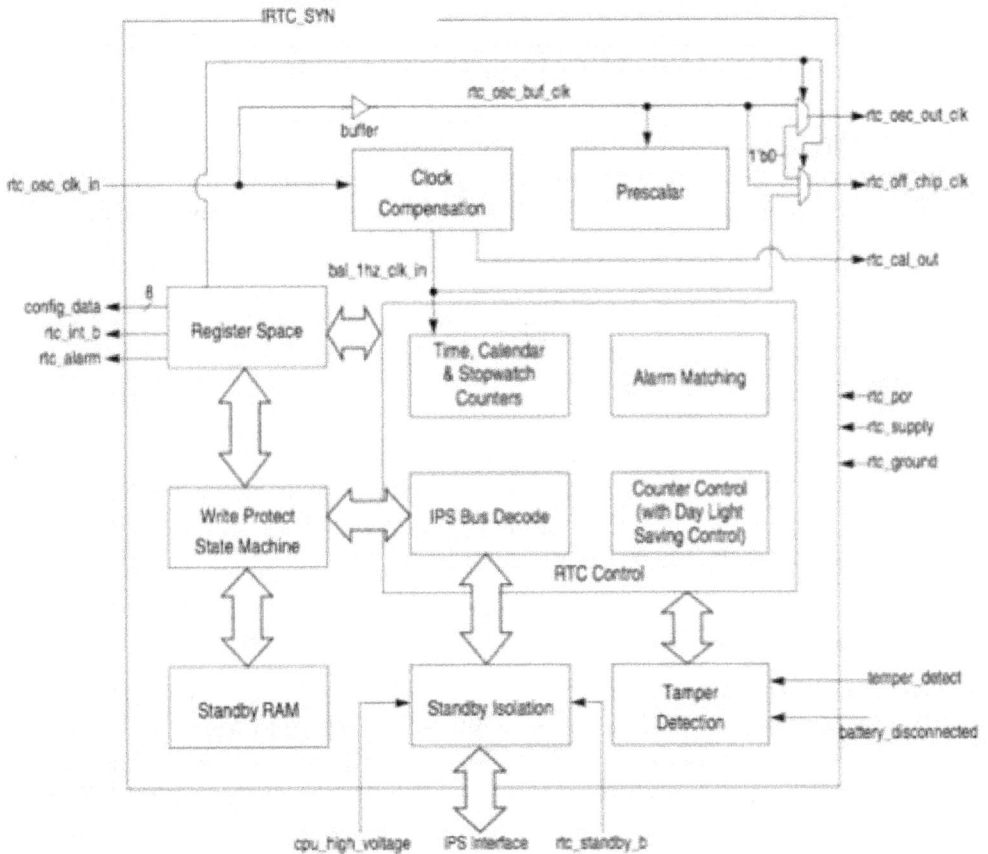

FIGURE 16.5. **IRTC Block Diagram**

## IRTC Overview

The Independent Real Time Clock (IRTC) provides time keeping functions with second, minute and hour registers, and calendaring functions with date, day-of-week, month and year registers, in either BCD or Binary format, including automatic adjustment for leap year and daylight saving.

The alarm is set for hour, minute and second, and on a match, the alarm flag is set, and if enabled an interrupt to CPU is generated. The alarm can also be configured to match days, months and year, and the alarm signal may be used for certain wakeup events. A minute countdown timer is included for time keeping, with separate enable, and interrupt. The IRTC module also includes seven sampling timer interrupts.

The frequency of the IRTC 32 kHz oscillator may vary due to crystal inaccuracy, board variations or changes in temperature. The frequency compensation module corrects for these variations and helps in the generation of an accurate 1 Hz clock.

The tamper detection includes two mechanisms, a tamper input for connecting external tamper sensors, and disconnection of the backup battery. A tamper event will generate an interrupt, and the time stamp of the even is captured. Tamper detection and its interrupt are enabled by default on reset.

IRTC also includes a 32-bit up-counter, that counts the number of writes to it's lowest 8-bits, and cannot be reset by software. This counter is intended as a secure backup count of energy units in metering applications.

The robust firmware update Flash array select bit is stored in bit 0 of the IRTC Configuration Data Register (IRTC_CFG_DATA), purely for being backed up by the IRTC backup battery, and is not part of any other IRTC functionality. Details of the robust firmware update are described in the Chapter 6.

## IRTC Clock and Calendar Functions

All of the clock, calendar, and alarm registers can be configured for either binary or Binary Coded Decimal (BCD) data format, with the BCD format Enable (BCDEN) control bit in the IRTC Control register (IRTC_CTRL). The calendar year is the one exception to this, as the year is stored as a 2's complement offset from the base year 2112. The user software must store the base year value, and perform the actual year calculation. The IRTC hardware does automatic leap year adjustment, including the 100 year leap year skip in year 2100. The use of any other value for the base year will result in incorrect leap year adjustments.This provides an available end value of the year from 1984 to 2239. For anyone wishing to extend this range, you could set the base year to 2212 for a range of 2084 to 2339 with correct leap year correction in hardware.

It should be noted that the IRTC hardware does not check any written time and clock values for being a valid value. If an invalid value is written into the time or calendar registers, the functionality of the IRTC will be indeterminate. The IRTC stores all values internally as binary, and performs the transformation between BCD and binary during read and write operations in the BCD mode.

Before and during accessing any of the time or calendar registers, the Invalid Time bit (INVAL), bit 0 of the IRTC Status Register (IRTC_STATUS), must be checked. This bit indicates that the time and calendar values are about to change, it is asserted 1 oscillator clock cycle before and after the 1 Hz (seconds') boundary, and is not cleared by a read.

Write accesses to time and calendar registers gets nullified and no register value gets changed while the INVAL bit is asserted, but will terminate normally. Reads while the INVAL bit is asserted return 16'hFFFF, and no Transfer Error is asserted.

In the calendar, the month value has the range 1 to 12, with 1 representing January. The Day of Week (DAY_OF_WEEK) value has a value range of 0, representing Sunday, to 6, representing Saturday. Value 7 is reserved and should not be used. Time is counted in the 0 to 23 hours format, if the 1 to 12 AM and PM format is required, the conversion will have to be done in the user software.

The alarm is set for hour, minute and second, and optionally for year and months. The Alarm Match (ALARM MATCH) field in the IRTC Control register (IRTC_CTRL) selects which values are compared by the alarm function:

- "00" – Seconds, Minutes and Hours matched
- "01" – Seconds, Minutes, Hours and Days matched
- "10" – Seconds, Minutes, Hours, Days and Months matched
- "11" – Seconds, Minutes, Hours, Days, Months and Year (offset) matched

It is possible to do fine adjustments to the clock with one second increments and decrements. A single second increment or decrement is achieved by writing a 1 to the Increment Seconds (INC_S) or Decrement Seconds (DEC_S) bit respectively, in the IRTC Seconds Alarm Register (IRTC_ALM_SEC). The value of the control bit is reset back to 0 once the adjustment takes place. This adjustment may be used for ongoing clock adjustments if an outside reference, such as network time, is available to the system. It may also be used for leap second adjustments, but be aware that the leap second adjustments process is currently under review, and adjustments may be suspended.

The IRTC includes the option of automatic Daylight Saving Time adjustment. Since the date of the change varies across regions and countries, and is subject to possible future changes, this has been made programmable. The Daylight Saving Time Enable (DSTEN) bit in the IRTC Control Register (IRTC_CTRL) enables Daylight Savings time adjustment when set to 1. The months for both starting and ending Daylight Savings Time is programmed in the IRTC Daylight Saving Month Register (IRTC_DST_MNTH), fields Daylight Saving Time Start Month (DST_START_MONTH) and Daylight Saving Time End Month (DST_END_MONTH) respectively. The day is programmed as the day of the month in IRTC Daylight Saving Day Register (IRTC_DST_DAY) fields Daylight Saving Time Start Day (DST_START_DAY) and Daylight Saving Time End Day (DST_END_DAY) for the starting and ending days respectively. Please note that the day for Daylight Savings Time changes is typically specified for a specific week day within a month, such as first Sunday in March. The user software will have to calculate the correct day of month for each year, and write this into the IRTC Daylight Saving Day Register (IRTC_DST_DAY). Similarly, the hour for starting and ending Daylight Savings Time is programmed in the IRTC Daylight Saving Hour Register (IRTC_DST_HOUR), fields Daylight Saving Time Start Hour (DST_START_HOUR) and Daylight Saving Time End Hour (DST_END_MONTH) in the 24 hour (0 to 23) format respectively.

The Software Reset (SWR) bit in the IRTC Control Register (IRTC_CTRL) may be used to clear the values of the alarm and interrupt status registers, by writing a 1 to it. This bit is self clearing following the clearing operation. Software Reset has no effect on the Tamper status and other non alarm functions.

The IRTC is protected against inadvertent writes, and has to unlocked with a specific unlock sequence to enable write. The IRTC is automatically unlocked for 14 to 15 seconds following power on reset, and for 1 to 2 seconds following a successful execution of the unlock sequence. The Write Enable (WE) bits in the IRTC Control register (IRTC_CTRL) are the only write accessible bits in the IRTC in the locked mode, and have to be written with the sequence 00 - 01 - 11 - 10. A single write of 10 will lock the IRTC. Reads of the WE field returns 00.

The IRTC includes a Minutes Countdown Timer, that can be programmed to generate an interrupt in 1 to 99 minutes. It is decremented by the clock's minute increment signal, so it's average variation is 0.5 minutes from the programmed countdown value. The Countdown Timer is programmed by writing the desired value to the IRTC Countdown (Minutes) Timer Register (IRTC_COUNT_DN), with the interrupt being generated when this reaches 0.

The IRTC Up Counter is not related to clock or calendar functions, and is described in "Tamper Detection" on page 340.

The IRTC offers the following Interrupts, all with separate enable bits in the IRTC Interrupt Enable Register (IRTC_IER), and status bits in the IRTC Interrupt Status Register (IRTC_ISR):

- Tamper Detect
- Countdown Timer has reached 0
- Alarm
- Day counter has been incremented
- Hour counter has incremented
- Minute counter has incremented
- 1 Hz flag, indicates that the seconds counter has incremented

- 2 Hz
- 4 Hz
- 8 Hz
- 16 Hz
- 32 Hz
- 64 Hz
- 128 Hz
- 256 Hz
- 512 Hz

The IRTC Status Register (IRTC_Status) includes the following status fields:

- INVAL - Indicates invalid or changing clock and calendar data
- C_DON - Indicates status of oscillator compensation
- BERR - Indicates a read or write attempt while the INVAL bit was
- OCAL - This bit toggles at the start of every compensation period
- WPE - The Write Protect Enable indicates if the IRTC is in the locked state, preventing writes
- CLV - CPU Low Voltage, this bit indicates that the main system power has fallen below the safe operating threshold, and when this bit is asserted writes to the IRTC are automatically blocked

**Oscillator Frequency Compensation**

The frequency of the IRTC 32 kHz oscillator may vary due to crystal inaccuracy, board variations or changes in temperature, and the Compensation logic is designed to correct this. The compensation is carried out by adjusting the number of oscillator clocks during a compensation period. The number of compensation clocks ranges from -128 to +127, and the compensation interval is adjustable from 1 to 255 seconds, giving a wide overall adjustment range of 0.119 to 3906 PPM. The actual values for both the compensation value and period need to be calculated by user firmware. It is recommended that the room temperature calibration for the crystal and board variations is carried out during initial system configuration, and stored in Flash memory. Then, during the initialization routine following reset, this calibration value should be written into the compensation registers.

To compensate for temperature variations, a temperature compensation table needs to be developed for the target application, with the required target accuracy and temperature step resolution. The system temperature may then be monitored using the integrated temperature sensor in the Analog to Digital Converter (ADC), but for best results the temperature measurement should be carried out as soon as possible following a microcontroller standby interval. This minimizes the self heating effects of the microcontroller power dissipation, and provides a more accurate reading of ambient temperature. The measured temperature reading can then be used to calculate the temperature compensation from the stored temperature calibration table, and then write it into the compensation register.

The IRTC Compensation Register (IRTC_COMPEN) consists of two 8-bit fields, the Compensation Interval (COMPENSATION_INTERVAL) and the Compensation Value (COMPENSATION_VALUE).The compensation interval is the binary value in seconds, with a range of 1 to 255, and if programmed to 0 disables oscillator compensation. The compensation value is a 2's complement of the number of oscillator clocks to be added or subtracted during the compensation interval.

### Tamper Detection

Three types of tamper events are monitored and recorded:

- External tamper events, requires an external (to the microcontroller) sensor
- Backup battery tampering when microcontroller power is on
- Backup battery tampering when microcontroller power is off

The simplest form of an external sensor may be an enclosure switch, that signals to the microcontroller that a tamper is being attempted. A wide range of other tamper sensors may be implemented by the user in the system, according to the application requirements. The tamper input includes a digital filter, with a programmable sample period during which the tamper signal must remain stable to be recognized as a tamper event. The sample period is programmable from 1 to 15 clocks of the 32 kHz oscillator in the TAMPER DETECT DURATION field in the IRTC Control Register (IRTC_CTRL). If this control field is set to 0 any positive edge on the tamper input pin will trigger a tamper event.

When a valid tamper event is detected on the tamper input pin, the current time and date is captured in the IRTC Tamper Time Stamp Registers, as seconds, minutes, hours, day, month, and year (offset) (IRTC_TTSR_SEC, IRTC_TTSR_HM, IRTC_TTSR_DAY, IRTC_TTSR_MY). If enabled, a tamper interrupt is generated to the microcontroller CPU, and the tamper status flag (TMPR) in the IRTC Interrupt Status Register (IRTC_ISR) is as set. It should be noted that the tamper detection continues to operate from the backup battery when the main microcontroller power is off, the tamper status should be checked as part of the power on reset and standby wake-up routines.

If backup battery power is removed while the microcontroller power is on, this will result in a tamper event with time and calendar values capture, interrupt generation, and tamper status flag assertion as described above for an external tamper event. If backup battery power is removed while the microcontroller power is off

The tamper detection includes two mechanisms, a tamper input for connecting external tamper sensors, and disconnection of the backup battery. A tamper event will generate an interrupt, and the time stamp of the even is captured. Tamper detection and its interrupt are enabled by default on reset.

## *Carrier Modulator Timer (CMT)*

The Carrier Modulator Timer (CMT) is designed for generating the Infra-Red (IR) remote control command signals used to control most consumer electronics equipment, and other devices. The CMT module supports the most commonly used IR modulation forms, but in order to support the widest range of IR remote control signalling standards, the CMT module is designed to generate one transmission cycle, or effectively one transmitted bit. Software has to load the CMT data registers for each cycle. The CMT transfers the data from the registers to the modulator at the start of each transmission cycle, generates an interrupt to signal that the following cycle data may be loaded. This allows the software almost the full transmission cycle time to service the CMT and maintain a continuos transmission data stream. The Infra-Red Output (ORI) is capable of driving an infra-red LED up to 20mA directly, for applications requiring more power an external power driver has to be used. The CMT consists of three main functional blocks plus control:

- Carrier Generator
- Modulator
- Transmitter Output
- Control and Interfaces

And, supports four modes of operation:

- Time with independent control of high and low times (gated carrier)
- Baseband
- Frequency shift key (FSK)
- Direct software control of IRO pin

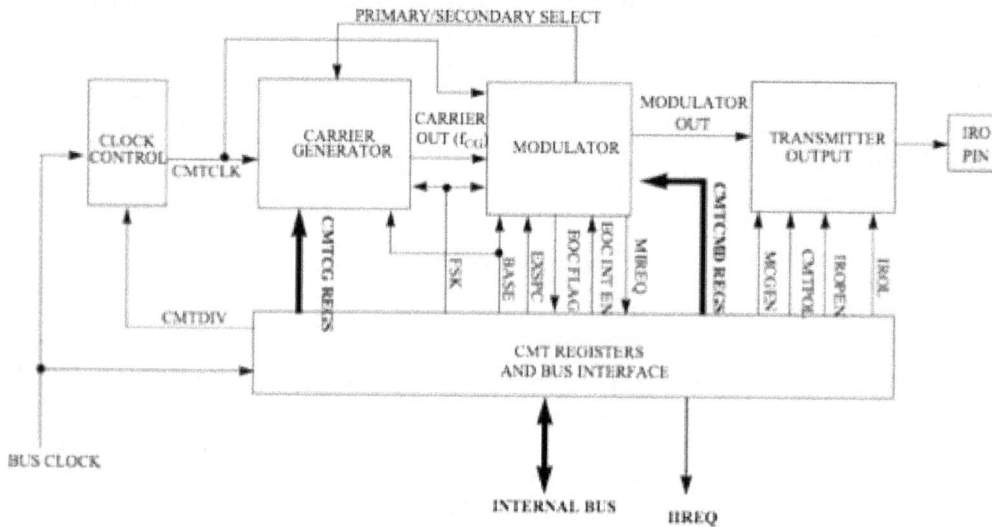

**FIGURE 16.6. CMT Block Diagram**

### Carrier Generator

The Carrier Generator has to able to generate one or two carrier frequencies, with programmable duty cycles. The generator uses an 8-bit up counter and two comparators to generate the low and high times of the carrier output waveform, this provides a wide range of flexibility for configuring the duty cycle and the period. The second output frequency is used for the Frequency Shift Key (FSK) modulation, and is implemented with an alternate

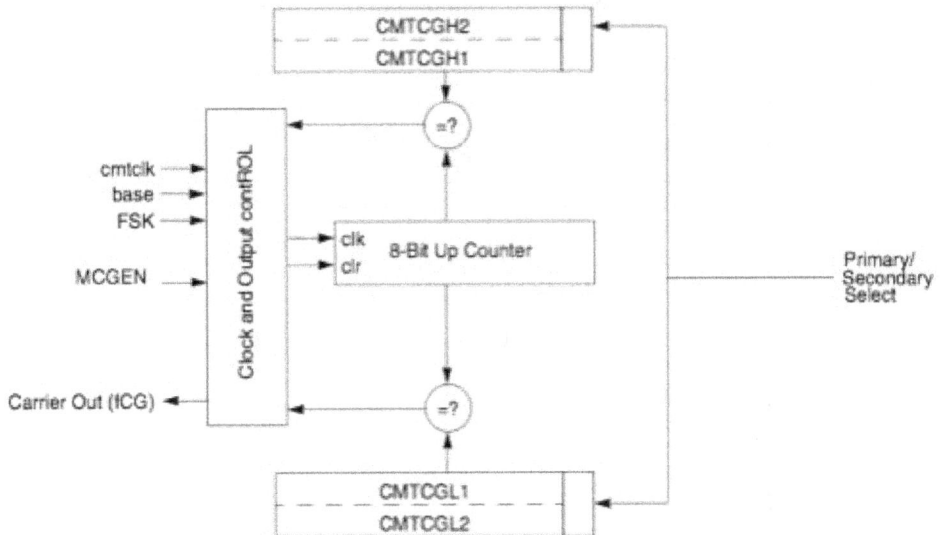

**FIGURE 16.7. CMT Carrier Generator**

set of registers for the high and low times, so that it may be independently configured for both frequency and duty cycle.

At the start of the cycle the 8-bit counter starts counting up from 0x01 and the output of the carrier out signal is driven high. The counter continues to count up until it reaches the value in the High count register (CMTCGH1), then the carrier out signal is driven low, the counter is reset to 0x01, and starts to count up until it reaches the value in the Low count register (CMTCGL1). The cycle is repeated automatically. For Frequency Shift Key (FSK) modulation a control signal from the modulator selects the alternate High and Low count registers (CMTGH2 and CMTGL2) as required for generating the alternate carrier frequency.

```
Carrier Generator Frequency = Frequency of CMT Clock/ (Highcount + Low-
count) Hz
```

### Modulator and Transmitter Output

The modulator generates the modulator output, either directly in the Baseband mode, or by modulating the carrier frequency in the Time and FSK modes. The main function of the modulator is to generate the mark and space periods of the output waveform.This is accomplished with 17-bit down counter which counts from a positive value to a negative value, with the mark period defined by the counter being positive, and the space period

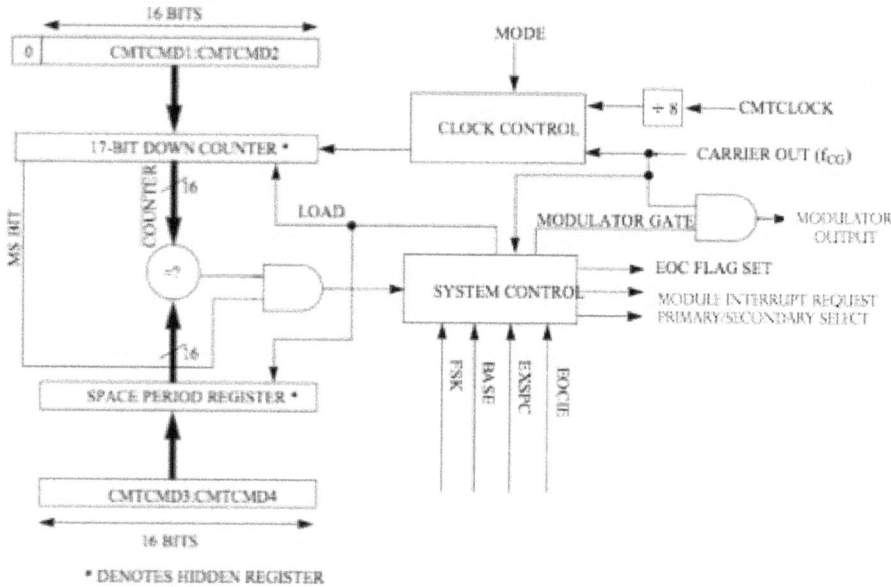

FIGURE 16.8. CMT Modulator

defined by the counter being negative. At the end of the space period, when the counter reaches the maximum negative value, the counter is reloaded with the positive value representing the mark period, and the cycle repeats. The down counter is decremented by the CMT clock divided by eight in Baseband and Time modes, and by the carrier out signal in FSK mode. The Carrier Modulation Timer Carrier Modulator Data (CMTCMDx) registers are user programmable, with CMTCMD1:CMTCMD2 holding the mark period, and CMTCMD3:CMTCMD4 holding the space period.

$$t_{mark} = (CMTCMD1:CMTCMD2 + 1) \div (f_{CMTCLK} \div 8)$$
$$t_{space} = CMTCMD3:CMTCMD4 \div (f_{CMTCLK} \div 8)$$

## Baseband Mode

In the Baseband mode the Mark and Space is the modulator output waveform, with Mark being the low output and Space the high output, as shown in Figure FIGURE 16.9. CMT Baseband Mode.

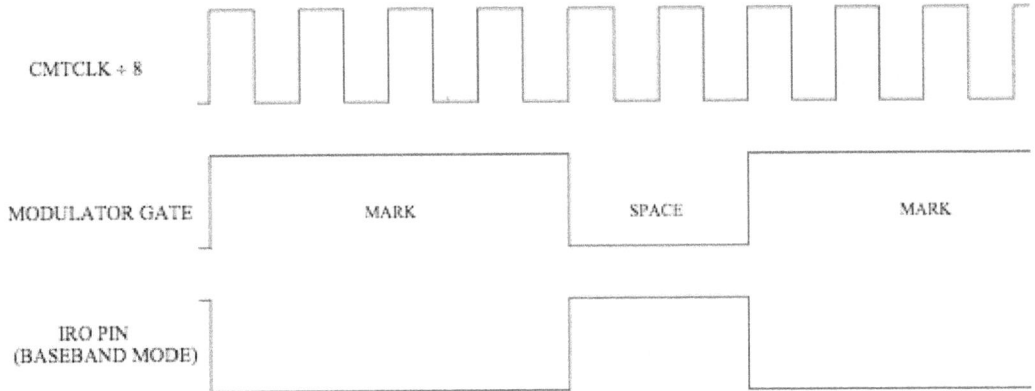

FIGURE 16.9. CMT Baseband Mode

## Time Mode

In the time mode the modulator Mark Space output is used to gate the carrier generator output, so that the modulator output is the carrier signal during the Mark period, and a steady high output during the Space period, as shown in Figure FIGURE 16.10. CMT Time Mode.

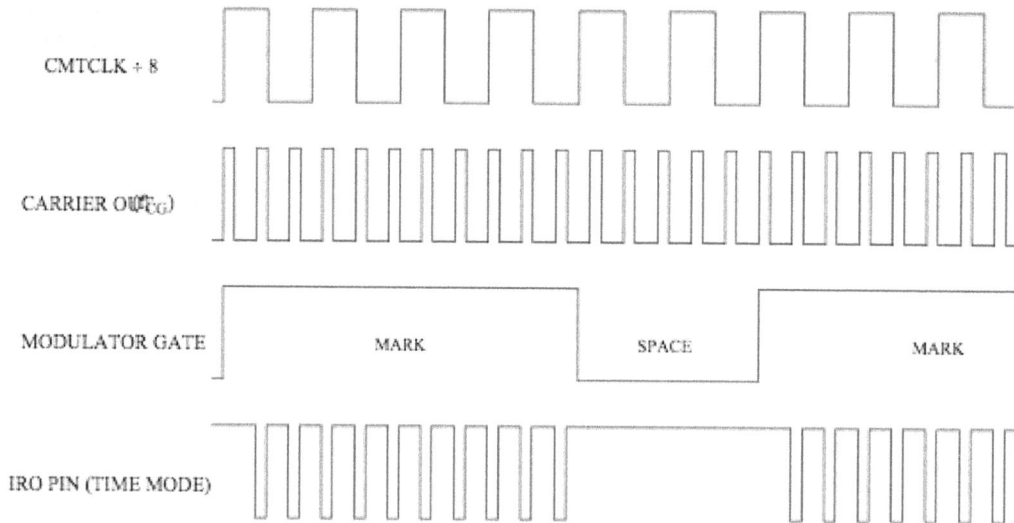

**FIGURE 16.10. CMT Time Mode**

### Frequency Shift Key (FSK) Mode

In the FSK mode an integer number of carrier clocks is output during the Mark period, with primary and alternate carrier frequency clocks being used in successive Mark periods, and a steady output during the Space periods. The use of the carrier output signal as the source clock to decrement the modulator down counter inherently ensures that both Mark and Space periods are an integer number of carrier output clocks, and coherent. It is permissible for the Space period to be zero. Please note that each Mark Space set is a modulation cycle, and the CMT must be serviced by software to setup the next, alternate, Mark

Space cycle. The switching of the carrier output frequency between primary and alternate is automatic in FSK mode.

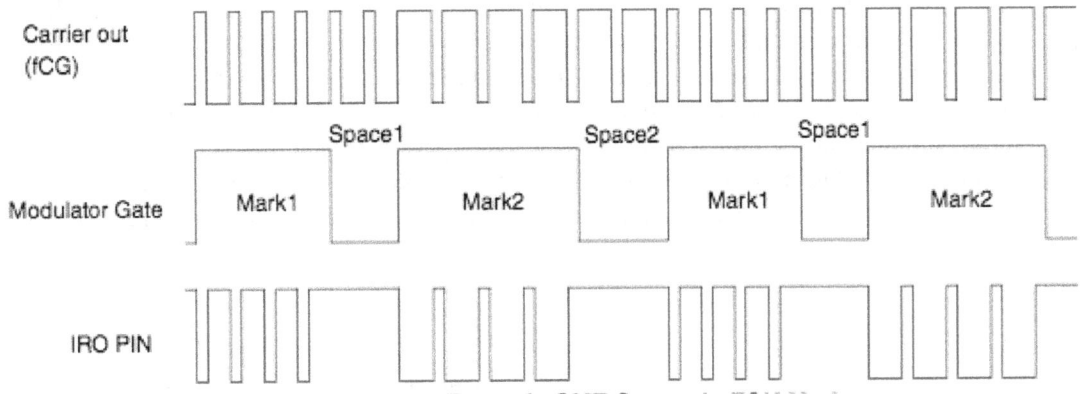

FIGURE 16.11. **CMT FSK Mode**

### Extended Space Operation

It is possible to extend the Space period beyond the maximum possible value in the Space period register (CMTCMD3:CMTCMD4) by setting the Extend Space (EXTSPC) bit in the CMT Modulator Status and Control Register (CMTMSC).This results in the modulator output being that of the Space state for the entire next Mark Space cycle, as shown in Figure FIGURE 16.12. Extended Space Operation.

FIGURE 16.12. **Extended Space Operation**

Note that in FSK mode the Mark and Space periods continue to be generated using the carrier generator primary and alternate output frequency clocks, and the user software must correctly track the FSK cycle and the extended space period generated.

The CMT Output Polarity (CMTPOL) bit in the CMT Output Control (CMTOC) register may be used to invert the polarity of the modulator output on the Infra Red Output (IRO) pin.

## PWM

The PWM timer provides the choice of two types of outputs: left- or center-aligned. In left-aligned output mode, the 8-bit counter is configured as an up counter only. It compares to two registers, a duty register and a period register. When the PWM counter matches the duty register, the output flip-flop changes state generating the PWM waveform. A match between the PWM counter and the period register resets the counter and the output flip-flop, and performs a load from the double buffer period and duty register to the associated registers. The counter counts from 0 to the value in the period register minus 1.

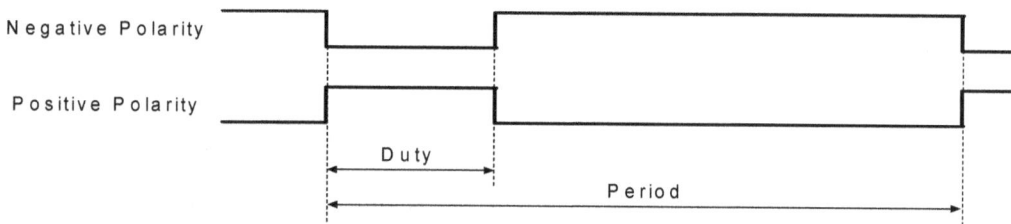

FIGURE 16.13. **Left Aligned PWM**

For center-aligned output mode the 8-bit counter operates as an up/down counter, and is set to count up whenever the counter is equal to 0x00. The counter is compared with two registers, a duty register and a period register. When the PWM counter matches the duty register, the output flip-flop changes state, generating the PWM waveform. A match between the PWM counter and the period register changes the counter direction from an up-count to a down-count. When the PWM counter decrements and matches the duty register again, the output flip-flop changes state causing the PWM output to also change state. When the PWM counter decrements and reaches zero, the counter direction changes from a down-count back to an up-count, and a load from the double buffer period and duty registers to the associated registers is performed. The counter counts from 0 up to the value in the period register and then back down to 0.

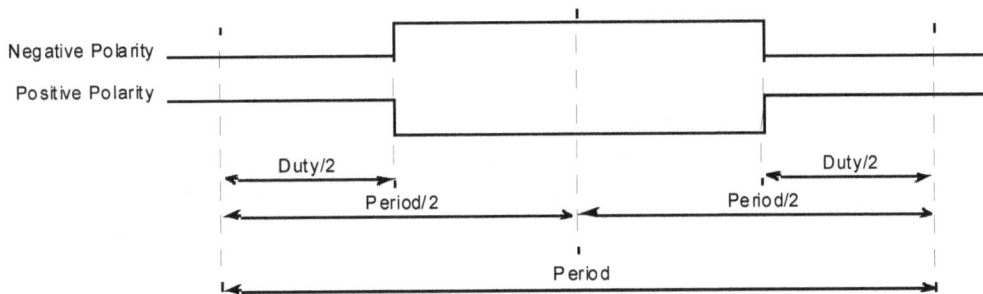

FIGURE 16.14. **Center Aligned PWM**

**PWM 16-Bit Functions.** The PWM timer has the option of eight 8-bit channels or four 16-bit channels for greater PWM resolution. The 16-bit channel option is achieved through the concatenation of two 8-bit channels.

The PWM control register contains four concatenation control bits, each used to concatenate a pair of PWM channels into one 16-bit channel. When channels are concatenated, the even numbered channel registers become the high order bytes of the double byte channel. When using the 16-bit concatenated mode, the clock source is determined by the low order 8-bit channel clock select control bits (the odd numbered channel). The resulting PWM is generated on the pins of the corresponding low order 8-bit channel. The polarity of the resulting PWM output is controlled by the corresponding low order 8-bit channel control registers.

Once concatenated mode is enabled, enabling/disabling the corresponding 16-bit PWM channel is controlled by the low order channel control bits. In this case, the high order bytes control bits have no effect, and their corresponding PWM output is disabled. In concatenated mode, writes to the 16-bit counter by using a 16-bit write will reset the 16-bit counter. Reads of the 16-bit counter must be made by 16-bit access to maintain data coherency.

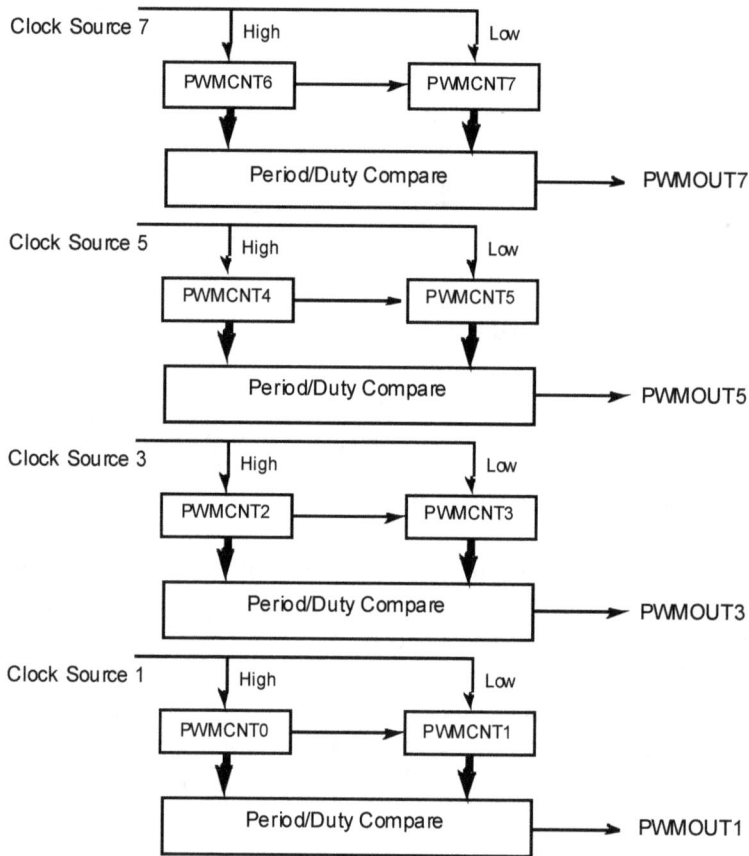

FIGURE 16.15. 16-bit PWM Mode

Either left- or center-aligned output mode can be used in concatenated mode and is controlled by the low order control bit.

# CHAPTER 17

# *Analog Systems*

## *Introduction to Analog Systems*

Most embedded control applications need to monitor analog values because the real world is primarily analog in nature. A real world value, such as temperature, is converted to an electrical value, either voltage or current, by a sensor. The electrical value then needs to be converted to a digital value usable by the microcontroller. This is done using an Analog to Digital Converter. Current Flexis and ColdFire® V1 products offer the following analog functionality modules:

- 12-bit resolution Analog-to-Digital Converter (ADC)
- 16-bit resolution Analog-to-Digital Converter (ADC16)
- Programmable Delay Block (PDB)
- 12-bit Digital to Analog Converter (DAC)
- Programmable Analog Comparator (PRACMP)
- Voltage Reference (VREF)
- General purpose Operation Amplifier (OPAMP)
- Trans-Impedance Amplifier (TRIAMP)

An Analog comparator compares two input voltages in the analog domain, but the output is digital indicating if one input voltage is greater than the other, or not.A Digital to Analog Converter (DAC) converts a digital value to an analog voltage or current, typically used to control external loads such as solenoids or other electromagnetic actuators. In most cases a Pulse Width Modulation (PWM) timer is used with an external low pass filter to generate an analog output, but in some applications it is highly desirable to be able to generate a steady analog output voltage without any additional external components.This need is met by an integrated Digital to Analog Converter Module (DAC). Operational Amplifiers (OPAMP) are used for analog signal conditioning and shifting, such scaling and shifting the analog output from a sensor to the input range of the microcontroller ADC for best overall system accuracy. A Trans-Impedance Amplifier (TRIAMP) is used to convert input current into voltages that can be measured with an ADC. Most of the analog systems require an accurate voltage reference to function. This reference is generated by the internal Voltage Reference (VREF) unit.

Since an analog signal may represent an infinite number of values, and may change very quickly, the conversion process invariably leads to some loss of information referred to as the conversion error. The conversion error has a number of components, quantization error, offset, gain, linearity, and noise. These all add up to an overall accuracy of the converter. The resolution of a converter depends on the number of binary bits used to represent the analog value, and is determined by the fundamental design of the converter.

In addition to these errors it is important to consider if a converter is monotonic. A monotonic converter will always follow the value direction of the analog signal including all errors. That means that if the analog value increases, the converted result will also have a greater value. A non-monotonic converter could return a converted value that is actually lower even when the analog value increased. Non-monotocity is typically caused by noise in the converter system. The reason monotocity is important in a control system is that quite often, for small changes in the analog value, it is critical for the control system to know the direction of the change rather than the absolute value. The ADC and ADC16, like all Freescale analog to digital converters, are monotonic.

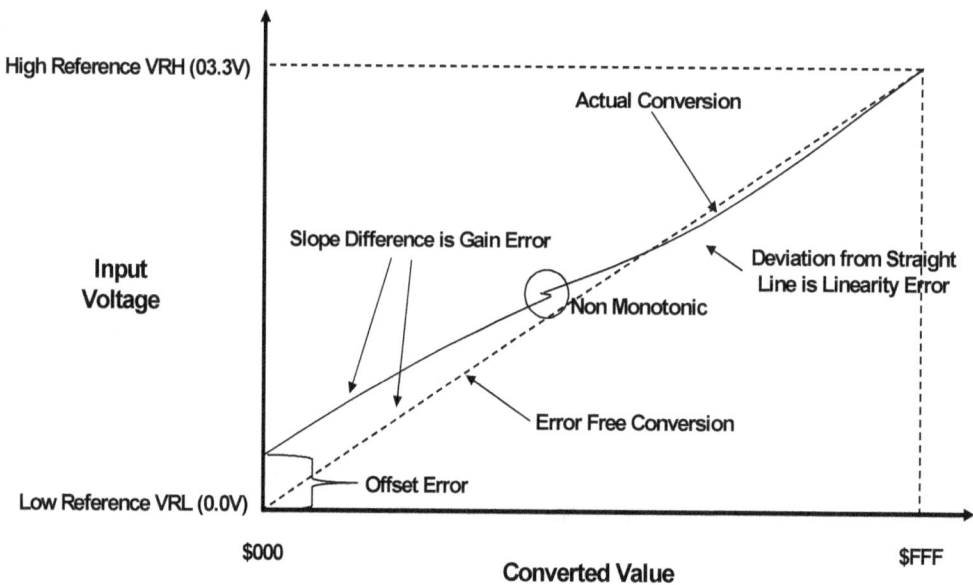

FIGURE 17.1. ADC Conversion Errors

Other parameters to be considered with analog to digital conversion systems is the sampling frequency, the sample time, and the sampling impedance of the converter and analog source. The sampling frequency determines the maximum frequency of the incoming signal to be sampled, with the sampling frequency needing to be at least twice the highest frequency of the analog signal. The sample time and impedance directly impact conversion accuracy. During sampling the converter connects a sampling capacitor to the analog voltage source. After the sampling time the sample capacitor is disconnected from the analog source and the voltage on the sample capacitor is converted to a digital value.

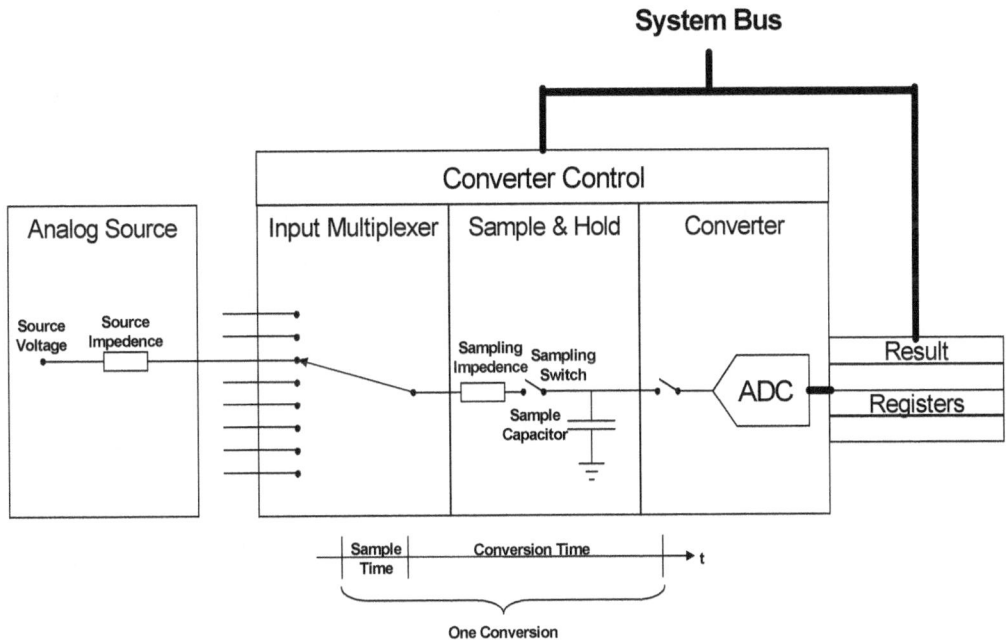

FIGURE 17.2. Simplified ADC Block

The greater the impedance of the analog source, the longer the sampling time needed to fully charge the sample capacitor for an accurate conversion. Analog source impedance is typically not an issue with active sensors such a semiconductor temperature sensor, but could be a problem with a passive thermistor sensor. Similar considerations apply when using resistive divider networks to bring the range of an analog voltage into the input range of the converter. The input range of the converter is set by applying the appropriate low and high voltages to the converter reference signals. For best conversion accuracy the voltage range between the two reference inputs should be as high as possible, and must be highly stable. The accuracy of the converter is specified with the maximum reference voltage differential and zero noise on the reference.

## *12-bit resolution Analog to Digital Converter (ADC)*

All Flexis and ColdFire V1 microcontrollers include either the 12-bit ADC or the 16-bit ADC16 Analog to Digital Converter, or both (see **Table xx for details**). The ADC consists of six main functional blocks:

- Input multiplexer, allows the selection of one out of up to 28 analog inputs (The MCF51QE128, for example, includes 24 analog inputs)
- Sample and hold unit, captures the analog voltage to be converted. This allows a precise sample to be taken and prevents the value from changing during the conversion process
- The converter unit, converts the analog voltage into a digital value using successive approximation
- Temperature sensor, to facilitate temperature calibration for maximum accuracy
- Result register(s), hold the converted value and allow this to be accessed by the system CPU
- Control unit, allows the selection of the required conversion mode

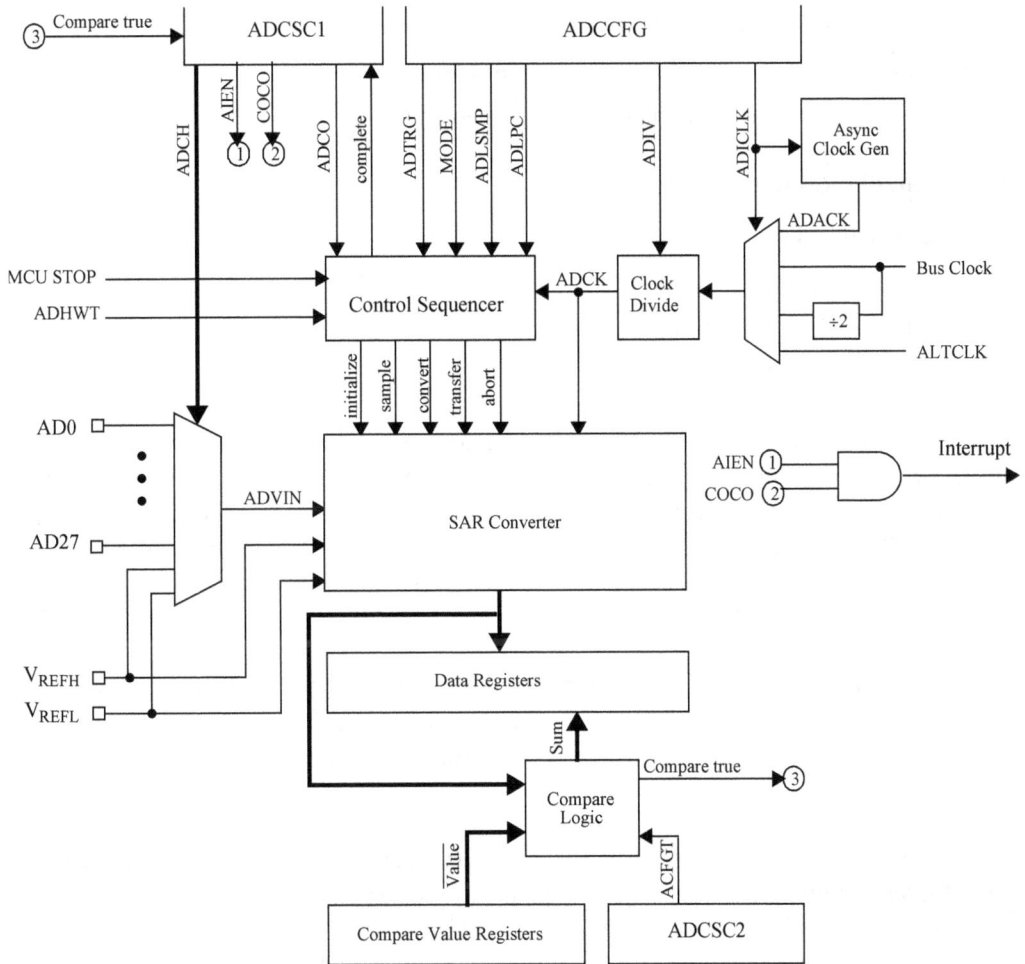

FIGURE 17.3. ADC12 Functional Block Diagram

The ADC offers a number of options in it's operation, that together provide a highly flexible system that is configureable for many varied applications. The options include:

- Programmable resolution, 12, 10, or 8 bits
- Conversion initiation, software initiated, externally triggered, and continuous
- Several module clocking options
- On chip temperature sensor
- Configureable sample time

- Configureable conversion time
- Conversion complete flag and interrupt
- Operation during low power wait and stop modes
- Automatic compare of result against a pre-selected value

**Programmable Resolution**

The ADC includes three programmable resolutions:

- 12-bit unsigned, right justified
- 10-bit unsigned, right justified
- 8-bit unsigned, right justified

Because the ADC uses successive approximation to convert the analog value to digital, the number of resolution bits has a direct impact on the time required to complete a conversion. The ADC's converter actually only employs 12-bit and 9-bit conversion sequences, so when a 10-bit resolution is selected the 12-it conversion result is rounded to 10-bits before being written to the result registers. Similarly, when 8-bit resolution is selected, the 9-bit conversion result is rounded to 8-bits.

The actual conversion time depends on the resolution selected, the frequency of the ADC clock, sample time, and whether single or continuous conversion mode is selected. The relevant product user manual includes the details for the actual conversion time calculation.

The integrated temperature sensor was included specifically to allow a system to be calibrated for temperature effects, which may then be used to compensate for temperature in the application and effectively raise the overall analog system accuracy. It should be remembered that this temperature sensor measures the microprocessors die temperature, not the ambient temperature. It is also possible to use this sensor to monitor the operating temperature of the microprocessor, and if necessary control system power consumption so as to keep the system operational, possibly with reduced functionality, at temperatures outside the normal operational range. In applications with regular standby mode operation, it may also be possible to make an approximate ambient temperature measurement if carried out immediately after exiting the standby mode.

**Conversion Initiation**

The ADC supports three method for conversion initiation:

- Software, initiated by writing to the ADC Status and Control register 1 (ADCSC1)

- An external hardware trigger signal (or internal timer trigger signal on some products)
- Continuous conversions, in this mode a new conversion will be started immediately following the completion of a conversion. Continuous conversion may be initiated by software or the hardware trigger

The ADC conversion system may be turned off by witting all 1's to the Input channel select field (ADCH) of the ADC Status and Control register 1. A conversion is completed when the result is written into the result registers (ADCRH and ADCRL) and the conversion complete flag (COCO) in the ADC Status and Control register 1 is set. A write to the ADC Status and Control register 1 (ADCSC1) while a conversion is in progress, will abort the current conversion and initiate the new conversion.

### ADC Clock

The ADC has two external clock inputs used to drive two clock domains within the ADC module.

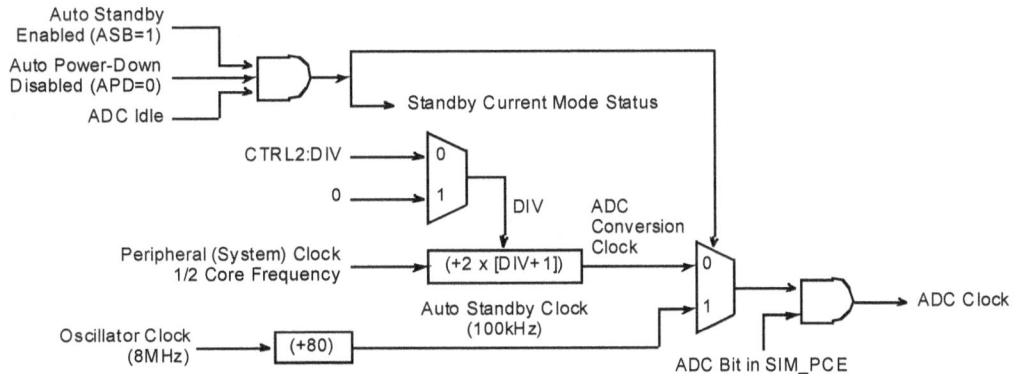

FIGURE 17.4. **ADC Clock Generation**

### Interrupts

Interrupts are available and separately enabled or disabled for the following events:

- End Of Scan reached
- Zero Crossing detected (positive, negative, or both)
- Low Limit detected
- High Limit detected

## *16-bit resolution Analog-to-Digital Converter (ADC16)*

The 16-bit ADC16 Analog to Digital Converter was designed for applications requiring very high analog conversion accuracy, or a high dynamic range, such as current measurement in electricity meters, or in medical applications. It is available in products specifically intended for these applications. (see **Table xx for details**). Like the 12-bit ADC, the ADC16 is a successive approximation converter. For applications where several analog signals have to be measured concurrently, such as multi-phase electricity meters, several ADC16 converters may be included in the microcontroller. The MCF51EM for example, includes four ADC16 converters. The Programmable Delay Block (PDB) facilitates flexible synchronous or phase shifted scheduling of the analog input sampling. The ADC16 consists of six main functional blocks:

- Input multiplexer with support up to 4 differential and up to 24 single ended input channels
- Sample and hold unit, captures the analog voltage to be converted. This allows a precise sample to be taken and prevents the value from changing during the conversion process
- The converter unit, converts the analog voltage into a digital value using successive approximation
- Temperature sensor and calibration unit
- Result register(s), hold the converted value and allow this to be accessed by the system CPU
- Control and trigger control units

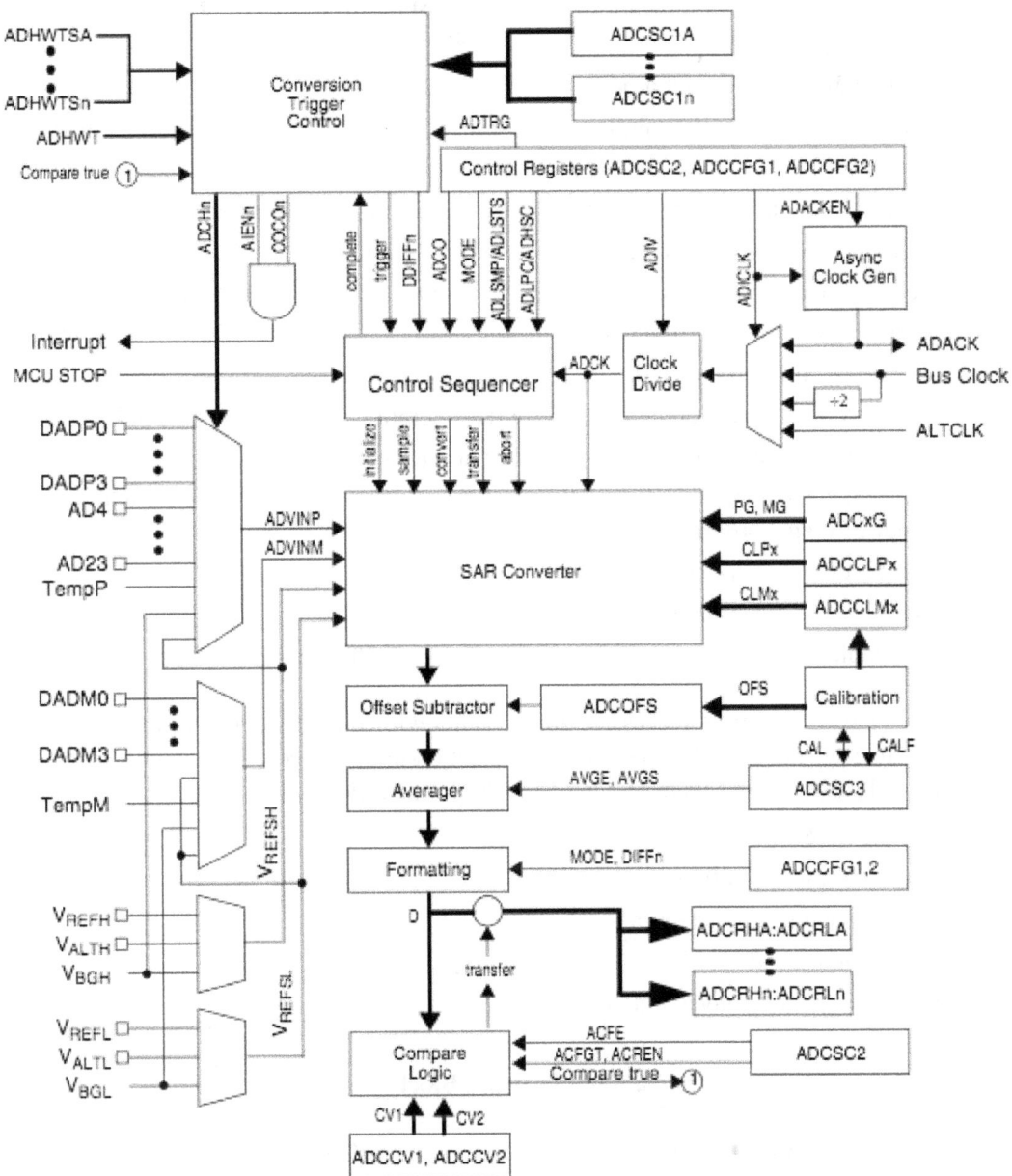

**FIGURE 17.5. ADC16 Block Diagram**

The ADC16 offers a number of options in it's operation, that together provide a highly flexible system that is configureable for many varied applications. The options include:

- Up to 4 pairs of differential and 24 single-ended external analog inputs
- Four Differential output modes; 16-bit, 13-bit, 11-bit and 9-bit
- Four single-ended output modes; 16-bit, 12-bit, 10-bit and 8-bit
- Output formatted in 2's complement 16-bit sign extended for differential modes
- Output in right-justified unsigned format for single-ended
- Single or continuous conversion (automatic return to idle after single conversion)
- Configureable sample time and conversion speed/power
- Conversion complete / Hardware average complete flag and interrupt
- Input clock selectable from up to four sources
- Operation in wait or stop3 modes for lower noise operation
- Asynchronous clock source for lower noise operation with option to output the clock
- Selectable asynchronous hardware conversion trigger with hardware channel select
- Automatic compare with interrupt for less-than, greater-than or equal-to, within range, or out-of-range, programmable value
- Temperature sensor and Self Calibration unit
- Hardware average function
- Selectable voltage reference, Internal, External, or Alternate

Because the ADC16 uses successive approximation to convert the analog value to digital, the number of resolution bits has a direct impact on the time required to complete a conversion. The ADC16 conversion time depends on the following configuration:

- Sample time
- Microprocessor bus frequency
- Conversion mode, differential or single ended
- Conversion speed
- Frequency of the conversion clock

**Conversion Initiation, Programmable Delay Block (PDB)**

The ADC16 supports three method for conversion initiation:

- Software, initiated by writing to the ADC Status and Control register 1 (ADCSC1A)
- Hardware trigger signal (selected in ADHWT), generated by the Programmable Delay Block (PDB)

- Continuous conversions, in this mode a new conversion will be started immediately following the completion of a conversion. Continuous conversion may be initiated by software or the hardware trigger

The ADC16 module remains in its idle state until a conversion is initiated. If ADACK is selected as the conversion clock source but the asynchronous clock output is disabled (ADACKEN=0), the ADACK clock generator will also remain in its idle state (disabled) until a conversion is initiated. If the asynchronous clock output is enabled (ADACKEN=1), it will remain active regardless of the state of the ADC16 or the micro-controller power mode. Power consumption of the active ADC16 can be reduced by using the ADC16 internal oscillator by setting the ADLPC bit. This selects a lower maximum value for ADC16 conversion clock.

### ADC16 Clock

The ADC16 has four possible clock sources:

- The bus clock. This is the default selection following reset
- The bus clock divided by two. For higher bus clock rates, this allows a maximum divide by 16 of the bus clock with using the ADIV bits
- Alternate clock (ALTCLK), as defined according to the microprocessor being used
- The asynchronous clock (ADACK). This clock is generated from a clock source within the ADC module. Conversions are possible using ADACK as the input clock source while the MCU is in stop3 mode

Whichever clock is selected, its frequency must fall within the specified frequency range for the ADC16 clock (ADCK). If the available clocks are too slow, the ADC16 may not perform according to specifications. If the available clocks are too fast, the clock must be divided to the appropriate frequency. This divider is specified by the ADIV bits and can be divide-by 1, 2, 4, or 8.

### Interrupts

An interrupt per ADC16 module is available to signal the completion of a conversion in addition to the Conversion Complete (COCO) flag being set. The interrupt is enabled by setting the ADC16 Interrupt Enable bit (AIENn)

*LCD Controller*

### *Introduction to the LCD Controller*

The LCD controller is designed to generate the appropriate waveforms to drive multi-plexed numeric, alpha-numeric, or custom LCD panels. The generic name used here to cover all forms of displays is "glass". Depending on LCD controller hardware and soft-ware configuration, the LCD glass can be either 3V or 5V. The LCD controller also has several timing and control settings that can be software configured depending on the appli-cations requirements.

While the LCD controller architecture supports up to 64 LCD drive pins, the maximum number of pins available on current microcontrollers is less than that, and varies with package type. Moreover, the number of available pins, in combination with the ratio of front and back planes, dictates the maximum number of LCD segments that are supported. Table 18.1 shows example options available in the MCF51EM256 series of devices, for different package types and the number of backplanes required by the LCD glass.

**TABLE 18.1 MCF51EM256 Example Configurations by Package Type**

| 100 pin package 44 LCD Pins | 80 pin package 37 LCD Pins | Backplanes | Frontplanes | Segments |
|:---:|:---:|:---:|:---:|:---:|
| ✔ | | 8 | 36 | 288 |
| ✔ | | 2 | 40 | 80 |
| ✔ | ✔ | 8 | 25 | 200 |

Figure 18.1 shows a typical 3 digit, 7 segment LCD. By taking advantage of the fact that an LCD segment responds to the rms value of an alternating voltage, it is possible to time division multiplex drive waveforms to each segment using a combination of frontplanes and backplanes provided by the LCD Controller. The front and back planes are configured in a matrix that ensures every segment is connected to unique front and back plane pairs. The value of the rms voltage across each pair determines whether the LCD segment is on

(high rms voltage) or off (low rms voltage). This minimizes the number of physical pins needed compared with statically driving each segment from a separate pin. In the example shown in Figure 18.1 each digit in the LCD has two groups of segments labelled FGED and ABC. Each group of segments is controlled by a separate front plane. The back plane is common across a different set of segments; D, EC, GB and FA.

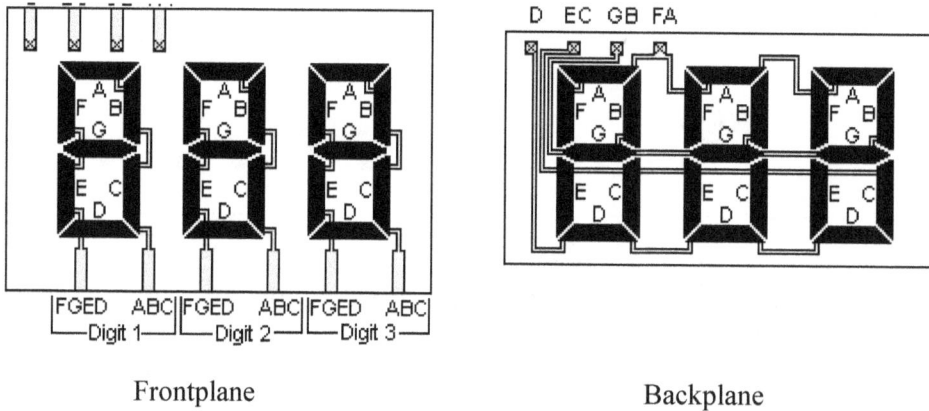

Frontplane                                              Backplane

FIGURE 18.1.  Front and Back Planes of 3 digit, 7 segment LCD Display

So to turn on, say segment E in any digit, it is necessary to apply a high rms voltage across the front and back plane pair FGED and EC. Of course, because more than one segment is normally turned on, multiple front and back planes must combine to generate "on" voltages at different time slots in the waveform, resulting in complex waveforms, similar to that shown in Figure 18.2, created by using multi-level voltages from a bias generator. This LCD Controller uses a 1/3 bias that produces a seven step differential voltage range from VLL3 to -VLL3.

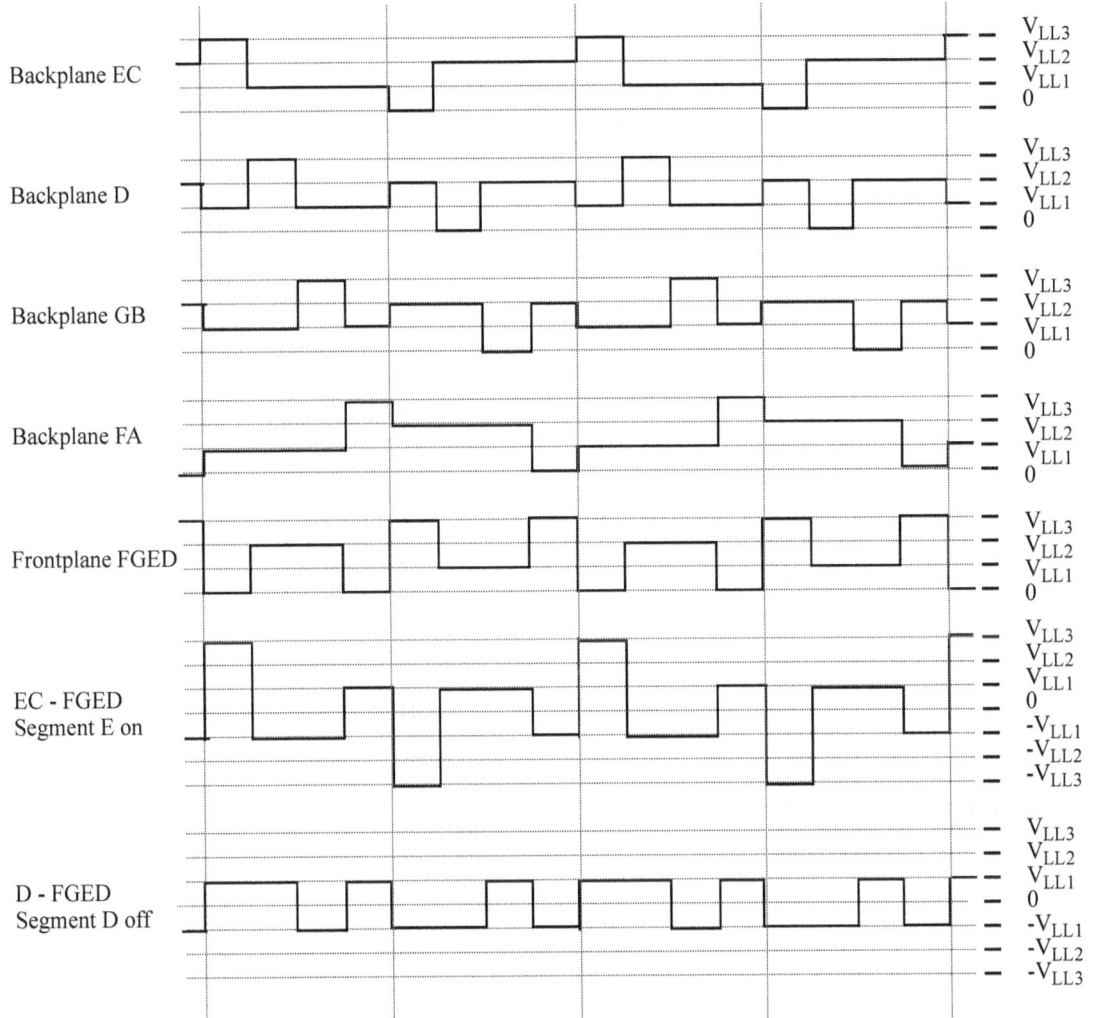

**FIGURE 18.2. Front and Backplane Waveforms to activate segment E in the example**

One consequence of these waveforms is that it is never possible to reduce the "off" rms voltage to zero, as shown for segment D. Another point to note is the applied voltage should contain a zero average DC value, since DC offsets cause long term degradation of the LCD segments.

Table 18.2 shows the relationship between on-off voltages, expressed as ratios for various combinations of backplanes and bias voltages. The values in italics are the optimum for a particular combination of backplane and bias voltage. The table indicates that as the number of backplanes increases, there comes a point when the optimum bias ratio must be increased too. The choice of 1/3 bias voltage for this LCD controller gives an optimum on-off ratio for 2 to 6 backplanes, and an almost ideal ratio for 7 and 8 backplanes.

TABLE 18.2 On-Off Ratio for Different Numbers of Backplanes and Bias Ratios

| Number of Backplanes | Bias Ratio | | | | | | |
|---|---|---|---|---|---|---|---|
| | 2 | 3 | 4 | 5 | 6 | 7 | 8 |
| 2 | 2.236 | *2.236* | 1.844 | 1.612 | 1.475 | 1.387 | 1.325 |
| 3 | 1.732 | *1.915* | 1.732 | 1.567 | 1.453 | 1.374 | 1.318 |
| 4 | 1.528 | *1.732* | 1.648 | 1.528 | 1.433 | 1.363 | 1.311 |
| 5 | 1.414 | *1.612* | 1.581 | 1.494 | 1.414 | 1.352 | 1.304 |
| 6 | 1.342 | *1.528* | 1.528 | 1.464 | 1.397 | 1.342 | 1.297 |
| 7 | 1.291 | 1.464 | *1.483* | 1.438 | 1.382 | 1.332 | 1.291 |
| 8 | 1.254 | 1.414 | *1.446* | 1.414 | 1.367 | 1.323 | 1.285 |
| 9 | 1.225 | 1.374 | *1.414* | 1.393 | 1.354 | 1.314 | 1.279 |
| 10 | 1.202 | 1.342 | *1.387* | 1.374 | 1.342 | 1.306 | 1.274 |

The main features of the LCD controller include:

- LCD waveforms functional in low-power run, wait and stop modes
- Up to 64 LCD pins with selectable frontplane and backplane configuration
- Generate up to 63 frontplane signals
- Generate up to 8 backplanes signals
- Programmable LCD frame frequency
- Programmable blink modes and frequency
- All segments blank during blink period
- Alternate display for each LCD segment in 4 or less backplane mode
- Blink operation in low-power modes
- Programmable LCD power supply switch, making it an ideal solution for battery-powered and board-level applications
- Internal LCD power using $V_{DD}$ (1.8V to 3.6 V)
- Internal $V_{IREG}$ regulated power supply option for 3V or 5V LCD glass
- External $V_{LL3}$ power supply option (3V)
- Internal regulated voltage source with a 4-bit trim register to apply contrast control
- Integrated charge pump for generating LCD bias voltages
- On-chip generation of bias voltages

- Backplane reassignment to assist in vertical scrolling on dot-matrix displays
- Software configurable LCD frame frequency interrupt
- An internal ADC channel is connected to VLL1 to allow software to monitor its value. This feature allows software to adjust the contrast.

## Architecture of the LCD Controller

Figure 18.3 shows the block diagram of the LCD Controller. The controller is driven by LCDCLK, derived from one of two clock sources. Typically, one of the sources is the OSCOUT1 crystal, while the alternate clock is the OSCOUT2. LCDCLK is applied through various prescalers to provide the blink frequency, LCD frame frequency and charge pump clock. The blink and charge pump clocks are described later in this chapter. The frame frequency is calculated from the LCLK and DUTY bits in the LCDC0 register and a constant Y derived from the DUTY value, which specifies the duty cycle of LCD the drive signals. The frame frequency *FrameFrequency* is given by:

$$FrameFrequency = \frac{LCDCLK}{(DUTY + 1) \times 8 \times (4 + LCLK) \times Y}$$

The Y value depends on the selected duty cycle, given by the three bit field DUTY, as shown in Table 18.3.

TABLE 18.3 Frame Frequency Y values for required Duty Cycle

| Duty Cycle | 1/1 | 1/2 | 1/3 | 1/4 | 1/5 | 1/6 | 1/7 | 1/8 |
|---|---|---|---|---|---|---|---|---|
| DUTY | 000 | 001 | 010 | 011 | 100 | 101 | 110 | 111 |
| Y | 16 | 8 | 5 | 4 | 3 | 3 | 2 | 2 |

The controller requires a voltage supply that can be derived from an external input to the VLL3 pin or from one of 2 internal supplies, VDD or VIREG.. The front and back plane pin allocation is defined by a set of up to 64 8 bit LCD waveform registers. The front and back plane allocation is defined by the LDCDPEN and LCDBPEN registers. Note that the number of waveform registers and allocation bits that can be programmed is equal to the number of physical pins on the particular device. The reader should refer to the Freescale product documentation before programming the controller. Various combinations of the bias voltages VLL1, VLL2 and VLL3 may be generated by either the on-chip charge pump or on-chip resistor ladder. The backplane sequencer generates up to 8 backplane phases. Other logic controls the display modes that are described in more detail later in this chapter. An interrupt can be enabled to occur at the frame rate of the waveform generator.

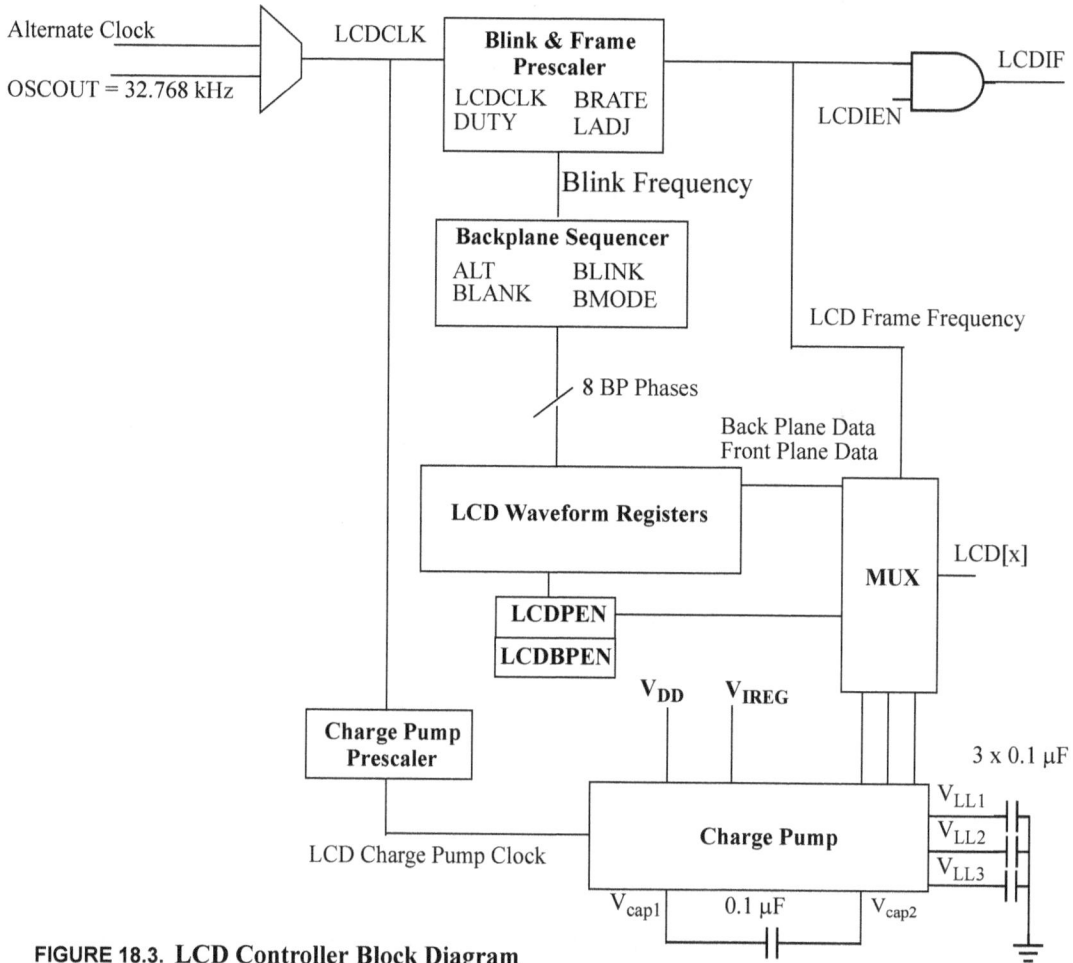

**FIGURE 18.3. LCD Controller Block Diagram**

## Pin Connections for the LCD Controller

The LCD module has several external pins dedicated to powering the LCD and providing frontplane and backplane signaling. These pins are:

- LCD[n:0]

  The value n depends on the particular device and package. When LCD functionality is enabled by the PEN bits in the LCDPEN registers, the corresponding LCD pin will generate a frontplane or backplane waveform depending on the configuration of the backplane-enable bit field, BPEN[n:0]. There is an optional configuration, controlled

**Chapter 18**

by the LCDC1[FCPDEN] bit that provides GPIO functionality on the LCD pin, if it is not used for waveform generation.

- VLL1, VLL2, VLL3

    VLL1, VLL2, and VLL3 are bias voltages for the LCD module driver waveforms which can be generated using the internal charge pump or the internal resistive divider network. The charge pump can be optionally configured to generate VLL2 and VLL1 from an external voltage applied to VLL3. In this case VLL3 should never be set to a voltage other than VDD. When using the charge pump, each of these pins should have a ceramic 0.1uf capacitor attached to ground. The capacitors may be retained when changing between the charge pump and internal resistive divider modes. However, if the application is being designed to operate only in resistive divider mode, the capacitors may be eliminated to minimize cost and board layout.

- Vcap1, Vcap2

    The charge pump capacitor is used to transfer charge from the input supply to the regulated output. A ceramic 0.1uf capacitor should be used.

### Waveform Generators

Three registers must be configured to enable LCD waveform signals on the device pins. They are the LCDPEN, BPEN and LCDWF registers.

Each bit in the LCDPEN registers maps into a single physical pin available on the device, and must be set to enable the pin for waveform generation. Figure 18.4 shows the bit allocations in the LCDPEN register for the first eight LCD pins.

FIGURE 18.4. LCDPEN register for LCD Pins 0 to 7

| | 7 | 6 | 5 | 4 | 3 | 2 | 1 | 0 |
|---|---|---|---|---|---|---|---|---|
| LCDPEN0 | PEN7 | PEN6 | PEN5 | PEN4 | PEN3 | PEN2 | PEN1 | PEN0 |

Each LCDBPEN register bit, which also maps into a single physical pin available on the device, selects whether the corresponding LCD pin operates as an LCD backplane (bit is set) or LCD frontplane (bit is cleared). Because the LCD waveform generator hardware supports up to eight backplanes, no more than eight BPEN bits should be set. These registers should be initialized before enabling the LCD controller. Figure 18.5 shows the bit allocations in the LCDBPEN register for the first eight LCD pins.

FIGURE 18.5. LCDBPEN register for LCD Pins 0 to 7

| | 7 | 6 | 5 | 4 | 3 | 2 | 1 | 0 |
|---|---|---|---|---|---|---|---|---|
| LCDBPEN0 | BPEN7 | BPEN6 | BPEN5 | BPEN4 | BPEN3 | BPEN2 | BPEN1 | BPEN0 |

**LCD Controller**                                                                                       **369**

Each LCD glass segment, as explained in the introduction to this chapter, is associated with a unique pair of frontplnae and backplane signals. For an LCD pin configured as a frontplane its LCDWF register controls the on/off state for all segments connected to that particular pin. Each of the eight bits in the register correspond to a separate backplane phase. Setting a bit in the register causes the segment associated with that frontplane and backplane to turn on.

For an LCD pin configured as a backplane, an LCDWF register defines which of eight backplane phases is assigned to the LCD pin. Figure 18.6 shows the bit allocations in the LCDWF register for the first LCD pin 0.

FIGURE 18.6. **LCDWF register for LCD Pins 0 to 7**

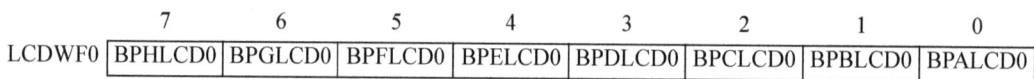

| | 7 | 6 | 5 | 4 | 3 | 2 | 1 | 0 |
|---|---|---|---|---|---|---|---|---|
| LCDWF0 | BPHLCD0 | BPGLCD0 | BPFLCD0 | BPELCD0 | BPDLCD0 | BPCLCD0 | BPBLCD0 | BPALCD0 |

Note that the actual number of backplane phases generated is controlled by the DUTY bit field in the LCDC0 register. Table 18.4 shows the relationship between the DUTY values and the phase sequence for the backplanes, designated A to H.

TABLE 18.4 **LCD Controller Duty Cycle Selection**

| Duty | LCDC0 Register | | | Number of Backplanes | Phase Sequence |
|---|---|---|---|---|---|
| | DUTY2 | DUTY1 | DUTY0 | | |
| 1/1 | 0 | 0 | 0 | 1 | A |
| 1/2 | 0 | 0 | 1 | 2 | A B |
| 1/3 | 0 | 1 | 0 | 3 | A B C |
| 1/4 | 0 | 1 | 1 | 4 | A B C D |
| 1/5 | 1 | 0 | 0 | 5 | A B C D E |
| 1/6 | 1 | 0 | 1 | 6 | A B C D E F |
| 1/7 | 1 | 1 | 0 | 7 | A B C D E F G |
| 1/8 | 1 | 1 | 1 | 8 | A B C D E F G H |

## Display Modes

The LCD module can be configured to implement several different display modes that contribute to lower power operation, or provide an alternative display output without CPU intervention.

The ALT bit in the LCDBCTL register enables the selection of an alternate display. The alternate display mode allows two separate frontplane values to be located in a single LCDWF register. The alternate values are located in the most significant nibble (4 bits) of the LCDWF frontplane register (corresponding to bit positions allocated to backplane phases E to H) while the normal segment values are in the least significant nibble (corresponding to blackplane phases A to D). The actual number of used bits in each nibble is equal to the required number of backplanes. Used bits must be right justified in the nibble, such that the order of use is from A to D for the normal display, and from E to H for the alternate display. This implementation restricts the use of the alternate display mode to a maximum of four backplanes. It is important to enable the alternate backplane on the same pin as the normal backplane in the appropriate LCDWF register.

The BLINK and BMODE bits in the LCDBCTL register allow alternating displays at a programmed frequency. If BLINK is clear, the LCD controller operates normally. If BLINK is set, the controller's operation is affected by BMODE, BLANK and ALT bits, as well as the selected LCD duty cycle. Table 18.5 shows the range of options available. Note that if BLINK is set when more than four backplanes are used, the alternate display cannot be viewed.

TABLE 18.5 Display Mode Interaction

| BLANK | ALT | BMODE | LCD Duty | BLINK = 1 | |
|-------|-----|-------|----------|-----------|------------|
| | | | | Normal Period | Blink Period |
| 0 | 0 | 0 | 1-4 | Normal Display | Blank Display |
| 0 | 0 | 1 | 1-4 | Normal Display | Alternate display |
| 0 | 1 | 0 | 1-4 | Alternate display | Blank Display |
| 0 | 1 | 1 | 1-4 | Alternate Display | Alternate display |
| 1 | X | 0 | 1-4 | Blank Display | Blank DIsplay |
| 1 | X | 1 | 1-4 | Blank Display | Alternate display |
| 0 | X | X | 5-8 | Normal Display | Blank Display |
| 1 | X | X | 5-8 | Blank Display | Blank Display |

The blink rate is defined by the LCDCLK frequency divided by the BRATE bits in the LCDBCTL register. The blink rate frequency *BlinkFrequency* can be computed from the following equation:

$$BlinkFrequency = (LCDCLK) \div 2^{(12 + BRATE)}$$

### Voltage Supplies and Charge Pump

The LCD controller contains a charge pump, resistive divider network, on-chip voltage regulator and power supply routing switches to support a range of power supply configurations and display voltage requirements.

Figure 18.7 shows the block diagram of the LCD power supply. The VSUPPLY control bits, in the LCDSUPPLY register, define where the LCD receives its power from. The choices are listed in Table 18.6. The VSUPPLY control bits should not be modified when the LCD controller is enabled (LCDEN=1).

**TABLE 18.6 LCD Power Supply Configuration Options**

| VSUPPLY[1:0] | Configuration |
|:---:|:---:|
| 00 | Drive $V_{LL2}$ from the internal $V_{DD}$ |
| 01 | Drive $V_{LL3}$ from the internal $V_{DD}$ |
| 11 | Drive $V_{LL3}$ from an external voltage<br>Or<br>Drive $V_{LL1}$ from the internal $V_{IREG}$ |

Setting the RVEN bit in the LCDRVC register enables the VIREG voltage regulator. Its voltage level may be adjusted in 1.5% increments by the RVTRIM bits in the same register. This feature allows software to control the contrast of the LCD glass. Additionally, HREFSEL, in the LCDSUPPLY register is used to set the VIREG to either 1V for 3V LCD glass, or 1.67V for 5V LCD glass applications.

The internal VDD may be used to power the charge pump, routed to either VLL2 or VLL3. The routing choices give the option of driving either 5V or 3V LCD glass from a nominally 3V device, a 3V LCD glass from a 2V device, or a 5V LCD glass from a 5V device.

When the charge pump is enabled, by setting CPSEL in the LCDSUPPLY register, the internal resistive divider is disabled. Control bits are provided in the LCDC1 register to allow the charge pump to continue operation during device wait and stop low power modes, if desired. The LCDSUPPLY register contains additional control bits to adjust the charge pump clock rate for different LCD glass capacitances, in four ranges up to 8000pf. Glass capacitance increases with increase in total area of the LCD segments. Setting the charge pump for higher capacitances will result in increased power consumption in the device. Applications that rely on low power operation should use the lowest capacitance setting possible.

If the charge pump is disabled and the resistive divider is used instead, the same control bits adjust the resistive divider to support LCD glass capacitance in two ranges up to 8000pf.

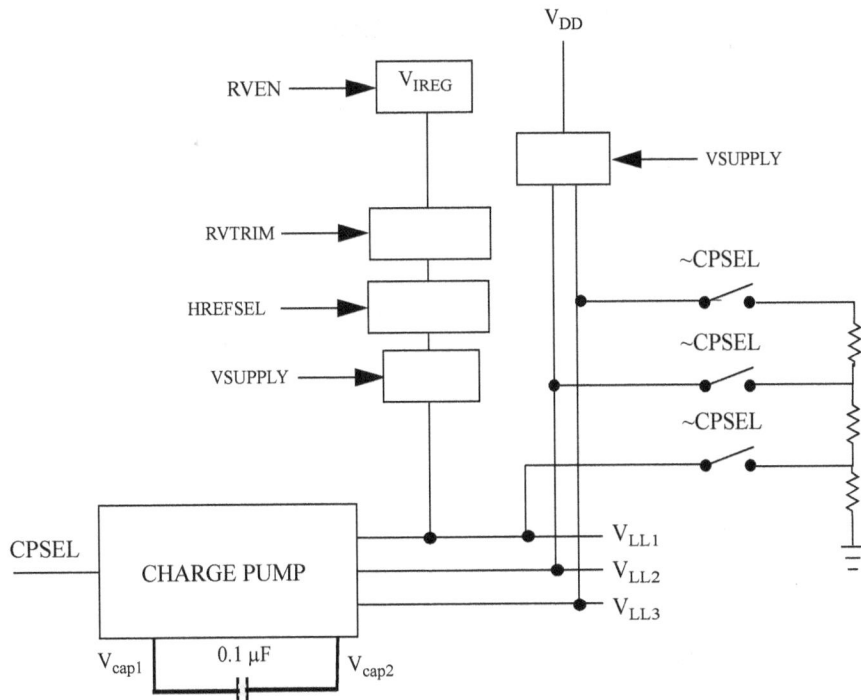

FIGURE 18.7. LCD Power Supply

The voltage applied to the VLL1 can be read internally from ADC channel 24, on devices that support the LCD controller.

The range of configuration options available to the power supply and charge pump allows system designers to trade-off power consumption against number of external components needed for a given application. Table 18.7 summarizes these trade-offs.

**TABLE 18.7 Power Supply Trade-offs**

| LCD Power Source Options | Benefits | Drawbacks | Power Consumption |
|---|---|---|---|
| External voltage applied to VLL3 | Least amount of external components required, since no capacitors are needed on charge pump or VLL2 and VLL1. Allows lower pin count devices. | Higher Power consumption, since internal resistive divider must be used. 3V devices cannot support 5V LCD glass. Degradation across battery life and temperature. | Low |
| Internal VIREG | Digital Contrast control No degradation of generated Voltage across battery life Support for 3V or 5V glass | Must support Charge pump operation with four 0.1uf external capacitors. | Lower |
| Internal VDD | Lowest Power | Must support charge pump operation with four 0.1uf external capacitors. | ULTRA LOW POWER |

## *Programming the LCD Controller*

After reset, the LCD controller has the following settings:

- LCDEN bit is cleared, thereby forcing all frontplane and backplane driver outputs to the high impedance state.
- 1/4 duty
- 1/3 bias
- LCLK[2:0], VSUPPLY[1:0], CPSEL, RVEN, and BRATE[2:0] revert to their reset values

There are some important caveats that should be considered when choosing the power source for the LCD controller and before programming its control registers.

- When the LCD controller power supply is configured in a mode that requires an external voltage applied to the VLL3 pin (with or without the charge pump enabled) the applied voltage must be equal to VDD. This means that for microcontrollers rated for the voltage range 1.8V to 3.6V, this power supply configuration cannot drive a 5V LCD display. If it is desired to drive a 5V display from a 1.8V to 3.6V rated microcontroller, use a mode that enables the internal VIREG supply.

- LCDWFx registers should be updated only with instructions that perform byte writes. Using instructions that perform word-write results in invalid data being stored. Because each write takes several clock cycles to complete, it is also important to separate each write with a non-write operation.

- During the first 16 timebase clock cycles after the LCDCPEN bit is set, all the LCD frontplane and backplane outputs are disabled, regardless of the state of the LCDEN bit.

- There are a number of additional restriction in modifying certain control bits when the controller is enabled. Refer to the Freescale documentation for a comprehensive description of these restrictions.

Software should initialize all LCDPEN, LCDBPEN, and LCDWF registers to the required values after a reset. The following registers should then be updated:

1. LCDC0 register

   Configure LCD clock source (SOURCE bit).

2. LCDRVC register (If the application uses the internally regulated voltage)

   Select 1.0 V or 1.67 V for 3 or 5 V glass (HREFSEL).

   Enable regulated voltage (RVEN).

   Trim the regulated voltage (RVTRIM).

3. LCDSUPPLY register

   Enable charge pump (CPSEL bit).

   Configure LCD module for doubler or tripler mode (HREFSEL bit).

   Configure charge pump clock (LADJ[1:0]).

   Configure op amp switch (BBYPASS bit).

   Configure LCD power supply (VSUPPLY[1:0]).

4. LCDC1 register

   Configure LCD frame frequency interrupt (LCDIEN bit).

   Configure LCD behavior in low power mode (LCDWAI and LCDSTP bits).

5. LCDC0 register

   Configure LCD duty cycle (DUTY[2:0]).

   Select and configure LCD frame frequency (LCLK[2:0]).

6. LCDBCTL register

   Configure display mode (ALT and BLANK bits).

   Configure blink mode (BMODE).

Configure blink frequency (BRATE[2:0]).

7. LCDPEN[7:0] register

Enable LCD module pins (PEN[43:0] bits).

8. LCDBPEN[7:0]

Enable LCD pins to operate as an LCD backplane (BPEN[43:0]).

9. LCDC0 register

Enable LCD module (LCDEN bit).

## 7-segment LCD example

A description of the connection between the LCD module and a seven segment LCD character is illustrated below in Figure 18.8, to provide a basic example for a 1/3 duty cycle LCD implementation. The example uses three backplane pins (LCD[0], LCD[1] and LCD[2]) and 3 frontplane pins (LCD[3], LCD[4], and LCD[5]). Also shown are the LCDWF contents required to display the value of 4 on the LCD segments.

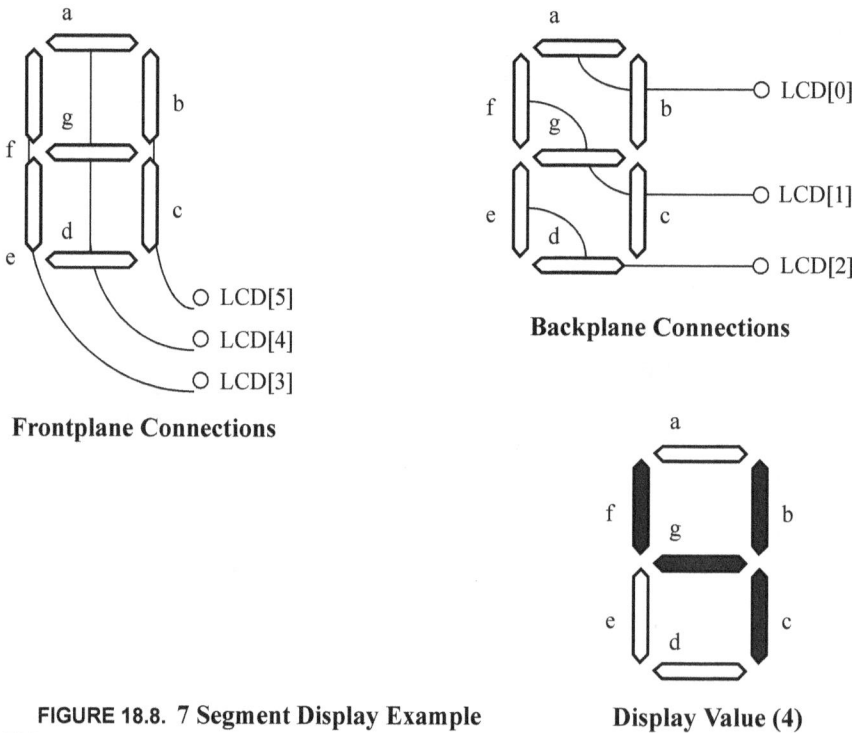

Frontplane Connections

Backplane Connections

Display Value (4)

FIGURE 18.8. 7 Segment Display Example

Figure 18.9 shows the backplane assignment register contents in LCDWF0-2 and the frontplane assignment register contents in LCDFW3-5 needed to display the digit "4".

FIGURE 18.9. **Backplane and Frontplane assignment registers**

| Backplane pins | - | - | - | - | - | LCD[2] | LCD[1] | LCD[0] |
|---|---|---|---|---|---|---|---|---|
| LCDWF0 | 0 | 0 | 0 | 0 | 0 | 0 | 0 | 1 |
| LCDWF1 | 0 | 0 | 0 | 0 | 0 | 0 | 1 | 0 |
| LCDWF2 | 0 | 0 | 0 | 0 | 0 | 1 | 0 | 0 |

| Frontplane Register and Pins | | | | | | | | |
|---|---|---|---|---|---|---|---|---|
| LCDWF3 (LCD[3]) | x | x | x | x | x | 0 (e) | 1 (f) | x |
| LCDWF4 (LCD[4]) | x | x | x | x | x | 0 (d) | 1 (g) | 0 (a) |
| LCDWF5 (LCD[5]) | x | x | x | x | x | x | 1 (c) | 1 (b) |

By first allocating each segment in the appropriate frontplane register and bit position that matches the intersection between front and backplanes pins, it is then a straightforward task to either set or clear the bit at that position to construct the desired displayed value.

So in this example, segment e is intersected by the frontplane on LCD[3], and the backplane on LCD[2]. Because the segment is not active, the corresponding LCDWF3 bit position for segment e is 0.

Segment f is intersected by frontplane on LCD[3], and the backplane on LCD[1]. Because the segment is active, the corresponding LCDWF3 bit position for segment f is 1.

The remaining segment bit values can be determined in an identical manner. Note that certain frontplane and backplane intersections are not implemented, so the values stored at those bit positions in the frontplane registers do not matter.

# CHAPTER 19

# *Tools and Software*

---

## *Introduction to Tools and Software*

Any embedded control system requires a number of supporting tools, and may require some standard software. Tools are required for the following stages of a project:

- Hardware and software development
- Hardware and software debug
- Application system calibration
- Production test
- Production support (e.g., calibration, Flash programming)
- Field support (e.g., performance analyses, product upgrades, fault evaluation)

Hardware development tools are general purpose and not processor architecture specific, and therefore outside the scope of this book. Software development tools provide support at different levels to meet a range of requirements of system complexity, development time, ease of development, and cost. Software development tools are broadly divided into the following levels of support, from lowest to highest:

- Assembly level
- Higher level languages (typically C, C++)
- Graphical level

### Assembly Level

Assembly level programming is at the level of the processor instructions, but using an English based representation of each instruction to ease programming, instead of the direct binary code. Each line of assembler code represents one processor instruction, so the assembler simply translates the assembly language instructions into the corresponding binary code. This gives a very high level of processor control, but becomes increasingly

---

inefficient with complex applications and multi register 32-bit architectures, such as Cold-Fire®.

Assembler language is now mainly used for critical real time code development. Assembler level is also very useful during debugging, as it allows instruction by instruction analyses of processor execution. Assembly level programs are very architecture specific, and generally not easily ported to a different processor architecture. Assembly level programs and are not directly migrateable even within the Flexis family products, and at the time of writing this book the authors are not aware of the availability of any assembler translators for Flexis. That said, it would not be a major problem to manually translate an HCS08 assembler routine to ColdFire assembler, due to the many more available CPU registers compared to HCS08 accumulators and index registers, and superset feel of the ColdFire instruction set compared to that of the HCS08. Translation in the other direction would be far more problematic.

### Higher Level Languages

The most widely used higher level programming languages in embedded microcotroller systems are C and C++. The use of a higher level programming language necessitates the use of a compiler to translate each language level instruction (or string of instructions) into the appropriate string of assembly level instructions. Although other higher level languages are available, most embedded system programs are written in C or C++. These higher level languages make it much easier to develop large and complex programs, and are independent of the processor architecture used. Although a high level language program can easily be recompiled to another processor architecture, any parts of the program that deal with processor specific functions such as system initialization, interrupt handling, peripheral servicing is unlikely to function correctly. The size and performance of the compiled code may also be very dependent on the Compiler used, and the options selected. The Flexis family products were specifically designed for compatibility between the 8-bit HCS08 and ColdFire V1 versions, with a large part of this compatibility being based on a common C/C++ compiler part of the CodeWarrior Development Studio. Other critical parts of the Flexis compatibility are the same peripherals and hardware pin compatibility. It is literally possible to compile a full application program for the HCS08 device, run it in the application, then recompile for the compatible ColdFire V1 device, swap the two devices in the application, and then run on the ColdFire V1. Although most available tool suites offer debug support at the higher level language, it may be necessary to go down to the assembler level to fully understand system behavior.

### Graphical level

Graphical level programming tools are becoming increasingly popular, especially for the development of complex control algorithms. These tools allow domain experts to develop control algorithms with little or no programming expertise. These tools also typically support system modeling and rapid prototyping, leading to much shorter development times and high quality software. The graphical programming tool then generates the equivalent lower level program, typically in C. A couple of popular examples of Graphical level programming tools are: LabVIEW from National Instruments (http://www.ni.com/), and MATLAB® and Simulink® from MathWorks (http://www.mathworks.com/). The main drawback of graphical level tools has been code efficiency, especially for microcontroller systems, although this continues to improve, graphical level programming tools are unlikely to be suitable for development with Flexis and ColdFire V1 microprocessors.

## Debug and Software

The debug portion of software development is dealt with in detail in Chapter 8. Since many tools are implemented purely in software, they are sometimes referred to simply as "Software". This can lead to confusion between software tools and run time software. To avoid this confusion we will refer to all tools simply as "Tools" and use the term "Software" to refer to run time software (unless otherwise specified). Run time software means any software code that is run on the microprocessor in the application.

Although most software is developed by the application developer, there is a growing number of commonly required functions that are available as software. This software is available in the following forms:

- Free software functions provided by the microcontroller supplier or other sources
- Open Source software developed by the Open Source community
- Proprietary software licensed by software vendors for a payment (with many different business models)

The commonly supported software functions are:

- Real Time Operating Systems (RTOS)
- Communication Protocol Stacks, such as for Ethernet or USB
- Standard functions (e.g. real time clock, floating point library, MP3 decoder)
- Application Program Interface (API)
- Initialization routines (typically generated by a configuration tool)
- Flash memory programming

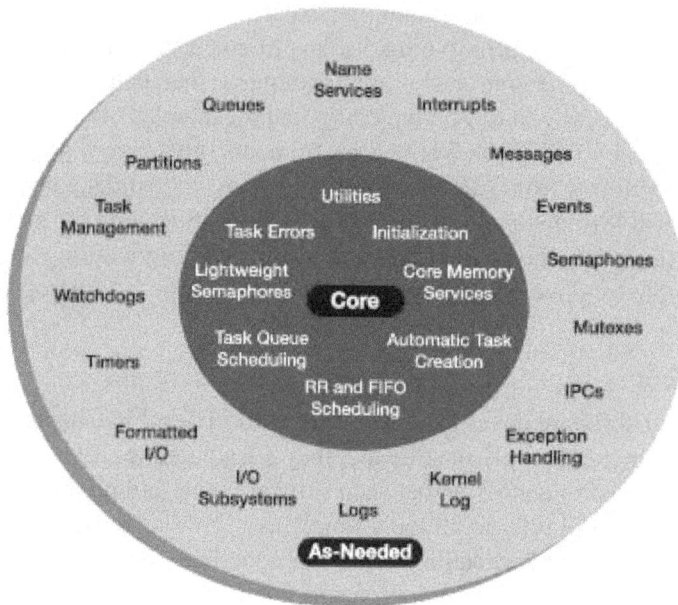

FIGURE 19.1. MQX RTOS

## MQX Real Time Operating System (RTOS)

Freescale provides the MQX Real Time Operating System (RTOS) including many protocol stacks and other software free for use on most ColdFire processors and microcontrollers.This includes the ColdFire V1 based microcontrollers, and is particularly valuable for use with the MCF51CN128 due to the included Ethernet protocol stack, and the MCF51JM due to the included USB protocol stacks. Due to the smaller memory systems on most of the HSC08 versions of Flexis products, MQX is currently (at the time of writing of this book) not available for HCS08 microcontrollers.

The Freescale MQX Real-Time Operating System (RTOS) provides real-time performance within a small, configured footprint. This RTOS is designed for configureability and balance of code size with performance requirements. It includes an easy-to-use API to allow first-time RTOS users to start developing their application on the day the software is installed. For experienced RTOS users, it is easy to migrate legacy application code to a Freescale MQX-based platform. The RTOS is tightly integrated with ColdFire® processors from Freescale and includes commonly used device drivers. Powerful Design and Development Tools are integrated with CodeWarrior™ tools to provide additional profiling and debugging capability. The MQX RTOS is designed with a modern, component-based micro-kernel architecture, allowing for customization by feature, size, and speed by selecting the components developers wish to include, while meeting the tight memory constraints of microcontroller systems.

## Hardware Tools

Freescale supports all Flexis and ColdFire V1 microcontrollers with hardware development and evaluation tools. Most current devices are supported either by a low cost Demo system or an Evaluation Board (EVB). Differences between the two are slight, with the EVBs typically including some additional support or connectivity. The EVBQE128 for example, includes a socket and both the MC9S08QE128 and MCF51QE128 microcontrollers, thus supporting development, evaluation and comparison of both available CPU architectures. Although there are no hard and fast rules, generally the HSC08 microcontrollers are supported by Demo systems, and ColdFire V1 microcontrollers by EVBs. The latest and future Flexis and ColdFire V1 products are supported by the new Tower System, designed to extend the traditional function of Demo systems and EVBs to include a level of rapid prototyping (see xxx below).

All of the Freescale hardware tools include an on board Debug interface to the microcontroller, requiring only a USB connection between the host PC and the board. To interface directly to a microcontroller in the application hardware a separate Debug and Programming interface is required, such as the USB BDM MULTILINK supplied by P&E Micro (www.pemicro.com), and also available from Freescale.

Most of the Freescale hardware tools include additional functionality, such as various sensors, to help demonstrate the microcontroller's capabilities, and potentially useful in many applications.

### Tower System

Freescale's Tower System was developed to save development time and costs through modular reconfigureable hardware that allows the user to start fast, finish early and move to the next project without starting over from scratch.

The Tower System is a simple concept. Take basic hardware modules, connect them together and start designing. There are two types of hardware modules, MCU/MPU and peripheral (i.e. serial, memory, LCD, etc.), which plug into backplane "elevator" boards. Through the use of open software and a standardized form factor and signal list, it is possible to continually design new modules to meet new design requirements. New modules are continuously being added to the system, so check http://www.freescale.com/tower for the latest catalog and information.

The Tower elevator boards have high-density, high-performance but low-cost PCI Express card-edge connectors for up to four modular boards. In addition, the elevator boards provide:

- Common serial and expansion bus signals
- Power regulation circuitry
- Standardized signal assignments
- Two expansion connectors for outward facing cards, such as an LCD module

The MCU/MPU modules are designed to work standalone or as part of the Tower System and feature:

- Basic user interfaces (buttons, DIP switches, LEDs, potentiometers, etc.)
- Sensor or other useful demonstration circuitry
- Debug interface via new OSBDM
- Power and debug through a single USB connector

Peripheral modules have also been developed for serial connectors, memory options, graphical LCDs and others. These modules and elevator boards combine to make a constantly evolving development system that will serve additional controllers, processors and peripherals to save users time and money through rapid prototyping and tool re-use for any number of projects.

## *CodeWarrior Development Studio for Microcontrollers*

To cover all of the features of CodeWarrior Development Studio for Microcontrollers and their detail use would require an entire book, so the following section is an introduction, but should be sufficient even for a new user to start a new project get it running. At the time of writing this book Freescale is introducing a major version (v10.x) that includes support Flexis and ColdFire V1 microcontrollers. Unlike previous version of CodeWarrior Development Studio version 10.x supports multiple processor architectures from 8 to 32-bit, and will support all Freescale microcontroller architectures. Most important for the readers of this book is, that it already supports both the HCS08 and ColdFire architectures.

The CodeWarrior Development Studio for Microcontrollers v10.x includes the following components:

- Eclipse Integrated Development Environment (IDE)
- Basic build tools, debug and flash support
- Processor Expert Configuration and Initialization Tool
- Welcome Screen

- New Project and MCU Change Wizards
- Tutorials
- Project importer
- Kernel aware debug (MQX, Professional Edition only)
- Host platform support

CodeWarrior is available in several editions:

- Evaluation Edition - Fully functional time limited, FREE
- Special Edition - Time unlimited, Code size and other limitations, FREE
- Standard Edition - Time and Code size unlimited, from $995
- Professional Edition - Time and Code size unlimited including Kernel aware debug, from $1995

It is worth noting that the Special Edition supports code sizes sufficient for many micro-controller applications.


### Getting Started with CodeWarrior Development Studio

For the latest detailed information on CodeWarrior refer to Freescale's web site at www.freescale.com/codewarrior. Most tools may be purchased directly on the Freescale web site using the "Buy Direct" button, or from a Freescale distributor using the "Distributor" button. After downloading or purchasing CodeWarrior follow the Quick Start Guide (QSG) instructions for installation, registration and activation. After you launch CodeWarrior use the Welcome Screen access the development resources.

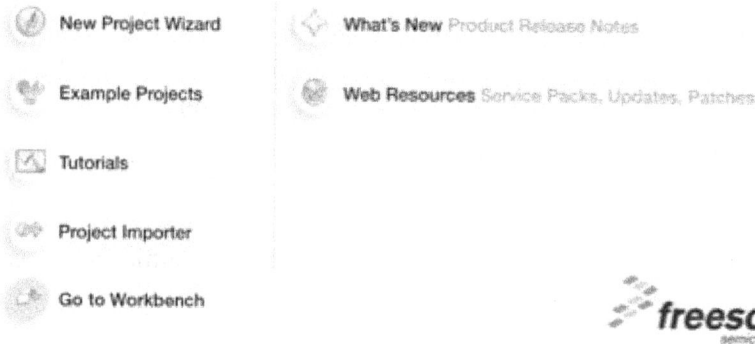

# CodeWarrior
## Development Studio

- New Project Wizard
- Example Projects
- Tutorials
- Project Importer
- Go to Workbench

- What's New Product Release Notes
- Web Resources Service Packs, Updates, Patches

*freescale*
semiconductor

FIGURE 19.2. CodeWarrior Welcome Screen

## Create a New CodeWarrior Project

From the Welcome Screen select the "New Project Wizard", and follow these easy steps

1. Enter the Project name, click "Next".
2. Select the either Flexis of ColdFire V1, the product family, and the target microcontroller, click "Next".
3. Select your programming language, click "Next".
4. Select your programming language options, click "Next".
5. Select Debug connection you are using (for Freescale Demo, EVB, and Tower boards select "P&E USB BDM Multilink"), click "Finish".

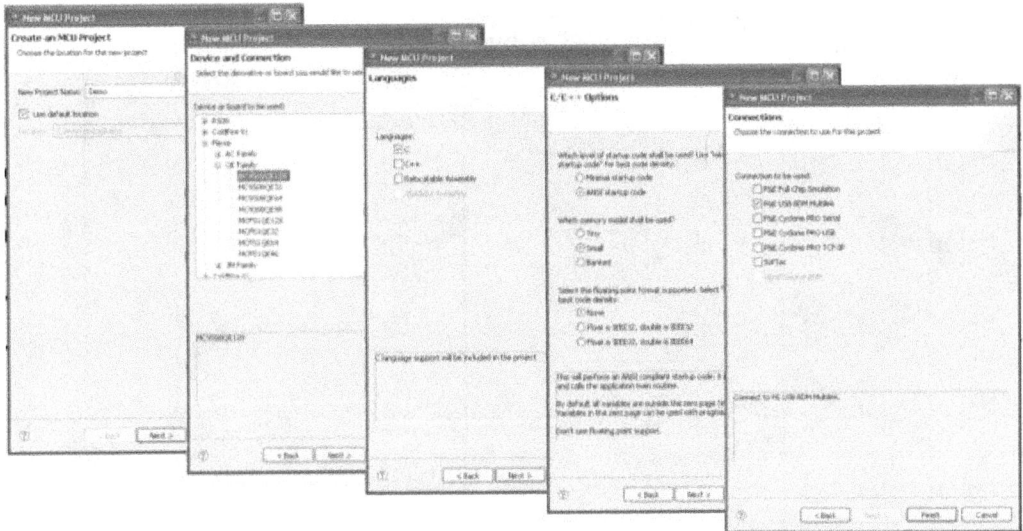

FIGURE 19.3. CodeWarrior Project Wizard

If you need to change the target microcontroller for any reason, such as requiring less or more memory, then use the MCU Change Wizard. This wizard may also be used to change the Debug connection, and the project name.

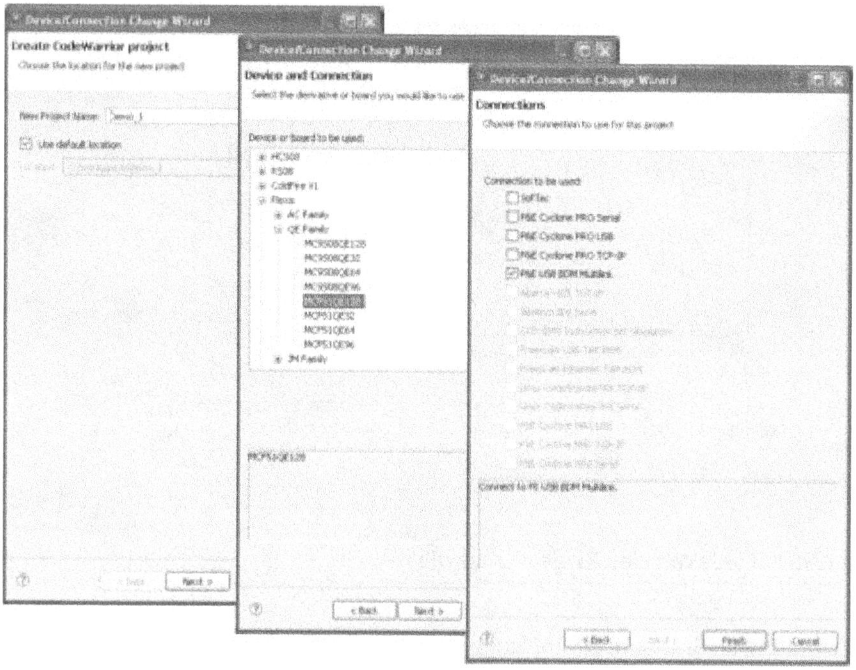

FIGURE 19.4. CodeWarrior MCU Change Wizard

If you have developed previous projects with CodeWarrior Classic (v6.x or v7.x), and wish to import these into CodeWarrior v10.x, there is the Project Importer for that.For additional guidance and details on using CodeWarrior use the included step by step Tutorials accessible from the Welcome Screen.

# Device Matrix

Table A.1 shows a matrix of Flexis and ColdFire® V1 devices that contain some of the modules covered in this book. Please refer to Freescale product documentation for complete details of product capabilities.

Column groups: CPU (Core, MAC, CRC, CAU) · Memory (Flash K bytes, SRAM K bytes) · Timers (TPM, FTM, RTC, iRTC, TOD, PDB, CMT, MTIM) · Analog (ADC (12-bit), ADC16, VREF, DAC, ACMP, PRACMP, TRIAMP) · Connectivity (UART/SCI, SPI, I2C, MSCAN, USB full speed)

| Order | Product Family | Flexis Product | Key Functionality/Application | Core | MAC | CRC | CAU | Flash K bytes | SRAM K bytes | MiniFlexBus | TPM | FTM | RTC | iRTC | TOD | PDB | CMT | MTIM | ADC (12-bit) | ADC16 | VREF | DAC | ACMP | PRACMP | TRIAMP | UART/SCI | SPI | I2C | MSCAN | USB full speed | 10/100 Ethernet | RGPIO | Segment LCD |
|---|---|---|---|---|---|---|---|---|---|---|---|---|---|---|---|---|---|---|---|---|---|---|---|---|---|---|---|---|---|---|---|---|---|
| 1 | MC9S08QE | X | Low Power | HCS08 | | | | 4 to 128 | 0.256 to 8 | | X | X | | | | | | X | | | | X | | | | X | X | X | | | | | |
| 2 | MCF51QE | X | Low Power | ColdFire V1 | | | | 32 to 128 | 4 to 8 | | X | X | | | | | | X | | | | X | | | | X | X | X | | | | X | |
| 3 | MC9S08JM | X | USB device | HCS08 | | | | 8 to 60 | 1 to 4 | | X | X | | | | | | X | | | | X | | | | X | X | X | | Device | | | |
| 4 | MCF51JM | X | USB otg | ColdFire V1 | | | X | 64 to 128 | 8 to 16 | | X | X | | | | | X | X | | | | X | | | | X | X | X | X | Host/ Device (otg) | | X | |
| 5 | MC9S08AC | X | Motor Control | HCS08 | | X | | 8 to 128 | 0.7 to 8 | | X | | | | | | | X | | | | X | | | | X | X | X | | | | | |
| 6 | MCF51AC | X | Motor Control | ColdFire V1 | | X | | 128 to 256 | 16 to 32 | | X | X | | | | | | X | | | | | X | | | X | X | X | X | | | X | |
| 7 | MC9S08JE | X | Low Power | HCS08 | | | | 32 to 128 | 4 to 12 | | X | | | X | X | X | | X | X | | X | X | X | | | X | X | X | | Device | | | |
| 8 | MCF51JE | X | Low Power | ColdFire V1 | X | X | | 128 to 256 | 32 | X | X | | | X | X | X | | X | X | | X | X | X | | | X | X | X | | Host/ Device (otg) | | X | |
| 9 | MC9S08MM | X | Medical | HCS08 | | | | 32 to 128 | 4 to 12 | | X | | | X | X | X | | X | X | X | X | X | X | X | X | X | X | X | | Device | | | |
| 10 | MCF51MM | X | Medical | ColdFire V1 | X | X | | 128 to 256 | 32 | X | X | | | X | X | X | | X | X | X | X | X | X | X | X | X | X | X | X | Host/ Device (otg) | | X | |
| 11 | MCF51CN | X | Ethernet | ColdFire V1 | | | | 128 | 24 | X | X | X | X | | | | | X | X | | | | | | | X | X | X | | | X | X | |
| 12 | MCF51EM | X | Electricity Meters | ColdFire V1 | X | X | | 128 to 256 | 8 to 16 | | X | | | X | | X | | X | X | | X | | X | | | X | X | X | | | | X | X |

**TABLE A.1 Modules used in Flexis and Coldfire V1 devices**

# Index

# Index

www.ingramcontent.com/pod-product-compliance
Lightning Source LLC
Chambersburg PA
CBHW080649220326
41598CB00033B/5147